Grundlehren der mathematischen Wissenschaften 264

A Series of Comprehensive Studies in Mathematics

Jean-Pierre Aubin   Arrigo Cellina

Differential Inclusions

Set-Valued Maps and Viability Theory

With 29 Figures

Springer-Verlag
Berlin Heidelberg New York Tokyo 1984

Jean-Pierre Aubin
CEREMADE, Université de Paris-Dauphine
Place du Marechal de Latte de Tassigny
75775 Paris Cedex 16
France

Arrigo Cellina
S.I.S.S.A.
Strada Costiera 11
Trieste
Italy

AMS Subject Classification (1980): 34A60, 34D, 39A,B, 49A,E

ISBN 3-540-13105-1 Springer-Verlag Berlin Heidelberg New York Tokyo
ISBN 0-387-13105-1 Springer-Verlag New York Heidelberg Berlin Tokyo

Library of Congress Cataloging in Publication Data
Aubin, Jean-Pierre.
Differential inclusions.
(Grundlehren der mathematischen Wissenschaften; #264)
Bibliography: p.
Includes index.
1. Differential inclusions, 2. Set-valued maps. 3. Feedback control systems. I. Cellina, Arrigo, 1941- . II. Title. III. Title: Viability theory. IV. Series.
QA371.A93 1984 515.3'5 84-1327
ISBN 0-387-13105-1 (U.S.)

Printed in Germany

Typesetting, printing and bookbinding: Universitätsdruckerei H. Stürtz AG, Würzburg
2141/3140-543210

This book is dedicated to
Anne-Laure, Claudia and Francesca

Epigraph

Why a book on differential inclusions?

There is a great variety of motivations that led mathematicians to study dynamical systems having velocities not uniquely determined by the state of the system, but depending loosely upon it, i.e., to replace differential equations

$$x' = f(x)$$

by differential inclusions

$$x' \in F(x)$$

when F is the set-valued map that associates to the state x of the system the set of feasible velocities.

Each one of these motivations offered a partial and biased view of differential inclusions, but all together contributed to the creation of a wealth of problems to a subject whose vitality is at present beyond doubts.

The first purpose of this book is to report on the common tools and ideas which were devised and proposed by those who were attracted by this field either for its own intrinsic interest and beauty or for its potential for applications in different fields.

But, besides this array of mathematical and physical motivations, social and biological sciences should provide many instances of differential inclusions. Indeed, if deterministic models are quite convenient for describing systems that arise in physics, mechanics, engineering and even, in microeconomics, their use for explaining the evolution of what we shall call "macrosystems" does not take in account the *uncertainty* (which, in particular, involves the impossibility of a comprehensive description of the dynamics of the system), the absence of *controls* (or the ignorance of the laws relating the controls and the states of the system) and the *variety* of available dynamics. These are reasons why usual dynamical systems, or even controled dynamical systems, may not be suitable for describing the evolution of states of systems derived from economics, social and biological sciences. This is our hope to supply scientists of these fields with an adequate tool. To justify this hope, we shall provide in this book an application to economics.

What are the problems which arise?

Naturally, as always in mathematics, we shall begin by studying the existence of solutions to several classes of differential inclusions and study the properties of the set of trajectories.

We may expect this set of trajectories to be rather large: hence a second class of problems consists naturally in devising mechanisms for selecting special trajectories.

A first class of such mechanisms is provided by *Optimal Control Theory:* it consists in selecting trajectories that optimize a given criterion, a functional on the space of all such trajectories.

This implicitely requires that:

1) there exists one decision maker who "controls" the system

2) such a decision maker has a perfect knowledge of the future (which is involved in the definition of the criterion)

3) the optimal trajectories are chosen once and for all at the origin of the period of time.

These requirements are not satisfied by the "macrosystems" that evolve according to the laws of Darwinian evolution.

Such macrosystems appear to have neither aims nor targets nor desire to optimize some criterion. But they face a minimal requirement, called *viability*, which is to remain "alive" in the sense of satisfying given binding constraints.

For that, they use a policy, *opportunism*, that enables the system to conserve viable trajectories that its lack of determinism – the *availability of several feasible velocities* – allows to find.

This provides a mathematical metaphor of this deep intuition of Democritus, "Everything that exists in the universe is due to chance and necessity".

This second class of mechanisms is the object of *Viability Theory*, which is thouroughly investigated in the second half of this book. In particular, we shall apply Viability Theory in the framework of Control Theory for *regulating* systems through *feedback controls.*

Paris, Venezia,
in Spring 1984

J.-P. Aubin
A. Cellina

Acknowledgments

The authors are indebted to their institutions (CEREMADE, Université de Paris-Dauphine, Paris and International School for Advanced Studies, Trieste), as well as the Mathematics Research Center of University of Wisconsin, Madison, the University of Southern California, Los Angeles, and the International Institute for Applied Systems Analysis, Laxenburg, Austria. They provided us with both facilities and the intellectual environment which was needed for the completion of this book. The colleagues we met there gave an important - even though hidden - contribution to this work.

The etching with the "putto" inscribing $\frac{dP}{dy}$ is taken from a 1787 edition of Euler's *Institutiones Calculi Differentiali.*

Table of Contents

Introduction 1

Chapter 0. Background Notes 9

Introduction 9
1. Continuous Partitions of Unity 9
2. Absolutely Continuous Functions 12
3. Some Compactness Theorems 13
4. Weak Convergence and Asymptotic Center of Bounded Sequences . 15
5. Closed Convex Hulls and the Mean-Value Theorem 18
6. Lower Semicontinuous Convex Functions and Projections of Best Approximation 21
7. A Concise Introduction to Convex Analysis 29

Chapter 1. Set-Valued Maps 37

Introduction 37
1. Set-Valued Maps and Continuity Concepts 39
2. Examples of Set-Valued Maps 46
3. Continuity Properties of Maps with Closed Convex Graph 54
4. Upper Hemicontinuous Maps and the Convergence Theorem . . . 59
5. Hausdorff Topology 65
6. The Selection Problem 68
7. The Minimal Selection 70
8. Chebishev Selection 73
9. The Barycentric Selection 77
10. Selection Theorems for Locally Selectionable Maps 80
11. Michael's Selection Theorem 82
12. The Approximate Selection Theorem and Kakutani's Fixed Point Theorem 84
13. σ-Selectionable Maps 86
14. Measurable Selections 90

Chapter 2. Existence of Solutions to Differential Inclusions 93

Introduction 93
1. Convex Valued Differential Inclusions 96

2. Qualitative Properties of the Set of Trajectories of Convex-Valued Differential Inclusions . . . 103
3. Nonconvex-Valued Differential Inclusions . . . 111
4. Differential Inclusions with Lipschitzean Maps and the Relaxation Theorem . . . 119
5. The Fixed-Point Approach . . . 127
6. The Lower Semicontinuous Case . . . 134

Chapter 3. Differential Inclusions with Maximal Monotone Maps . . . 139

Introduction . . . 139
1. Maximal Monotone Maps . . . 140
2. Existence and Uniqueness of Solutions to Differential Inclusions with Maximal Monotone Maps . . . 147
3. Asymptotic Behavior of Trajectories and the Ergodic Theorem . . . 151
4. Gradient Inclusions . . . 158
5. Application: Gradient Methods for Constrained Minimization Problems . . . 163

Chapter 4. Viability Theory: The Nonconvex Case . . . 172

Introduction . . . 172
1. Bouligand's Contingent Cone . . . 176
2. Viable and Monotone Trajectories . . . 179
3. Contingent Derivative of a Set-Valued Map . . . 188
4. The Time Dependent Case . . . 191
5. A Continuous Version of Newton's Method . . . 195
6. A Viability Theorem for Continuous Maps with Nonconvex Images . 198
7. Differential Inclusions with Memory . . . 204

Chapter 5. Viability Theory and Regulation of Controled Systems: The Convex Case . . . 213

Introduction . . . 213
1. Tangent Cones and Normal Cones to Convex Sets . . . 218
2. Viability Implies the Existence of an Equilibrium . . . 228
3. Viability Implies the Existence of Periodic Trajectories . . . 235
4. Regulation of Controled Systems Through Viability . . . 238
5. Walras Equilibria and Dynamical Price Decentralization . . . 245
6. Differential Variational Inequalities . . . 264
7. Rate Equations and Inclusions . . . 274

Chapter 6. Liapunov Functions . . . 281

Introduction . . . 281
1. Upper Contingent Derivative of a Real-Valued Function . . . 284
2. Liapunov Functions and Existence of Equilibria . . . 290

3. Monotone Trajectories of a Differential Inclusion 293
4. Construction of Liapunov Functions 305
5. Stability and Asymptotic Behavior of Trajectories 309

Comments 322

Bibliography 328

Index 341

Introduction

A great impetus to study differential inclusions came from the development of Control Theory, i.e. of dynamical systems

$$(*) \qquad x'(t)=f(t,x(t),u(t)), \qquad x(0)=x_0$$

"controlled" by parameters $u(t)$ (the "controls"). Indeed, if we introduce the set-valued map

$$F(t,x) \doteq \{f(t,x,u)\}_{u\in U}$$

then solutions to the differential equations $(*)$ are solutions to the "differential inclusion"

$$(**) \qquad x'(t)\in F(t,x(t)), \qquad x(0)=x_0$$

in which the controls do not appear explicitely.

Systems Theory provides dynamical systems of the form

$$x'(t)=A(x(t))\frac{d}{dt}(B(x(t))+C(x(t)); \qquad x(0)=x_0$$

in which the velocity of the state of the system depends not only upon the state $x(t)$ of the system at time t, but also on *variations of observations* $B(x(t))$ of the state.

This is a particular case of an *implicit differential equation*

$$f(t,x(t),x'(t))=0$$

which can be regarded as a differential inclusion $(**)$, where the right-hand side F is defined by

$$F(t,x) \doteq \{v \mid f(t,x,v)=0\}.$$

During the 60's and 70's, a special class of differential inclusions was thoroughly investigated: those of the form

$$x'(t)\in -A(x(t)), \qquad x(0)=x_0$$

where A is a "maximal monotone" map.

This class of inclusions contains the class of "gradient inclusions" which generalize the usual gradient equations

$$x'(t)=-\nabla V(x(t)), \qquad x(0)=x_0$$

when V is a differentiable "potential".

There are many instances when potential functions are not differentiable. This is the case when the potential function V "equals $+\infty$" outside a given closed subset K: in other words, the state of the system must belong to K (a first occurence of a viability problem).

When the potential function V is a lower semicontinuous convex function, we replace $\nabla V(x)$ by a "generalized gradient", also called "subdifferential", $\partial V(x)$, which associates to any point x a set of "subgradients". The gradient inclusions:

$$x'(t) \in -\partial V(x(t)), \quad x(0)=x_0$$

enjoy an important property: if the state $\bar{x}$ minimizes the potential V, then the trajectories $x(t)$ of the gradient inclusion do converge to such minimizers.

Also, differential inclusions provide a mathematical tool for studying differential equations

$$x'(t)=f(t,x(t)), \quad x(0)=x_0$$

with *discontinuous* right-hand side, by embedding $f(t,x)$ into a set-valued map $F(t,x)$ which, as a set-valued map, enjoys enough regularity to have trajectories closely related to the trajectories of the original differential equation.

Differential variational inequalities do form a special class of differential inclusions. They are differential inclusions of the form

$$(***) \qquad \begin{cases} \text{i)} \ \forall t \geq 0, \ x(t) \in K, \\ \text{ii)} \ \sup\limits_{y \in K} \langle x'(t) - f(x(t)), \ x(t) - y \rangle = 0, \\ \text{iii)} \ x(0) = x_0, \end{cases}$$

when K is a *closed convex subset*. We observe that whenever $x(t)$ belongs to the interior of K, equation (***) ii) reduces to the differential equation

$$x'(t)=f(x(t)).$$

So, differential variational inequalities "modify" the differential equation $x'(t)=f(x(t))$ only at boundary points of K. The reason for this, as we shall see, is to "force" the trajectories to remain in K, as it is stated in (***) i).

Naturally, as often in mathematics, we shall begin by studying the existence (local or global) of solutions to a differential inclusion

$$x'(t) \in F(t,x(t)), \quad x(0)=x_0$$

and by investigating the topological properties of the set $\mathcal{T}(x_0)$ of such solutions, as well as the nature of its dependence upon the initial state x_0. In doing so, we shall have to overcome difficulties which do not exist for ordinary differential equations.

Since we may expect a differential inclusion to have a rather "large" set of trajectories, a second class of problems consists in devising mechanisms for selecting special trajectories.

The first example of a very noticeable trajectory is provided by *equilibria* – stationary states –, which are the constant trajectories $\bar{x}$, solution to

$$0 \in F(\bar{x}).$$

We may also ask for the existence of "periodic" trajectories. If we denote by $\mathscr{A}_T(x_0)$ the values of the trajectories at time T, this amounts to finding fixed points of the set-valued map $x_0 \to \mathscr{A}_T(x_0)$. This is an issue that we shall tackle.

Another method of selection is provided by Optimal Control Theory.

Since $\mathscr{T}(x_0)$ is a set of functions defined on an interval $[0, T]$ of the real line, we can use a continuous functional $\underset{\sim}{W}$ associating to each trajectory $x(\cdot) \in \mathscr{T}(x_0)$ a cost $\underset{\sim}{W}(x(\cdot))$ for singling out "optimal" trajectories $\bar{x}(\cdot) \in \mathscr{T}(x_0)$ minimizing the functional $\underset{\sim}{W}$ over the set $\mathscr{T}(x_0)$ of trajectories. We can even devise more sophisticated selection procedures using game theoretic approaches. We shall not study these problems in this book. A first reason – justification – is that many excellent books cover this topic (however, this book provides several technical tools useful for this theory, for instance the concept of contingent derivatives for setting the Hamilton-Jacobi-Carathéodory-Bellman equation).

We shall study quite thoroughly the selection mechanism provided by Viability Theory.

We begin with the *viability problem:* we select the trajectories which are "viable" in the sense that they always satisfy given constraints. We can summarize this by saying that a trajectory $x(\cdot)$ is viable when

$$\forall t, \quad x(t) \in K(t)$$

where $K(t)$ is the *viability subset* at time t, which is at least closed, and often compact. The "smaller" is the set, the more restrictive is this selection procedure.

More generally, we can select "monotone" trajectories: a preorder $\succcurlyeq$ is given, and we say that a trajectory is monotone with respect to this preorder when

$$\forall t \geq s, \quad x(t) \succcurlyeq x(s).$$

This selection procedure is consistent with the behavioral assumption of "limited rationality" proposed by March and Simon, where optimality is replaced by mere satisfaction.

The concept of Bouligand's contingent cone will play a crucial role, as a pioneering Theorem, proved in 1942 by Nagumo, shows: let f be a bounded continuous function from a closed subset K of R^n to R^n. Nagumo's theorem states that a necessary and sufficient condition for the differential equation

$$x'(t) = f(x(t)), \quad x(0) = x_0$$

to have viable trajectories for all initial state $x_0 \in K$ is that

$$\forall x \in K, \quad f(x) \text{ belongs to the contingent cone to } K \text{ at } x.$$

When K is a smooth manifold embedded in R^n, the contingent cone to K at x coincides with the tangent space to K at x, so that the above condition states only that

$$\forall x \in K, \quad f(x) \text{ is a vector field}$$

and the Nagumo Theorem yields the existence of solutions to differential equations on manifolds.

However, this approach does not require any smoothness on K. This is important because many viability problems use subsets K such as intersections of half-spaces, as in economics where the "simplest" subsets are defined by linear inequality constraints.

So, the purpose of the second part of this book is to adapt the Nagumo Theorem to differential inclusions and to investigate many of its consequences.

We can describe at this stage what are some of its consequences in the framework of *regulation* of controled dynamical systems, i.e. of finding controls yielding viable trajectories of

$$x'(t) = f(x(t), u(t)), \quad x(0) = x_0.$$

We shall introduce the feedback map associating to each state x the subset of controls

$$C(x) \doteq \{u \in U \mid f(x,u) \text{ belongs to the contingent cone to } K \text{ at } x\}.$$

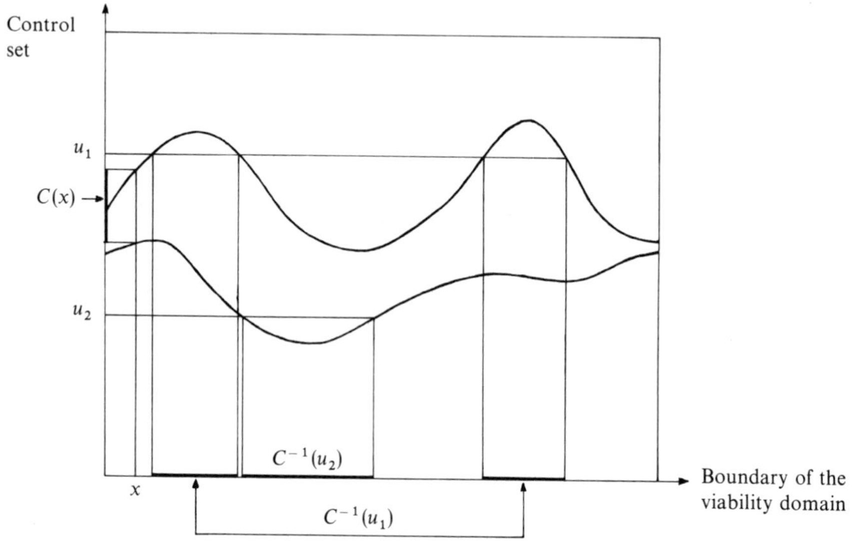

Fig. 1. Graph of the "Feedback" map C. The Feedback map reveals a division of the viability domain into "cells"; each cell is the subset of viable states which can be regulated by a given control. To pass from one cell to another requires the control to be changed. The boundaries of these cells signal the need for "structural change"

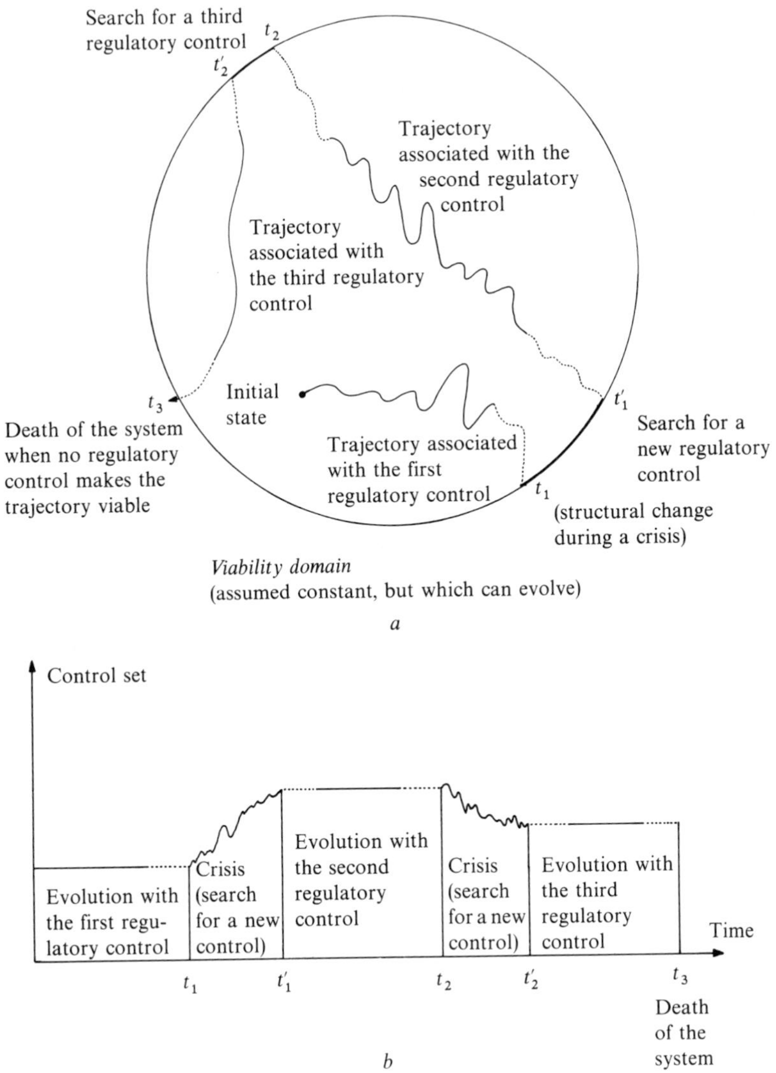

Fig. 2. *a* Evolution of the state (in the state space); *b* evolution of the control. The dotted lines represent the parts of the trajectories close to the boundary of the viability domain which disappear after a certain period of time

Under suitable assumptions, we shall prove that the controled system can be regulated – i.e. that there exists a viable trajectory – if and only if the subsets $C(x)$ are nonempty for every state x. Furthermore, the controls $u(t)$ yielding a viable trajectory $x(\cdot)$ do satisfy the feedback relation

$$\forall t \geq 0, \quad u(t) \in C(x(t)).$$

In other words, the controls depend upon the time through the state of the system.

One subsequent problem is to find single-valued feedback controls, i.e. "selections" $\underset{\sim}{u}$ of the map C, satisfying $\underset{\sim}{u}(x) \in C(x)$ for all x. For such feedback controls, the dynamical system

$$x'(t) = f(x(t), \underset{\sim}{u}(x(t))), \qquad x(0) = x_0$$

has viable trajectories.

Another problem is to find explicit and/or convenient sufficient conditions implying that the subsets $C(x)$ are nonempty.

We shall point out quite fascinating relations between the viability problem and the problem of finding equilibria.

Under convexity assumptions, we shall prove that the *necessary and sufficient condition for viability implies the existence of equilibria.*

We shall continue by studying monotone trajectories, and in particular, by characterizing the differential inclusions

$$x'(t) \in F(x(t)), \qquad x(0) = x_0$$

which have trajectories "monotone" in the sense that:

$$\forall t \geq s, \qquad V(x(t)) - V(x(s)) + \int_s^t W(x(\tau), x'(\tau))\, d\tau \leq 0$$

where V: $\mathrm{Dom}\, F \to R_+$ and W: $\mathrm{graph}(F) \to R_+$ are nonnegative functions.

Such trajectories do satisfy:

$$t \to V(x(t)) \text{ is nonincreasing}$$

and we can infer from the properties of the function W several informations on the asymptotic behavior of the trajectory when $t \to \infty$.

For that purpose, we shall adapt Lyapunov method for studying the stability of trajectories. The concept of "upper contingent derivative" of any function will allow us to use *non differentiable Lyapunov functions.*

At this point we shall make our brief connection with optimal control theory.

We recall that $\mathscr{T}(x_0)$ denotes the subset of trajectories of the differential inclusion $x'(t) \in F(x(t))$ issued from x_0. We denote by

$$V_F(x_0) \doteq \inf_{x(\cdot) \in \mathscr{T}(x_0)} \int_0^\infty W(x(\tau), x'(\tau))\, d\tau$$

the *value function.* As we shall observe, monotone trajectories for such a function V_F are the optimal trajectories of this problem of optimal control and the function $x_0 \to V_F(x_0)$ satisfies the Hamilton-Jacobi-Carathéodory equation when we replace the usual derivative (which may not exist) by the upper contingent derivative (which always does exist).

We shall show how most of these results can be extended to differential difference inclusions, Volterra inclusions and, more generally, to the so-called "functional differential inclusions" or "differential inclusions with memory", where the velocity depends upon the past history of the trajectory, and not only upon its current state.

We devote a part of this book to an application to Economic Theory. We apply viability theory to built a "dynamical" analog to the static concept of Walras equilibrium. In other words, we regard the price system as a "control" which is used by each consumer $i=1,\ldots,n$ to govern the evolution of his consumption of a commodity bundle $x_i(t)$ according to a differential equation

$$x_i'(t)=d_i(x_i(t), p(t)), \qquad x_i(0)=x_i^0.$$

The viability constraint in this framework is the requirement that the sum $\sum_{i=1}^{n} x_i(t)$ of the consumed commodity bundles lies in the set of available goods. We shall prove that financial laws (- it is forbidden to spend more than it is earned -) guarantees the existence of price systems $p(t)$ yielding viable trajectories. We observe that this mechanism is decentralized, since each consumer does not take in account the consumptions of his fellow consumers in his differential equation. We shall argue that this dynamical approach retains the "good features" of the Walras model of general equilibrium, by letting aside the static concept of equilibrium, which is beeing highly critized nowadays because of its obvious shortcomings.

As the material listed so far indicates, this is not strictly speaking a technical book on differential inclusions, but is intended for readers that are not necessarily mathematicians, or, even more, mathematicians with a working interest in this area. Although the chapter on set-valued maps is well developed and presents more material than that is later used in the book, technical difficulties have been avoided whenever possible. In particular we have decided not to go into the investigation of maps $(t,x)\to F(t,x)$ that are assumed only measurable with respect to t. Two are the reasons for this exclusion: on one hand the monograph of Castaing-Valadier is already available on the subject and on the other, we felt that this technical difficulty is not really intrinsic to the subject of differential inclusions.

We organized the book as follows.

Chapter 0 gathers heteroclite material which shall be used later in the book. The first chapter deals rather comprehensively with the set-valued maps which will be used in the right-hand side of differential inclusions. In Chapter 2, we study the existence of solutions to differential inclusions and the properties of the set of trajectories.

Chapter 3 is self-contained: it presents the basic results on maximal monotone maps and on the associated differential inclusions, including gradient inclusions.

The viability issue is the purpose of the fourth and fifth chapters and owes much to Georges Haddad. Chapter 6 is devoted to the study of monotone trajectories and to applications to stability theory.

Chapter 0. Background Notes

Introduction

We gather several known (or less known) results which are used in several distinct places of this book. We prove the existence of locally Lipschitzean partitions of unity subordinated to a locally finite open covering of a metric space and we recall the definition and the characterization of absolutely continuous functions.

We then recall the Ascoli-Arzelà Theorem characterizing compact subsets of continuous functions and the Alaoglu Theorem characterizing the weakly compact subsets of the dual of a Banach space; we use them for providing a Compactness Theorem which is of constant use in this book and we prove a compactness criterion in the space of bounded functions.

We then introduce Edelstein's concept of asymptotic center of a bounded sequence of elements of a Hilbert space, which is useful for studying the asymptotic behavior of trajectories of differential inclusions.

We also present several properties of projectors of best approximation onto closed convex subsets of Hilbert spaces, with a special emphasis to orthogonal projectors onto closed convex cones. Finally, we propose a concise introduction to convex analysis, by defining conjugate functions of lower semicontinuous convex functions and applying their properties to the support functions of convex subsets and by recalling the calculus of subdifferentials of convex functions.

1. Continuous Partitions of Unity

Let X be any space. A family of subsets of X, $\{\Omega_i\}_{i \in I}$, such that $X = \bigcup_{i \in I} \Omega_i$, is called a covering of X. In the case when X is a topological space and each $\{\Omega_i\}_{i \in I}$ is an open subset of X, we call $\{\Omega_i\}_{i \in I}$ an *open* covering.

Definition 1. Let $\{\Omega_i\}_{i \in I}$ and $\{\omega_j\}_{j \in J}$ be two coverings of X. $\{\Omega_i\}$ is a *refinement* of $\{\omega_j\}$ if for every $i \in I$ there exists a $j \in J$ such that $\Omega_i \subset \omega_j$. ▲

Definition 2. Let $\{\Omega_i\}_{i\in I}$ be a covering. Then if J is contained in I and $\{\Omega_j\}_{j\in J}$ is again a covering, it is called a *subcovering*. ▲

Definition 3. A covering $\{\Omega_i\}_{i\in I}$ of a topological space X is called *locally finite* if for every $x\in X$, there exists a neighborhood V of x such that $\Omega_i \cap V \neq \emptyset$ only for a finite number of indexes. ▲

Definition 4. A Hausdorff space is called *paracompact* if each open covering has a locally finite open refinement. ▲

We recall that:

Proposition 1. *A closed subset of a paracompact space is paracompact.*

and the following Theorem (see Bourbaki [1958]).

Theorem 1. *Every metric space is paracompact.* ▲

Definition 5. Let f be a single-valued map from the topological space X to R. The closure of $\{x \mid f(x)\neq 0\}$ is called the *support* of $f(\mathrm{supp}(f))$.

A family $\{\psi_i\}_{i\in I}$ is called a locally Lipschitz *partition of unity* if for all $i\in I$

a) ψ_i is locally Lipschitz and non negative

b) the supports of ψ_i are a closed locally finite convering of X

c) for each $x\in X$, $\sum_{i\in I}\psi_i(x)=1$.

We say that a partition of unity $\{\psi_i\}_{i\in I}$ is subordinated to a covering $\{\Omega_i\}_{i\in I}$ if $\forall i\in I$, $\mathrm{supp}(\psi_i)\subset\Omega_i$. ▲

Theorem 2. *Let X be a metric space. To any locally finite open covering $(\Omega_i)_{i\in I}$ of X, we can associate a locally Lipschitzean partition of unity subordinated to it.* ▲

Proof. Since X is paracompact, we know that there exists an open covering $(\omega_i)_{i\in I}$ of X such that $\bar{\omega}_i\subset\Omega_i$ for any $i\in I$. We notice that since $(\Omega_i)_{i\in I}$ is locally finite, the same is true for $(\omega_i)_{i\in I}$. (See Bourbaki, chapt 9, § 4.3.)

Let us define for any $i\in I$ the function $\varphi_i\colon X\to R_+$ by $\varphi_i(x)=d(x, X\backslash\omega_i)$ where $d(\cdot, X\backslash\omega_i)$ denotes the distance function to the subset $X\backslash\omega_i \doteq \mathrm{comp}(\omega_i)$.

Then each φ_i is obviously Lipschitzean and verifies

$$\mathrm{supp}(\varphi_i)=\overline{\{x\in X/\varphi_i(x)\neq 0\}}=\bar{\omega}_i\subset\Omega_i.$$

Let us introduce for any $i\in I$ the function $\psi_i\colon X\to R^+$ such that

$$\psi_i(x)=\frac{\varphi_i(x)}{\sum_{j\in I}\varphi_j(x)} \quad \text{for any } x\in X. \tag{1}$$

We verify that for any $x \in X$, $\sum_{j\in I} \varphi_j(x)$ is well defined since the covering $(\omega_i)_{i\in I}$ is locally finite and that $\sum_{j\in I} \varphi_j(x) > 0$ since for at least one index $j \in I$, $x \in \omega_j$, which gives $\varphi_j(x) = d(x, X\setminus\omega_j) > 0$. Furthermore each ψ_i is continuous on X and takes its values in the interval $[0, 1]$.

Moreover we verify that $\operatorname{supp}\psi_i = \operatorname{supp}\varphi_i = \bar{\omega}_i \subset \Omega_i$ and that $\sum_{i\in I} \psi_i(x) = 1$ for any $x \in X$.

It is left to prove that each function ψ_i is locally Lipschitz on X.

Indeed let $x \in X$ be given. Then there exists an open neighborhood $V(x)$ of x which meets only a finite number of ω_i, $i \in I$, say

$$\omega_{i_1}, \omega_{i_2}, \ldots, \omega_{i_p}.$$

By construction we easily verify that when $y \in V(x)$, $\psi_i(y) = 0$ if $i \notin \{i_1, \ldots, i_p\}$.

Consider ψ_{i_k} with $i_k \in \{i_1, \ldots, i_p\}$. Since

$$\sum_{i\in I} \varphi_i(x) = \sum_{i\in\{i_1,\ldots,i_p\}} \varphi_i(x) > 0,$$

by continuity, there exists a neighborhood $W(x) \subset V(x)$ of x and $m > 0$, $M > 0$ such that:

$$m \le \sum_{i\in I} \varphi_i(y) = \sum_{i\in\{i_1,\ldots,i_p\}} \varphi_i(y) \le M \qquad \text{for any } y \in W(x).$$

Then for any $y, z \in W(x)$ we get:

$$\begin{aligned}
|\psi_{i_k}(y) - \psi_{i_k}(z)| &= \left| \frac{\varphi_{i_k}(y)}{\sum\limits_{i\in\{i_1,\ldots,i_p\}} \varphi_i(y)} - \frac{\varphi_{i_k}(z)}{\sum\limits_{i\in\{i_1,\ldots,i_p\}} \varphi_i(z)} \right| \\
&= \frac{\left|\varphi_{i_k}(y) \sum\limits_{i\in\{i_1,\ldots,i_p\}} \varphi_i(z) - \varphi_{i_k}(z) \sum\limits_{i\in\{i_1,\ldots,i_p\}} \varphi_i(y)\right|}{\sum\limits_{i\in\{i_1,\ldots,i_p\}} \varphi_i(y) \sum\limits_{i\in\{i_1,\ldots,i_p\}} \varphi_i(z)} \\
&\le \frac{\left|\varphi_{i_k}(y) \sum\limits_{i\in\{i_1,\ldots,i_p\}} \varphi_i(z) - \varphi_{i_k}(z) \sum\limits_{i\in\{i_1,\ldots,i_p\}} \varphi_i(y)\right|}{m^2} \\
&\le \frac{1}{m^2} \sum_{i\in\{i_1,\ldots,i_p\}} |\varphi_{i_{k_i}}(y)\, \varphi_i(z) - \varphi_{i_k}(z)\, \varphi_i(y)| \\
&\le \frac{1}{m^2} \sum_{i\in\{i_1,\ldots,i_p\}} |\varphi_i(z)\, \varphi_{i_k}(z) - \varphi_{i_k}(z)\, \varphi_i(y)| \\
&\quad + \frac{1}{m^2} \sum_{i\in\{i_1,\ldots,i_p\}} |\varphi_i(z)\, \varphi_{i_k}(y) - \varphi_i(z)\, \varphi_{i_k}(z)| \\
&\le \frac{1}{m^2} \sum_{i\in\{i_1,\ldots,i_p\}} |\varphi_{i_k}(z)| \cdot |\varphi_i(z) - \varphi_i(y)| \\
&\quad + \frac{1}{m^2} \sum_{i\in\{i_1,\ldots,i_p\}} |\varphi_i(z)| \cdot |\varphi_{i_k}(y) - \varphi_{i_k}(z)| \\
&\le \frac{2Mp}{m^2}\, d(y, z), \qquad \text{since each } \varphi_i \text{ for } i \in \{i_1, \ldots, i_p\}
\end{aligned}$$

is obviously bounded above by M on $W(x)$ and Lipschitzean with constant equal to 1. Hence each ψ_{i_k} with $i_k \in \{i_1, \dots, i_p\}$ is Lipschitzean on $W(x)$. □

2. Absolutely Continuous Functions

In the theory of differential equations one often considers functions $x(\cdot)$, candidates to being solutions of a differential equations $x' = f(t, x)$, that are obtained as limit of some sequence $\{x_n(\cdot)\}$. To handle the passing to the limit one is lead to consider, instead of the Cauchy problem in differential form, its integral analogue $x(t) = x^0 + \int_0^t f(s, x(s))\, ds$. The study of the equivalence of these two formulations is the purpose of this paragraph.

Definition 1. A function $x\colon [\alpha, \beta] \to R^n$ is called absolutely continuous if $\forall \varepsilon > 0$, $\exists \delta$ such that, for any countable collection of disjoint subintervals $[\alpha_k, \beta_k]$ of $[\alpha, \beta]$ such that

(1) $$\sum (\beta_k - \alpha_k) < \delta,$$

we have

(2) $$\sum |x(\beta_k) - x(\alpha_k)| < \varepsilon.$$ ▲

An absolutely continuous function is at once continuous and of bounded variation (the converse is false). Any lipschitzean function is absolutely continuous.

As it is well known a function of bounded variation, hence a fortiori an absolutely continuous function, has a finite derivative except at most on a set of measure zero. However for a continuous function x, even of bounded variation, having a finite derivative x' except at most on a set of measure zero, it need not be true that

(3) $$x(\beta) - x(\alpha) = \int_\alpha^\beta x'(s)\, ds.$$

In fact the exceptional set E of points where the derivative need not exist, being of measure zero, has no influence on the value of the integral on the right-hand side of (3), in the sense that, even if the derivative were to exist on such a set, nothing would change in the integral. However the behavior of x' (hence of x) on a set of measure zero can very well have an influence on the left-hand side of (3). For it not to occurr we should have that our function x maps subsets of measure zero of (α, β) onto subsets of measure zero of R^n. This is not true in general for continuous functions of bounded variation. A counterexample is the celebrated function of Vitati (see 1943). However this is exactly what is expressed by condition (3). This remark makes plausible the following

Theorem 1. *A continuous function is the integral of its derivative if and only if it is an absolutely continuous function.* ▲

3. Some Compactness Theorems

We begin by recalling the Ascoli-Arzelà Theorem.

Theorem 1. *Let X be a Banach space and I be an interval of R. Let $\mathscr{C}(I, X)$ denote the space of continuous functions from I to X supplied with the topology of uniform convergence on compact subintervals of I. Then a subset $\mathscr{H}$ of $\mathscr{C}(I, X)$ satisfying*

$$(1)\qquad \begin{cases} \text{i) } \mathscr{H} \text{ is equicontinuous,} \\ \text{ii) } \forall t \in I, \ \mathscr{H}(t) \doteq \{x(t)\}_{x(\cdot)\in\mathscr{H}} \text{ is precompact} \end{cases}$$

is precompact in $\mathscr{C}(I, X)$. ▲

Another useful Compactness Theorem is provided by the following consequence of Banach-Steinhaus's Theorem.

Theorem 2. *Let X be a Banach space and K be a subset of its dual X^*, weakly bounded in the sense that*

$$(2)\qquad \forall x \in X, \quad \sup_{p\in K} \langle p, x\rangle < +\infty.$$

Then K is weakly relatively compact. ▲

As a consequence, we obtain the Alaoglu Theorem.

Theorem 3. *Let X be a Banach space. The unit ball of its dual X^* is weakly* compact.* ▲

We shall use the following consequence of the Ascoli-Arzelà and Alaoglu Theorems.

Theorem 4. *Let us consider a sequence of absolutely continuous functions $x_k(\cdot)$ from an interval I of R to a Banach space X satisfying*

$$(3)\qquad \begin{cases} \text{i) } \forall t \in I, \ \{x_k(t)\}_k \text{ is a relatively compact subset of } X, \\ \text{ii) there exists a positive function } c(\cdot) \in L^1(I) \text{ such that,} \\ \quad \text{for almost all } t \in I, \ \|x_k'(t)\| \le c(t). \end{cases}$$

Then there exists a subsequence (again denoted by) $x_k(\cdot)$ converging to an absolutely continuous function $x(\cdot)$ from I to X in the sense that

(4) $\begin{cases} \text{i) } x_k(\cdot) \textit{ converges uniformly to } x(\cdot) \textit{ over compact subsets of } I, \\ \text{ii) } x'_k(\cdot) \textit{ converges weakly to } x'(\cdot) \textit{ in } L^1(I,X). \end{cases}$ ▲

Proof. a) Assumption (3) ii) implies that the sequence $\{x_k(\cdot)\}$ is equicontinuous: for any $\varepsilon>0$, for any $t\geq 0$, there exists η such that

$$\int_{t-\eta}^{t+\eta} c(\tau)\,d\tau \leq \varepsilon.$$

Hence, if

$$|t-s|\leq\eta, \qquad \|x_k(t)-x_k(s)\| \leq \int_t^s \|x'(\tau)\|\,d\tau \leq \int_t^s c(\tau)\,d\tau \leq \varepsilon \qquad \text{for all } k.$$

Since $\{x_k(\cdot)\}_k$ lies in a relatively compact subset by assumption (3) i), the Ascoli-Arzelà theorem implies that $\{x_k(\cdot)\}_k$ is a relatively compact subset of $\mathscr{C}(I,X)$ supplied with the topology of uniform convergence over compact intervals. Therefore, there exists a subsequence (again denoted) $x_k(\cdot)$ converging to a continuous function $x(\cdot)\in\mathscr{C}(I,X)$ uniformly over compact intervals.

b) Since $\|x'_k(t)\|\leq c(t)$, we deduce that the sequence of functions $w_k(\cdot)$ defined by $w_k(t) \doteq x'_k(t)/c(t)$ belongs to the unit ball of $L^\infty(I,X)$ which is weakly* compact by Alaoglu's Theorem (for $L^\infty(I,X)$ is the dual of $L^1(I,X)$). Therefore, a subsequence (again denoted) $w_k(\cdot)$ converges weakly* to some function $w(\cdot)$ in $L^\infty(I,X)$.

The operator of multiplication by the function c is continuous from $L^\infty(I,X)$ (supplied with the weak* topology $\sigma(L^\infty,L^1)$) to $L^1(I,X)$ (supplied with the weakened topology $\sigma(L^1,L^\infty)$). Indeed, for any $\phi\in L^\infty$, $\psi \doteq c\phi\in L^1(I,X)$ (for $c\in L^1(I)$) and $|\langle\phi,cx\rangle| = |\langle\psi,x\rangle|$. Then the sequence of functions $x'_k = cw_k$ converges weakly in $L^1(I,X)$ to the function $v(\cdot) = c(\cdot)\,w(\cdot)$.

Since $x_k(t)-x_k(s) = \int_s^t x'_k(\tau)\,d\tau$, we deduce that $x(t)-x(s) = \int_s^t v(\tau)\,d\tau$. Hence, for almost all $t\in I$, $x'(t) = v(t)$. □

We shall need also the following compactness criterion in the space of bounded functions.

Let $\mathscr{B}(I,X)$ denote the Banach space of bounded functions from an interval I of R to the Hilbert space X, supplied with the sup norm.

We denote by $\mathscr{R}(I,X)$ the closure of the vector space of step functions. Functions of this Banach subspace, which are uniform limits of step functions, are called *regulated functions*.

We define the *oscillation* of a bounded function $x(\cdot)$ on an interval J by

(5) $$\omega_J(x) \doteq \sup_{t_1,t_2\in J} \|x(t_1)-x(t_2)\|.$$

We adapt the definition of an equicontinuous family of continuous functions by saying that a family $\mathscr{H}$ of functions $x(\cdot)\in\mathscr{B}(I,X)$ is "*equioscillating*" if

(6) $\forall \varepsilon>0$, $\exists$ a finite partition of I into subintervals J_k $(1 \leq k \leq r)$ such that, $\forall x(\cdot) \in \mathscr{H}$, $\forall k=1, \ldots, r$, $\omega_{J_k}(x) \leq \varepsilon$.

A proof analogous to the Ascoli-Arzelà Theorem yields the following compactness criterion.

Theorem 5. *Assume that a subset $\mathscr{H} \subset \mathscr{B}(I, X)$ of bounded functions satisfies*

$$\text{(7)} \quad \begin{cases} \text{i) } \mathscr{H} \text{ is equioscillating,} \\ \text{ii) } \forall t \in I, \ \mathscr{H}(t) \doteq \{x(t)\}_{x(\cdot) \in \mathscr{H}} \text{ is precompact.} \end{cases}$$

Then $\mathscr{H}$ is precompact in $\mathscr{B}(I, X)$. ▲

Proof. Let $\varepsilon>0$ be fixed. We consider the partition of I into r intervals J_k on which the oscillations of the functions of $\mathscr{H}$ are bounded by $\varepsilon/3$. Take $t_k \in J_k$. By (7) ii), the image L of $\mathscr{H}$ by the map $x(\cdot) \rightarrow (x(t_1), \ldots, x(t_r)) \in X^r$ is precompact. Hence there exist n functions $x_1(\cdot), \ldots, x_n(\cdot)$ of $\mathscr{H}$ such that

$$\text{(8)} \qquad \inf_{1 \leq i \leq n} \sup_{1 \leq k \leq r} \|x(t_k) - x_i(t_k)\| \leq \frac{\varepsilon}{3} \quad \text{for all } x(\cdot) \in \mathscr{H}.$$

Since $\mathscr{H}$ is equioscillating, we deduce that $\mathscr{H}$ is covered by the n balls $x_i(\cdot) + \varepsilon B$ in $\mathscr{B}(I, X)$. □

4. Weak Convergence and Asymptotic Center of Bounded Sequences

We shall associate with a bounded sequence its "asymptotic center" introduced by Edelstein, which *always exists and is unique*, and which coincides with the weak limit whenever it exists. It could also have been called the "shadow limit" or "virtual limit". In this section, X is a Hilbert space.

For simplicity, we consider D to be either N or R_+ and bounded sequences $T \in D \rightarrow x_T \in X$. We associate with the *bounded sequence* of elements $x_T \in X$ the function ϕ defined on X by

$$\text{(1)} \qquad \phi(y) \doteq \limsup_{T \rightarrow \infty} \|x_T - y\|^2 = \inf_{S \in D} \sup_{T>S} \|x_T - y\|^2.$$

It is finite and nonnegative since the sequence is bounded.

Lemma 1. *The function ϕ is locally Lipschitzean, strictly convex and satisfies*

$$\text{(2)} \qquad \lim_{\|y\| \rightarrow \infty} \phi(y) = \infty. \qquad ▲$$

Proof. It is obvious that ϕ is locally Lipschitzean. The strict convexity follows from the identity

$$\|x-\alpha y-\beta z\|^2=\alpha\|x-y\|^2+\beta\|x-z\|^2-2\alpha\beta\|y-z\|^2$$

when $\alpha+\beta=1$. The last property is also obvious. □

Therefore, *the function ϕ has a unique minimum.*

Definition 1. The unique point $x_\infty\in X$ that minimizes ϕ on X is called the *asymptotic center* of the bounded sequence of elements x_T. ▲

This definition is motivated by the following result:

Proposition 1. *The asymptotic center of a bounded sequence of elements $x_T\in X$ belongs to the closed convex hull of its weak cluster points.*

Moreover, if the bounded sequence of elements x_T converges weakly, then the limit coincides with the asymptotic center. ▲

Proof. Let x_∞ be the asymptotic center of the bounded sequence of elements x_T and y_∞ be the projection of x_∞ onto the closed convex hull C of the weak cluster points of the sequence $\{x_T\}_T$. It is non-empty because any bounded sequence is weakly relatively compact. There exists a subsequence of elements $x_{T'}$, such that

$$\phi(y_\infty)=\lim_{T'\to\infty}\|x_{T'}-y_\infty\|^2.$$

Hence

$$\limsup_{T'\to\infty}\|x_{T'}-x_\infty\|^2\leq\phi(x_\infty).$$

Also there exists a subsequence of elements $x_{T''}$ that converges to some $z\in C$.

Therefore,

$$\limsup_{T''\to\infty}\langle y_\infty-x_\infty, y_\infty-x_{T''}\rangle=\langle y_\infty-x_\infty, y_\infty-z\rangle\leq 0 \tag{3}$$

because y_∞ is the projection of x_∞ to C.

From the identity

$$\|x_T-x_\infty\|^2=\|x_T-y_\infty\|^2+\|y_\infty-x_\infty\|^2+2\langle x_T-y_\infty, y_\infty-x_\infty\rangle$$

we deduce that

$$\phi(x_\infty)\geq\phi(y_\infty)+\|x_\infty-y_\infty\|^2\geq\phi(y_\infty).$$

From the uniqueness of the minimum of ϕ, it follows that $x_\infty=y_\infty\in C$. The second part of the proposition is an immediate consequence of the first. □

Now, we make more precise the above result.

Proposition 2. *Let us consider a bounded sequence of elements* x_T. *We denote by* C *the closed convex hull of its weak cluster points and by* N *the subset defined by*

$$N \doteq \{y \in X \mid \phi(y) = \lim_{T \to \infty} \|x_T - y\|^2\}. \tag{4}$$

Then, if $N \cap C \neq \emptyset$, *this intersection reduces to the asymptotic center* x_∞:

$$N \cap C = \{x_\infty\}. \tag{5}$$ ▲

Proof. Let $y \in N \cap C$; we shall prove that $y = x_\infty$ and, for that purpose, we shall check that $\phi(z) \geq \phi(y)$ for all $z \in X$ [in such a way that y is the unique minimum x_∞ of ϕ].

Let w be any weak cluster point of $\{x_T\}$: Hence w is the weak limit of a subsequence $\{x_{T'}\}$ of $\{x_T\}$. Passing to the limit as $T' \to \infty$ in the identity

$$\|x_{T'} - z\|^2 = \|x_{T'} - y\|^2 + \|y - z\|^2 + 2\langle x_{T'} - y, y - z\rangle$$

we obtain, since $y \in N$,

$$\phi(z) \geq \phi(y) + \|y - z\|^2 + 2\langle w - y, y - z\rangle.$$

Since this inequality holds for all weak cluster points, it also holds for all $w \in C$. In particular, we can take $w = y$, for $y \in C$. Hence:

$$\phi(z) \geq \phi(y) + \|y - z\|^2 \geq \phi(y).$$ □

As a corollary, we obtain the following sufficient condition for weak convergence.

Proposition 3. *If the weak cluster points of a bounded sequence of elements* x_T *belong to the subset* N *defined by* (4), *then this sequence converges weakly to its asymptotic center.* ▲

Proof. Indeed, $C \cap N$ has a nonempty intersection.

Proposition 4. *Let us consider two bounded sequences of elements* x_T *and* y_T. *Let* N *be the subset defined by* (4) *and* C *be the closed convex hull of the weak cluster points of* $\{x_T\}_T$. *If the weak cluster points of* $\{y_T\}_T$ *belong to* $C \cap N$, *then* y_T *converges weakly to the asymptotic center of the sequence* $\{x_T\}_T$. ▲

Proof. Since the weak cluster points of $\{y_T\}$ belong to $C \cap N$, this set is nonempty and, by *Proposition 2*, reduces to x_∞, the asymptotic center of $\{x_T\}_T$. Since the sequence $\{y_T\}_T$ is relatively compact and since it has a unique cluster point, x_∞, then y_T converges weakly to x_∞ as $T \to \infty$. □

5. Closed Convex Hulls and the Mean-Value Theorem

The convex subsets of a vector space play a fundamental role in analysis.

Definition 1. Let X be a real vector space. A subset K of X is called "convex" if for every $x, y \in K$ and every $\lambda \in [0,1]$, $\lambda x + (1-\lambda) y \in K$. A subset K of X is a "convex cone" with vertex 0 if for every $x, y \in K$ and every $\lambda, \mu \geq 0$, $\lambda x + \mu y \in K$. We say that $x = \sum_{i=1}^{n} \lambda_i x_i$ is a "convex combination of elements x_i" if $\lambda_i \geq 0$ and $\sum_{i=1}^{n} \lambda_i = 1$. ▲

We agree to say that the empty set is convex.

The following proposition summarizes the elementary properties of convex sets.

Proposition 1. a) *A subset $K \subset X$ is convex if and only if every convex combination of elements $x_i \in K$ belongs to K.*

b) *Under a linear mapping, the image and the inverse image of a convex set are convex.*

c) *The intersection of any family of convex sets is convex.*

d) *Every product of convex sets is convex.*

e) *If K and L are convex sets, $K+L$ is convex.* ▲

We shall now give some topological properties of convex sets.

Theorem 1. *Let K be a convex subset of a normed space. Suppose that $x_0 \in \mathring{K}$ and $x_1 \in \overline{K}$. Then*

$$\forall \lambda \in]0,1], \quad \lambda x_0 + (1-\lambda) x_1 \in \mathring{K}. \tag{1}$$ ▲

Proof. Let $\lambda \in]0,1]$ and $x_\lambda \doteq \lambda x_0 + (1-\lambda) x_1$. We are going to show that x_λ belongs to an open set $\tilde{B}$ contained in K. Since $x_0 \in \mathring{K}$, there exists $\varepsilon > 0$ such that $\mathring{B}(x_0, \varepsilon) \subset K$. Since $x_1 \in \overline{K}$, there exists $y \in K$ such that $y \in \mathring{B}(x_1, \varepsilon\lambda/(1-\lambda))$. Consider the open ball

$$\tilde{B} = \mathring{B}(\lambda x_0 + (1-\lambda) y, \lambda \varepsilon). \tag{2}$$

Then x_λ belongs to $\tilde{B}$ since

$$\|x_\lambda - \lambda x_0 - (1-\lambda) y\| = (1-\lambda) \|x_1 - y\| < \frac{(1-\lambda)\varepsilon\lambda}{(1-\lambda)} = \varepsilon\lambda.$$

Moreover, $\tilde{B} \subset K$: Indeed, if $z \in \tilde{B}$, we conclude that

$$\left\| \left(\frac{1}{\lambda}\right)(z - (1-\lambda) y) - x_0 \right\| < \frac{\lambda\varepsilon}{\lambda} = \varepsilon,$$

and, consequently, that $(1/\lambda)(z-(1-\lambda)\,y) \in \mathring{B}(x_0, \varepsilon) \subset K$. Since $y \in K$ and since K is convex, we obtain that

$$\lambda\left(\frac{1}{\lambda}\right)(z-(1-\lambda)\,y)+(1-\lambda)\,y = z \in K. \qquad \square$$

Theorem 2. *Let K be a convex subset of a normed space X. The closure $\overline{K}$ of K and the interior $\mathring{K}$ of K are convex. Moreover, if $\mathring{K} \neq \emptyset$, $\overline{K}$ is the closure of $\mathring{K}$ and $\mathring{K}$ is the interior of $\overline{K}$.* ▲

Proof. a) Let us show that $\overline{K}$ is convex. Let λ be given in $]0,1[$, and take $x = \lim x_n$ and $y = \lim y_n$ in the closure $\overline{K}$ of K (where x_n and y_n belong to K). Since $\lambda x+(1-\lambda)\,y$ is the limit of $\lambda x_n+(1-\lambda)\,y_n$ (which belongs to K since K is convex), we conclude that $\lambda x+(1-\lambda)\,y \in \overline{K}$. Thus $\overline{K}$ is convex.

b) If x_0 and $x_1 \in \mathring{K}$, the elements $x_\lambda = \lambda x_0+(1-\lambda)\,x_1$ belong to $\mathring{K}$ when $\lambda \in]0,1[$ according to Theorem 1. Hence the interior $\mathring{K}$ of K is convex.

c) If $\mathring{K} \neq \emptyset$, $\overline{K}$ is the closure of $\mathring{K}$ since if $x_0 \in \mathring{K}$ and if $x_1 \in \overline{K}$, $x_1 = \lim_{n\to\infty} x_n$, where $x_n \doteq (1/n)\,x_0+(1-1/n)\,x_1 \in \check{K}$ according to Theorem 1.

d) Let us show that if $\mathring{K} \neq \emptyset$, then $\mathring{K}$ is the interior of $\overline{K}$. Since $\mathring{K} \subset \mathring{K}$ it is sufficient to show that if $x_0 \in \check{K}$, then $x_0 \in \check{K}$. By hypothesis, there exists $\varepsilon > 0$ such that $B(x_0, \varepsilon) \subset \overline{K}$. Moreover, since $x_0 \in \overline{K}$ and since $\mathring{K} \neq \emptyset$, then $\overline{K} = \overline{\mathring{K}}$ according to c) and there exists x_1 belonging to $\mathring{K} \cap B(x_0, \varepsilon)$. Thus, since $x_1 \in B(x_0, \varepsilon)$, $2x_0 - x_1$ belongs to $B(x_0, \varepsilon) \subset \overline{K}$ since $\|2x_0 - x_1 - x_0\| = \|x_0 - x_1\| < \varepsilon$. Thus $x_0 = (1/2)\,x_1 + (1/2)(2x_0 - x_1)$, where $x_1 \in K$ and $2x_0 - x_1 \in \overline{K}$, and, therefore, $x_0 \in \mathring{K}$ using Theorem 1. $\square$

Definition 2. Let K be a subset of a normed space X. We call the "convex hull" of K the intersection of all the convex sets containing K and the "closed convex hull" of K the intersection of all the closed convex sets containing K. ▲

In other words, the convex hull $\mathrm{co}(K)$ of K is the smallest convex set containing K, and the closed convex hull $\overline{\mathrm{co}}(K)$ of K is the smallest closed convex set containing K.

We state the following results.

Proposition 2. *Let K be a subset of a normed space X. Then the convex hull $\mathrm{co}(K)$ of K is the set of convex combinations $\sum \lambda_i x_i$ of elements x_i of K and the closed convex hull $\overline{\mathrm{co}}(K)$ is the closure of $\mathrm{co}(K)$.* ▲

Proposition 3. *Let us denote by*

$$(3) \qquad S^n \doteq \left\{\lambda = \{\lambda_i\}_{i=1,\dots,n} \in R^n \text{ such that } \lambda_i \geq 0,\ \forall\, i \text{ and } \sum_{i=1}^{n} \lambda_i = 1\right\}.$$

Consider a sequence of n convex sets K_i. Then the convex hull of the union of the K_i's is the set of convex combinations $\sum_{i=1}^{n} \lambda_i x_i$ as λ runs over S^n and x_i runs over K_i for all i:

(4) $$\operatorname{co}\left(\bigcup_{i=1}^{n} K_i\right) = \left\{\sum_{i=1}^{n} \lambda_i x_i\right\}_{\substack{\lambda \in S^n \\ x_i \in K_i}}. \qquad \blacktriangle$$

Proposition 4. *Let K_i $(1 \le i \le n)$ be n compact convex sets of a normed space X. Then the convex hull $\operatorname{co}\left(\bigcup_{i=1}^{n} K_i\right)$ of the union of the K_i's is compact (thus equal to the closed convex hull of the union).* ▲

Corollary 1. *The convex hull of a finite set is compact.* ▲

More generally, the convex hull of a compact subset is compact. To show this we need the following theorem of Carathéodory.

Proposition 5. *Let K be a subset of R^n. The convex hull* co(*K*) *of K is the set of elements of the form*

$$x = \sum_{i=1}^{n+1} \lambda_i x_i \quad \text{where } \lambda \in S^{n+1} \text{ and } x_i \in K \ (i = 1, \ldots, n+1). \qquad \blacktriangle$$

Proof. Every element $x \in \operatorname{co}(K)$ is of the form $x = \sum_{i=1}^{k} \lambda_i x_i$ where $\lambda_i \ge 0$ and $\sum_{i=1}^{k} \lambda_i = 1$. Let us show that if $k > n+1$, we can find among the representations of this type one for which $\lambda_i = 0$ for an index *i*. Indeed if $k > n+1$, the elements $x_1 - x_k, x_2 - x_k, \ldots, x_{k-1} - x_k$ are linearly dependent. Thus there exist $\mu_1, \ldots, \mu_{k-1}$ not all zero such that

$$\mu_1(x_1 - x_k) + \mu_2(x_2 - x_k) + \ldots + \mu_{k-1}(x_{k-1} - x_k) = 0.$$

Set $\mu_k = -\sum_{i=1}^{k-1} \mu_i$. Then we have that $\sum_{i=1}^{k} \mu_i x_i = 0$ and $\sum_{i=1}^{k} \mu_i = 0$. Let *I* be the (nonempty) set of indices *i* such that $\mu_i > 0$. Set

$$t \doteq \min_{i \in I} \frac{\lambda_i}{\mu_i} = \frac{\lambda_{i_0}}{\mu_{i_0}}, \qquad v_i = \lambda_i - t\,\mu_i.$$

The scalars v_i are positive or zero, $v_{i_0} = \lambda_{i_0} - t\,\mu_{i_0} = 0$ and

$$\sum_{i=1}^{k} v_i = \sum_{i=1}^{k} \lambda_i - t \sum_{i=1}^{k} \mu_i = 1 - 0 = 1.$$

Moreover,

$$x = \sum_{i=1}^{k} \lambda_i x_i - t \sum_{i=1}^{k} \mu_i x_i = \sum_{i=1}^{k} v_i x_i.$$

Consequently, we have shown that x is a convex combination of k points.

By repeating this procedure, we can successively write x as a convex combination of $k-1, k-2, \ldots, n+1$ points of K, which proves the proposition. □

Proposition 6. *The convex hull of a compact subset of a finite dimensional space is compact.* ▲

Proof. We consider an infinite sequence of elements $x^k \in \mathrm{co}(K)$. Applying Proposition 5 we can write $x^k = \sum_{i=1}^{n+1} \lambda_i^k x_i^k$ where $\lambda^k \in S^{n+1}$ and $x_i^k \in K$. The sets S^{n+1} and K being compact, we can extract from these sequences subsequences λ^{k_p} and $x_i^{k_p}$, which converge, respectively, to $\lambda \in S^{n+1}$ and $x_i \in K$. Thus the sequence of elements $x^{k_p} = \sum_{i=1}^{n+1} \lambda_i^{k_p} x_i^{k_p}$ converges to $x = \sum_{i=1}^{n+1} \lambda_i x_i \in \mathrm{co}(K)$. We have, therefore, shown that $\mathrm{co}(K)$ is compact. □

We now state and prove a very useful result.

Theorem 3 (Mean-value). *Let* $]a, b[$ *be an interval,* $v(\cdot)$ *a function of* $L^1_{\mathrm{loc}}(a, b; X)$ *with values in a subset* $K \subset X$. *Then for every* t_0, t_1 *in* $]a, b[$, *we have*

$$\int_{t_0}^{t_1} v(s)\, ds \in (t_1 - t_0)\, \overline{\mathrm{co}}(K). \tag{5}$$ ▲

Proof. The map $v(\cdot)$ can be approximated in L^1_{loc} by a sequence of simple functions $v_n(\cdot) \doteq \sum_i y_{n,i} \chi_{E_i}(\cdot)$, where χ_E denotes the characteristic function of a measurable set E. Fix t_1 and t_0 in $[a, b]$. Then it is clear that

$$\int_{t_0}^{t_1} v_n(t)\, dt \in (t_1 - t_0)\, \mathrm{co}(K). \tag{6}$$

By letting n go to infinity, the mean value property ensues. □

6. Lower Semicontinuous Convex Functions and Projections of Best Approximation

Let X be a Hilbert space. We shall, most of the time, use functions V from X to $R \cup \{+\infty\}$, i.e., real-valued functions which are allowed to take the value $+\infty$.

We define the domain of V by

$$\text{Dom}\, V \doteq \{x \in X \mid V(x) < +\infty\} \tag{1}$$

so that the restriction of V to its domain is real-valued. We say that V is a *proper* function if its domain is nonempty, i.e., if it is not the constant $+\infty$.

When V is a real-valued function on a subset K of X, we associate its extension V_K defined by

$$V_K(x) \doteq V(x) \quad \text{when } x \in K, \qquad V_K(x) \doteq \{+\infty\} \quad \text{when } x \notin K. \tag{2}$$

We say that the subset

$$\text{Ep}(V) \doteq \{(x, \lambda) \in X \times R \mid \lambda \geq V(x)\} \tag{3}$$

is the epigraph of V. We observe that

$$\text{Ep}(\sup_{i \in I} V_i) = \bigcap_{i \in I} \text{Ep}(V_i). \tag{4}$$

We recall that

(5) A function V from X to $R \cup \{+\infty\}$ is convex if and only if its epigraph is convex

and that

(6) A function V is lower semicontinuous if and only if its epigraph is closed.

The latter property is equivalent to either

$$\forall \lambda \in R, \quad \{x \in X \mid V(x) \leq \lambda\} \quad \text{is closed} \tag{7}$$

or to

$$\forall x_0 \in \text{Dom}\, V, \quad V(x_0) \leq \liminf_{x \to x_0} V(x). \tag{8}$$

We denote by ψ_K the *indicator* of K, defined by

$$\psi_K(x) \doteq 0 \quad \text{when } x \in K \quad \text{and} \quad \psi_K(x) \doteq +\infty \quad \text{when } x \notin K. \tag{9}$$

Theorem 1. *Let V be a proper lower semicontinuous convex function from a Hilbert space X to $R \cup \{+\infty\}$. Then, for every $x \in X$, there exists a unique $\bar{x} \in X$ satisfying*

$$V(\bar{x}) + \tfrac{1}{2}\|\bar{x} - x\|^2 = \inf_{y \in X}(V(y) + \tfrac{1}{2}\|y - x\|^2). \tag{10}$$

Furthermore, it is characterized by the inequalities

$$\forall y \in X, \quad V(\bar{x}) - V(y) + \langle \bar{x} - x, \bar{x} - y \rangle \leq 0. \tag{11}$$ ▲

Proof. Let us set

$$W(x) \doteq \inf_{y \in X}(V(y) + \tfrac{1}{2}\|y - x\|^2). \tag{12}$$

It is possible to prove that $W(x) > -\infty$.

a) Let us prove the existence of $\bar{x}$, solution to $W(x)=V(\bar{x})+\frac{1}{2}\|\bar{x}-x\|^2$. To this end, we consider a minimizing sequence of elements y^n, which satisfy: $V(y^n)+\frac{1}{2}\|y^n-x\|^2 \leq W(x)+\frac{1}{n}$. It is a Cauchy sequence:

$$\begin{aligned}\|y^n-y^m\|^2 &= 2\|y^n-x\|^2+2\|y^m-x\|^2-4\left\|\frac{y^n+y^m}{2}-x\right\|^2\\ &\leq 4\left[\frac{1}{n}+\frac{1}{m}+2W(x)-V(y^n)-V(y^m)\right]+8\left[V\left(\frac{y^n+y^m}{2}\right)-W(x)\right]\\ &\leq 4\left[\frac{1}{n}+\frac{1}{m}+2V\left(\frac{y^n+y^m}{2}\right)-V(y^n)-V(y^m)\right]\\ &\leq 4\left(\frac{1}{n}+\frac{1}{m}\right) \quad \text{(for } V \text{ is convex).}\end{aligned}$$

Hence y^n converges to an element $\bar{x}$. Since V is lower semicontinuous, it follows that

$$V(\bar{x})+\tfrac{1}{2}\|\bar{x}-x\|^2 \leq \liminf_{n\to\infty} V(y^n)+\lim_{n\to\infty}\tfrac{1}{2}\|y^n-x\|^2 \leq W(x).$$

b) Let $\bar{x}$ be a solution to the minimization problem (10). Then, by taking $y=\bar{x}+\theta(\bar{x}-z)$, we deduce that

$$\begin{aligned}&V(\bar{x})+\tfrac{1}{2}\|\bar{x}-x\|^2\\ &\quad\leq(1-\theta)V(\bar{x})+\theta V(z)+\tfrac{1}{2}(\|\bar{x}-x\|^2+2\theta\langle\bar{x}-x,\bar{x}-z\rangle+\theta^2\|\bar{x}-z\|^2).\end{aligned}$$

After simplification and division by $\theta>0$, we obtain

$$V(\bar{x})-V(z)\leq\langle(\bar{x}-x),\bar{x}-z\rangle+\frac{\theta}{2}\|\bar{x}-z\|^2.$$

By letting $\theta\to 0$, we have proved that (11) holds true.

c) Conversely, inequalities (11) imply that $\bar{x}$ minimizes $y\to V(y)+\frac{1}{2}\|x-y\|^2$, because $\frac{1}{2}\|\bar{x}-x\|^2-\frac{1}{2}\|y-x\|^2\leq\langle\bar{x}-x,\bar{x}-y\rangle$.

d) For proving uniqueness, let $\bar{\bar{x}}$ be another solution to inequalities

$$\text{(13)} \qquad \forall z\in X, \quad V(\bar{\bar{x}})-V(z)+\langle\bar{\bar{x}}-x,\bar{\bar{x}}-z\rangle\leq 0.$$

We take $y\doteq\bar{\bar{x}}$ in (11), $z\doteq\bar{x}$ in (13) and we add the two inequalities. We deduce that $\|\bar{x}-\bar{\bar{x}}\|^2\leq 0$, i.e., that $\bar{x}=\bar{\bar{x}}$. □

As a consequence, we obtain the Best-Approximation Theorem.

Corollary 1. *Let K be a closed convex subset of a Hilbert space X. We can associate to any $x\in X$ a unique element $\pi_K(x)\in K$ satisfying*

$$\text{(14)} \qquad \|x-\pi_K(x)\|=\min_{y\in K}\|x-y\|.$$

It is characterized by the following variational inequalities:

$$\text{(15)} \qquad \langle\pi_K(x)-x,\pi_K(x)-y\rangle\leq 0 \quad \forall y\in K.$$

▲

Proof. We take $V \doteq \psi_K$, the indicator of K. □

Definition 1. The map π_K is called the *projector* (of best approximation) onto K. It is non-expansive:

$$\|\pi_K(x) - \pi_K(y)\| \leq \|x - y\| \tag{16}$$

and monotone

$$\langle \pi_K(x) - \pi_K(y), x - y \rangle \geq 0. \tag{17}$$

▲

Definition 2. If K is a convex subset, we set

$$m(K) \doteq \pi_K(0) \tag{18}$$

which is the *element of K with minimal norm.* ▲

Proposition 1. *Let K be a nonempty closed convex set and d_K^2 the function defined on X by $d_K^2(x) \doteq \inf\{\|x-y\|^2 \mid y \in K\}$. Then d_K^2 is continuously differentiable and*

$$\nabla d_K^2(x) = 2(x - \pi_K(x)). \tag{19}$$

▲

Proof. The projector π_K is non-expansive. Hence, since $x \to \|x\|^2$ is continuously differentiable

$$\begin{aligned}
2\langle x - \pi_K(x), v \rangle &= \lim_{\theta \to 0+} \frac{\|x - \pi_K(x+\theta v) + \theta v\|^2 - \|x - \pi_K(x+\theta v)\|^2}{\theta} \\
&\leq \liminf_{\theta \to 0+} \frac{d_K^2(x+\theta v) - d_K^2(x)}{\theta} \leq \limsup_{\theta \to 0+} \frac{d_K^2(x+\theta v) - d_K^2(x)}{\theta} \\
&\leq \lim_{\theta \to 0+} \frac{\|x + \theta v - \pi_K(x)\|^2 - \|x - \pi_K(x)\|^2}{\theta} = 2\langle x - \pi_K(x), v \rangle.
\end{aligned}$$

□

We devote special attention to the case when K is a closed convex cone T.

Proposition 2. *If T is a closed convex cone of a Hilbert space X, then $\pi_T(x)$ is characterized by*

$$\begin{cases} \text{i)} & \pi_T(x) \in T, \\ \text{ii)} & \langle x - \pi_T(x), y \rangle \leq 0 \quad \forall y \in T, \\ \text{iii)} & \langle x - \pi_T(x), \pi_T(x) \rangle = 0. \end{cases} \tag{20}$$

The projector π_T satisfies

$$\begin{cases} \text{i)} & \pi_T(\lambda x) = \lambda \pi_T(x) \quad \forall x \in X, \ \forall \lambda > 0, \\ \text{ii)} & \|x\|^2 = \|\pi_T(x)\|^2 + \|(1 - \pi_T)(x)\|^2, \\ \text{iii)} & \|\pi_T(x)\| \leq \|x\|. \end{cases} \tag{21}$$

▲

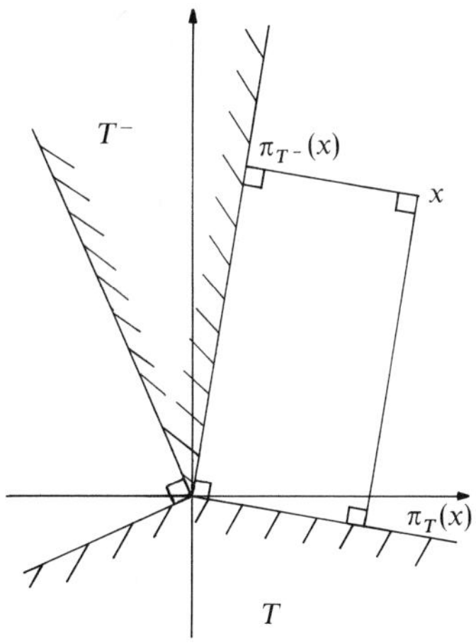

Fig. 3. Orthogonal projection onto polar closed convex cones T and T^-

Proof. a) It is clear that conditions (15) of *Corollary 1* imply conditions (20). Conversely, let us choose $\bar{x}\in T$ satisfying, for all $y\in T$, $\langle\bar{x}-x, \bar{x}-y\rangle\leq 0$. Since T is a cone, we can choose $y=0$ and $y=2\bar{x}$. This yields $\langle\bar{x}-x,\bar{x}\rangle=0$ and thus, $\langle\bar{x}-x, y\rangle\geq 0$ for all $y\in T$.

b) If $\lambda>0$, we can write

$$\left\langle x-\frac{1}{\lambda}\pi_T(\lambda x), \frac{1}{\lambda}\pi_T(\lambda x)\right\rangle=\frac{1}{\lambda^2}\langle\lambda x-\pi_T(\lambda x), \pi_T(\lambda x)\rangle=0$$

and, for all $y\in T$,

$$\left\langle x-\frac{1}{\lambda}\pi_T(\lambda x), y\right\rangle=\frac{1}{\lambda}\langle\lambda x-\pi_T(\lambda x), y\rangle\leq 0.$$

Therefore, by (20), $\frac{1}{\lambda}\pi_T(\lambda x)=\pi_T(x)$.

c) Consequently, we can develop the following:

$$\begin{aligned}\|x\|^2&=\|x-\pi_T(x)+\pi_T(x)\|^2\\&=\|x-\pi_T(x)\|^2+\|\pi_T(x)\|^2+2\langle x-\pi_T(x), \pi_T(x)\rangle\\&=\|x-\pi_T(x)\|^2+\|\pi_T(x)\|^2\quad\text{by (20) ii).}\end{aligned}$$

Hence (21) iii) follows. □

If T is a closed subspace of X, then the projector π_T is linear, with a norm equal to one.

Definition 3. When T is a closed convex cone or a closed subspace, the projector π_T is called the *orthogonal projector* onto T. If T is a cone, then

$$T^- \doteq \{p \in X^* \mid \langle p, x\rangle \le 0 \text{ for all } x \in T\}$$

is called the (negative) polar cone. ▲

The main fact is that $1-\pi_T$ is the orthogonal projector on T^-.

Proposition 3. *Let T be a closed convex cone and $N \doteq T^-$ be its negative polar cone. Then the projector π_N onto N equals $1-\pi_T$. Actually, we obtain:*

(22) *if $x=y+z$ where $y\in T$, $z\in N$ and $\langle y,z\rangle=0$, then $y=\pi_T(x)$ and $z=\pi_N(x)$.*

Furthermore,

(23) $$N=\pi_T^{-1}(0) \quad \text{and} \quad T=\pi_N^{-1}(0).$$ ▲

Proof. a) We consider the characterization (20) of the projector π_T. It can be written

(24) $$\begin{cases} \text{i)} \ \langle (1-\pi_N)(x), y\rangle \le 0 \quad \text{for all } y\in N,\\ \text{ii)} \ \pi_N(x)\in N=T^-,\\ \text{iii)} \ \langle \pi_N(x), (1-\pi_N)(x)\rangle=0.\end{cases}$$

Therefore, $\pi_N(x)$ is the projection of x onto N.

b) The elements $y=\pi_T(x)\in T$ and $z=\pi_N(x)\in N$ satisfy the conditions $x=y+z$ and $\langle y,z\rangle=0$. *Conversely*, if these conditions are satisfied, we obtain $y\in T$, $\langle x-y, v\rangle=\langle z,v\rangle\le 0$ for all $v\in T$ and $\langle x-y,y\rangle=\langle z,y\rangle=0$. Hence, by characterization (20), $y=\pi_T(x)$.

c) If $x\in N$, then $\pi_T(x)=x-\pi_N(x)=0$. *Conversely*, if $\pi_T(x)=0$, then $x=\pi_N(x)$ and thus, $x\in N$. □

The following proposition shall play an important role in the study of differential variational inequalities (Section 5.6).

Proposition 4. *Let F be a compact convex subset, T be a closed convex cone and $N=T^-$ be its negative polar cone. Then*

(25) $$\pi_T(F)\subset F-N.$$

The elements of minimal norm are equal

(26) $$m(\pi_T(F))=m(F-N)$$

and satisfy

(27) $$\sup_{z\in -F}(z, m(\pi_T(F)))+\|m(\pi_T(F))\|^2\le 0.$$

If $m(F)\in T$, then $m(F)=m(\pi_T(F))$. ▲

Proof. a) Since we have $\pi_T(x)=x-\pi_N(x)\in F-N$ when $x\in F$, we obtain the first inclusion.

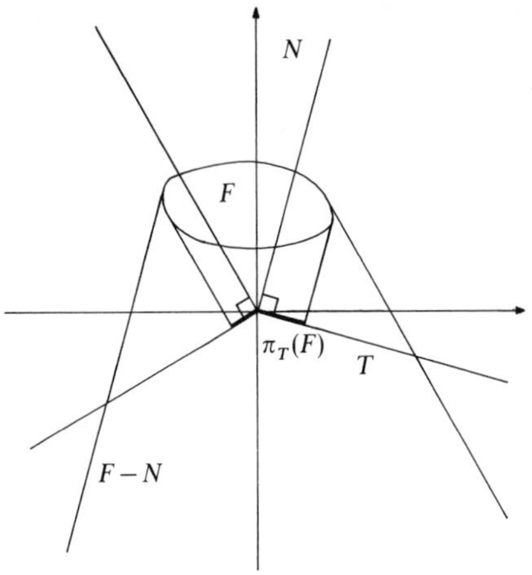

a

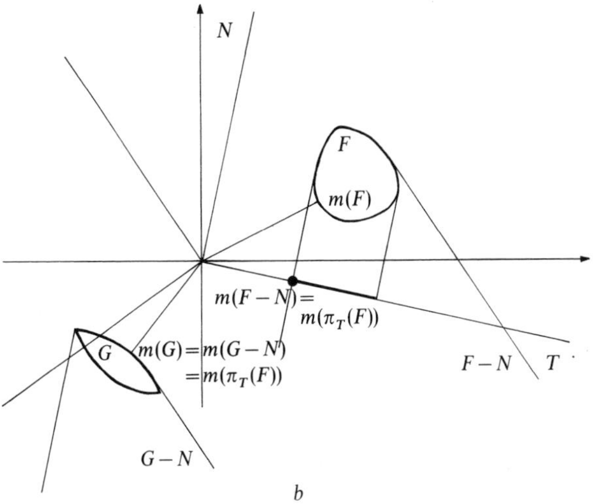

b

Fig. 4. *a* Projection of F onto T; *b* Elements of minimal norms of projections

b) Let us write $m(F-N)=x_0-y_0$ where $x_0 \in F$ and $y_0 \in N$. We have the following equalities:

$$\|x_0-y_0\| = \min_{x\in F}\min_{y\in N}\|x-y\| = \min_{x\in F}\|x-\pi_N(x)\|$$
$$= \min_{x\in F}\|\pi_T(x)\| = \min_{y\in N}\|x_0-y\| = \|\pi_T(x_0)\|.$$

Hence $\pi_T(x_0)=m(\pi_T(F))=m(F-N)$.

c) Since $m(\pi_T(F))$ is the projection of 0 onto $F-N$, we have

$$\sup_{z\in -F}(z, m(\pi_T(F))) + \|m(\pi_T(F))\|^2 = \sup_{y\in F}\langle m(\pi_T(F)), m(\pi_T(F))-y\rangle$$
$$= \sup_{y\in F,\, z\in N}\langle m(\pi_T(F)), m(\pi_T(F))-(y-z)\rangle \le 0$$

for $\langle m(\pi_T(F)), z\rangle \ge 0$ for all $z\in N$.

d) Since $m(F)=\pi_F(0)$, we know that

$$\|m(F)\|^2 \le \langle m(F), y\rangle = \langle m(F), \pi_T(y)\rangle + \langle m(F), \pi_N(y)\rangle \qquad \text{for all } y\in F.$$

Since $m(F)\in T$, we deduce that $\|m(F)\|^2 \le \langle m(F), \pi_T(y)\rangle \le \|m(F)\|\,\|\pi_T(y)\|$ for all $y\in F$. Hence $m(F)=m(\pi_T(F))$. □

Proposition 5. *Let X and Y be Hilbert spaces, $F\subset X$ and $G\subset Y$ be closed convex subsets and $B\in\mathscr{L}(Y,X)$ be an injective continuous linear operator with closed range. We denote by $B^- \doteq (B^*B)^{-1}B^*\in\mathscr{L}(X,Y)$ its orthogonal left inverse* (see for instance Aubin [1979a], Chapter 4, Proposition 5.3, p. 87) *and by π_G the projector onto G when Y is supplied with the "initial norm" $\|y\|_Y \doteq \|B^- y\|_X$. Then*

(28) $$m(F-BG) = m[(1-B\pi_G B^-)(F)]. \qquad \blacktriangle$$

Proof. Let us set $m(F-BG)\doteq \bar{x}-B\bar{y}$ where $\bar{x}\in F$ and $\bar{y}\in G$. Then

$$\|\bar{x}-B\bar{y}\| = \inf_{x\in F}\inf_{y\in G}\|x-By\|.$$

Let $\tilde{y}\in G$ achieve the minimum of $\|x-By\|$ over G. By taking $z=\tilde{y}+\theta(y-\tilde{y})$, writing that $\|x-Bz\|^2-\|x-B\tilde{y}\|^2\ge 0$, dividing by $\theta>0$ and letting θ converge to 0, we deduce that

(29) $$\langle B\tilde{y}-x, B\tilde{y}-By\rangle \le 0 \qquad \forall y\in G.$$

This can be written

(30) $$\langle B\tilde{y}-B(B^*B)^{-1}B^*x, B\tilde{y}-By\rangle \le 0 \qquad \forall y\in G$$

and thus, $\tilde{y}=\pi_G(B^*B)^{-1}B^*x)=\pi_G(B^-x)$. Hence, we obtain equality

(31) $$\|\bar{x}-B\bar{y}\| = \inf_{x\in F}\|x-B\pi_G(B^-x)\| = \|\bar{x}-B\pi_G(B^-\bar{x})\|,$$

from which we deduce our proposition. □

7. A Concise Introduction to Convex Analysis

Conjugate Functions

Definition 1. Let V be a function from a Hilbert space X to $R \cup \{+\infty\}$ whose domain Dom V is nonempty. The function V^*: $V^* \to R \cup \{+\infty\}$ associating to every $p \in X^*$:

(1) $$V^*(p) \doteq \sup_{x \in X} [\langle p, x \rangle - V(x)] \in]-\infty, +\infty]$$

is called the "conjugate function" of V. ▲

Note that V^* never takes the value $-\infty$ since if $x_0 \in \text{Dom } V$, then

$$-\infty < \langle p, x_0 \rangle - V(x_0) \leq V^*(p) \quad \text{for all } p \in X^*.$$

Obviously, V^* is a lower semicontinuous convex function. If we introduce the "biconjugate" V^{**}, we see that

(2) $$V(x) \geq V^{**}(x).$$

The importance of the concept of conjugate functions lies in the following theorem.

Theorem 1. *A function V is lower semicontinuous and convex if and only if $V = V^{**}$.* ▲

Proof. See for instance Aubin [1979a], Chapter 10, Theorem 1.1, p. 211.

Remark. We always have the following inequality

(3) $$\forall x \in X, \quad \forall p \in X^*, \quad \langle p, x \rangle \leq V(x) + V^*(p)$$
called the *Fenchel inequality.*

Example. Let $\|x\|$ be the norm of a Hilbert space, and V be the function defined by

(4) $$V(x) \doteq \frac{1}{\alpha} \|x\|^{\alpha}, \quad 1 < \alpha < +\infty.$$

Let $\|p\|_* = \sup_{\|x\| \leq 1} \langle p, x \rangle = \sigma_B(p)$ be the dual norm (B denotes the unit ball). Let $\alpha^* = \dfrac{\alpha}{\alpha - 1}$. Then we can prove that

(5) $$V^*(p) = \frac{1}{\alpha^*} \|p\|_*^{\alpha^*}.$$

Support Functions

Definition 2. Let ψ_K be the indicator of K. Its conjugate function is called the support function of K and is denoted by $\sigma_K(\cdot)$: For all $p \in X^*$,

$$\sigma_K(p) \doteq \psi_K^*(p) = \sup_{x \in K} \langle p, x \rangle \in R \cup \{+\infty\}. \tag{6}$$ ▲

It is a convex lower semicontinuous function that is positively homogeneous.

Its domain is called the *barrier cone* of K; it is a convex cone, not necessarily closed.

Definition 3. Let K be a nonempty subset of a Hilbert space X. We define its negative polar cone K^- by

$$K^- = \{p \in X^* \text{ such that } \langle p, x \rangle \leq 0 \text{ for all } x \in K\}. \tag{7}$$ ▲

We set $K^+ = -K^-$.

It is obvious that K^- is a *closed convex cone* of X^*.

Corollary 1. *Let $K \subset X$ be a closed convex subset of a Hilbert space X. Then*

$$K = \{x \in X \mid \forall p \in X^*, \ \langle p, x \rangle \leq \sigma_K(p)\}. \tag{8}$$ ▲

Proof. Since ψ_K is convex and lower semicontinuous, we write that $\psi_K = \sigma_K^*$. □

This result is very useful since it allows to characterize closed convex sets by a family of linear inequalities.

Corollary 2. *Let σ be a proper positively homogeneous lower semicontinuous convex function from X^* to $R \cup \{+\infty\}$. Then it is the support function of the closed convex subset K defined by*

$$K = \{x \in X \mid \forall p \in X^*, \ \langle p, x \rangle \leq \sigma(p)\}. \tag{9}$$ ▲

Proof. We observe that $\sigma^* = \psi_K$. Hence $\sigma = \sigma^{**} = \psi_K^* \doteq \sigma_K$.

We recall the main properties of support functions. □

Proposition 1. a) *Continuous linear functionals are the support functions of singletons.*

b) *If T is a cone, then*

$$\sigma_T(p) = \begin{cases} 0 & \text{if } p \in T^-, \\ +\infty & \text{if } p \notin T^-. \end{cases} \tag{10}$$

c) *If K and L are two non-empty subsets and* $\alpha, \beta > 0$, *then*

(11) $$\sigma_{\alpha K+\beta L}(p) = \alpha\,\sigma_K(p) + \beta\,\sigma_L(p).$$

d) *If A belongs to* $\mathscr{L}(X, Y)$ *and* $K \subset X$ *is a nonempty subset, then*

(12) $$\sigma_{A(K)}(q) = \sigma_K(A^* q).$$

If $K = \bigcup_{i \in I} K_i$, *then*

(13) $$\sigma_K(p) = \sup_{i \in I} \sigma_{K_i}(p).$$ ▲

Corollary 3. *Let K be a nonempty subset and N be a cone of a Hilbert space X. Then*

$$\sigma_{K+N}(p) = \begin{cases} \sigma_K(p) & \text{when } p \in N^- \\ +\infty & \text{when } p \notin N^- \end{cases}$$ *and consequently,*

(14) $$(K+N)^- = K^- \cap N^-.$$ ▲

Corollary 4. a) *If T is a non-empty subset of a Hilbert space X, then* T^{--} *is the closed convex cone spanned by T, in the sense that it is the smallest closed convex cone containing T.*

b) *If K is a nonempty subset and N is a cone, the closed convex cone spanned by* $K+N$ *is equal to* $(K^- \cap N^-)^-$.

c) *If K is a nonempty subset of X and if* $A \in \mathscr{L}(X, Y)$, *then*

(15) $$A(K)^- = A^{*-1}(K^-)$$

and consequently, the closed convex cone spanned by $A(K)$ *is equal to* $[A^{*-1}(K^-)]^-$. ▲

Differentiability and Sub-Differentiability

Let V be a function from X to $R \cup \{+\infty\}$.

Definition 4. We set, when $x \in \operatorname{Dom} V$,

(16) $$DV(x)(v) = \lim_{h \to 0+} \frac{V(x+hv) - V(x)}{h} \in [-\infty, +\infty].$$

We say that $DV(x)(v)$ is the *derivative from the right of V at x in the direction v.* If $DV(x)(v)$ exists for all directions v, we say that V is *differentiable from the right* at x. If moreover

(17) $$v \to DV(x)(v) = \langle \nabla V(x), v \rangle$$

is continuous and linear, we say that $\nabla V(x) \in X^*$ is the *gradient* of V at x and that V is "*Gâteaux differentiable*" (in short, "*differentiable*"). We say

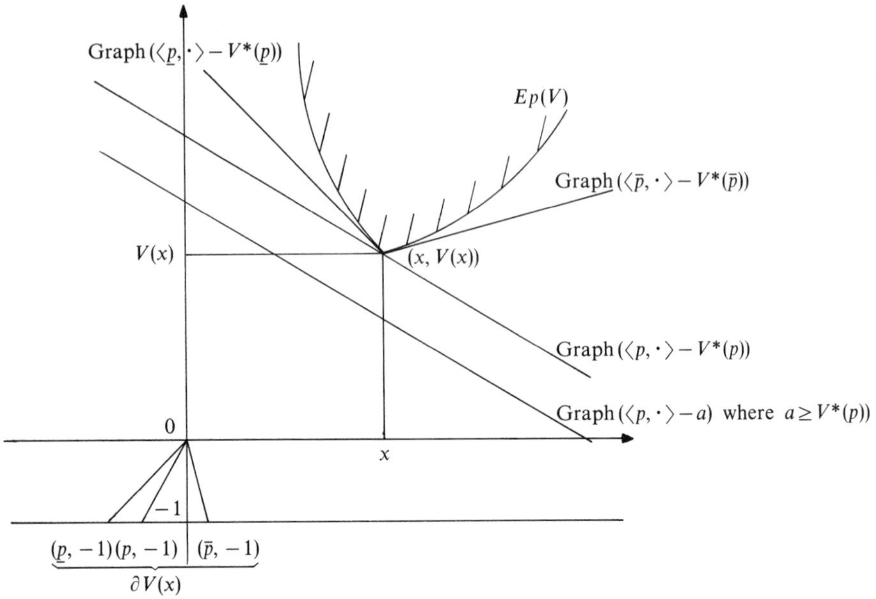

Fig. 5. Conjugate function and sub-differential

$$p\in\partial V(x)\Leftrightarrow(p,-1)\in N_{Ep(V)}(x, V(x))$$

that the closed convex subset (possibly empty)

(18) $$\partial V(x) \doteq \{p\in X^* \mid \forall v\in X,\ \langle p, v\rangle \le DV(x)(v)\}$$

is the "*sub-differential*" of V at x. The elements p of $\partial V(x)$ are called the "sub-gradients of V at x". ▲

Remark. If $v\to DV(x)(v)$ is convex and lower semicontinuous, then it is the support function of $\partial V(x)$.

Proposition 2. *If V is differentiable at x, then $\partial V(x)=\{\nabla V(x)\}$.* ▲

Proof. Indeed $p\in\partial V(x)$ if and only if $\langle p, v\rangle \le \langle \nabla V(x), v\rangle$ for all $v\in X$; this is equivalent to saying that $p=\nabla V(x)$. □

When V is a convex function, we obtain the following properties.

Proposition 3. *If V is a convex function from X to $R\cup\{+\infty\}$ and if $x\in\operatorname{Dom} V$, then $v\to DV(x)(v)$ is convex positively homogeneous and the following inequalities hold:*

(19) $$-\infty \le V(x)-V(x-v) \le DV(x)(v) \le V(x+v)-V(x) \le +\infty.$$

Furthermore

(20) $$\partial V(x)=\{p\in X^* \text{ such that } V(x)-V(y)\le\langle p,x-y\rangle \ \ \forall y\in X\}.$$ ▲

Proof. See Aubin [1979a], Chapter 10, § 2 and 3. □

Corollary 5. *Let V be a proper convex function from X to $R\cup\{+\infty\}$. The following propositions are equivalent*

(21) $$\begin{cases} \text{a) } \bar{x}\in X \quad \text{minimizes } V \text{ on } X, \\ \text{b) } 0\in\partial V(\bar{x}), \\ \text{c) } \forall v\in X, \quad 0\le DV(\bar{x})(v). \end{cases}$$ ▲

Corollary 6. *Let V be a convex lower semicontinuous function with nonempty domain. The following statements are equivalent:*

(22) $$\begin{cases} \text{a) } p\in\partial V(x), \\ \text{b) } x\in\partial V^*(p), \\ \text{c) } \langle p,x\rangle=V(x)+V^*(p). \end{cases}$$ ▲

The above proposition shows that the set-valued map $\partial V^*(\cdot)$ is the inverse of $\partial V(\cdot)$.

Example. Let K be a nonempty closed convex subset. Then

(23) $$\partial\psi_K(x)=\{p\in X^* \text{ such that } \langle p,x\rangle=\max_{y\in K}\langle p,y\rangle\}.$$

Definition 5. We set

(24) $$\partial\psi_K(x)\doteq N_K(x)$$

and we say that $N_K(x)$ is the *normal cone* to K at x. ▲

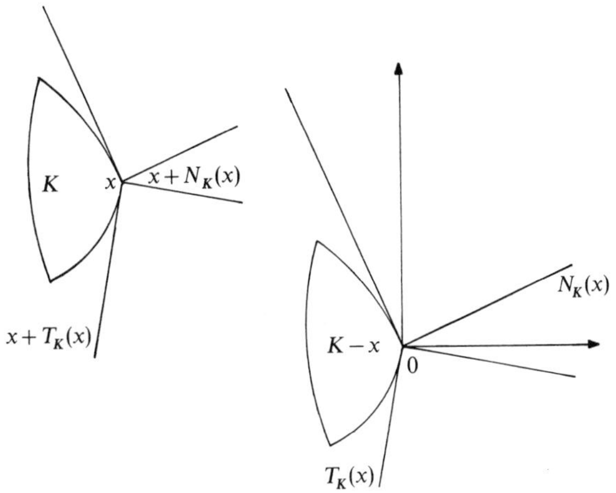

Fig. 6. Tangent and Normal cones to a closed convex subset K

The normal cones to closed convex subsets shall play an important role in viability theory, since their polar cones are *tangent cones* to K at x. See Section 5.1.

If V is continuous at x, we can prove that $\partial V(x)$ is nonempty and bounded.

Theorem 2. *Let us assume that V is convex and lower semicontinuous and that* $\mathrm{Int}(\mathrm{Dom}\, V)$ *is nonempty.*

a) *For any* $x \in \mathrm{Int}(\mathrm{Dom}\, V)$, *$\partial V(x)$ is nonempty and bounded and $DV(x)(\cdot)$ is its support function.*

b) *The map* $(x, v) \in \mathrm{Int}(\mathrm{Dom}\, V) \times X \to DV(x)(v)$ *is upper semicontinuous (and thus, $\partial V(\cdot)$ is upper semicontinuous on* $\mathrm{Int}(\mathrm{Dom}\, V)$).

c) *The following regularity property holds*

$$DV(x)(v) = \limsup_{\substack{y \to x \\ h \to 0+}} \frac{V(y+hv) - V(y)}{h}. \tag{25}$$ ▲

Proof. a) Since V is convex and lower semicontinuous on $\mathrm{Int}(\mathrm{Dom}\, V)$, it is continuous. (This is a consequence of Baire's theorem; see [Aubin 1979a], Chapter 5, § 3, Theorem 4, p. 189.)

b) If $x \in \mathrm{Int}(\mathrm{Dom}\, V)$ and if $v \in X$ satisfies $x + \theta v \in X$ for $|\theta| < 1$, then $\theta \to \dfrac{V(x+\theta v) - V(x)}{\theta}$ is increasing and satisfies

$$V(x) - V(x-v) \le \frac{V(x+\theta v) - V(x)}{\theta} \le V(x+v) - V(x).$$

Hence

$$DV(x)(v) = \inf_{\theta > 0} \frac{V(x+\theta v) - V(x)}{\theta}. \tag{26}$$

c) Since the functions $(x, v) \to V(x+\theta v) - V(x)$ are continuous, we deduce that $DV(x)(v)$ is upper semicontinuous.

d) Since V is continuous at x, there exists $\eta > 0$ such that $V(x+\eta v) - V(x) \le 1$ for all $v \in B$. Hence

$$DV(x)(v) \le \frac{V(x+\eta v) - V(x)}{\eta} \le \frac{1}{\eta} \|v\|.$$

This proves that $v \to DV(x)(v)$ is continuous and thus that it is the support function of $\partial V(x)$. This also implies that $\partial V(x) \subset \dfrac{1}{\eta} B$.

e) We prove property (25). Let $\varepsilon > 0$ and $\eta \in]0, 1[$ be fixed. Since $(\theta, y) \to \dfrac{V(y+\theta v) - V(y)}{\theta}$ is continuous at (η, x), there exist $\alpha, \beta > 0$ such that if $\theta < \eta + \alpha$ and $\|y - x\| < \beta$, we have

$$(V(y+\theta v) - V(y))/\theta \le (V(x+\eta v) - V(x))/\eta + \varepsilon.$$

Therefore

$$\limsup_{\substack{y\to x\\ \theta\to 0+}} \frac{V(y+\theta v)-V(y)}{\theta} = \inf_{\substack{\alpha,\eta>0\\ \beta>0}} \sup_{\substack{\theta\le\alpha+\eta\\ \|y-x\|<\beta}} \frac{V(y+\theta v)-V(y)}{\theta}$$

$$\le \inf_{\eta>0} \frac{V(x+\eta v)-V(x)}{\eta} + \varepsilon = DV(x)(v)+\varepsilon.$$

The result follows by letting $\varepsilon\to 0$. □

In general, even when the interior of the domain of V is empty, we can still prove that the set of points where V is subdifferentiable is dense in the domain of V.

Theorem 3. *Let V be a proper lower semicontinuous convex function from a Hilbert space X to $R\cup\{+\infty\}$. Then*

a) *The map $x\to x+\partial V(x)$ is surjective.*

b) *The set of points where V is subdifferentiable is dense in the domain of V.* ▲

Proof. a) The first statement follows from Theorem 5.1 which states that for all $x\in X$, there exists a unique $\bar{x}\in X$ such that $x-\bar{x}$ belongs to $\partial V(\bar{x})$, i.e., that $\bar{x}\in X$ is a solution to the inclusion $x\in\bar{x}+\partial V(\bar{x})$.

b) Let $\lambda>0$ converge to 0. By replacing V by λV in Theorem 5.1, we deduce that there exists a unique x_λ solution to the inclusion $x\in x_\lambda+\partial V(x_\lambda)$. We check that when $x\in \operatorname{Dom} V$, x_λ converges to x when λ converges to 0. Take p in $\operatorname{Dom} V^*$, which is nonempty by *Theorem 1*. Hence, by definition of x_λ,

$$\lambda V(x_\lambda)+\tfrac{1}{2}\|x_\lambda-x\|^2 \le \lambda V(x)+\tfrac{1}{2}\|x-x\|^2=\lambda V(x).$$

Since $\langle p,x_\lambda\rangle \le V(x_\lambda)+V^*(p)$, we deduce that

$$\begin{aligned}\tfrac{1}{2}\|x_\lambda-x\|^2 &\le \lambda(V(x)-V(x_\lambda))\\ &\le \lambda(V(x)+V^*(p)-\langle p,x\rangle+\langle p,x-x_\lambda\rangle)\\ &\le \lambda(V(x)+V^*(p)-\langle p,x\rangle+\lambda\|p\|^2)+\tfrac{1}{4}\|x_\lambda-x\|^2\end{aligned}$$

because $ab\le a^2/4\lambda+b^2\lambda$. Hence, when λ converges to 0, so does $\|x_\lambda-x\|^2$ because

$$\tfrac{1}{4}\|x_\lambda-x\|^2 \le \lambda(V(x)+V^*(p)-\langle p,x\rangle+\lambda\|p\|^2).$$ □

Subdifferentials of convex functions do enjoy enough properties. We summarize some of them in the following statement.

Theorem 4. *Let A be a continuous linear operator from a Hilbert space X to a Hilbert space Y and $U\colon X\to R\cup\{+\infty\}$, $V\colon Y\to R\cup\{+\infty\}$ be proper lower* *semicontinuous convex functions. We assume that*

(27) $$0 \in \operatorname{Int}(A \operatorname{Dom} U - \operatorname{Dom} V).$$

Then, for all $x \in \operatorname{Dom} U \cap A^{-1} \operatorname{Dom} V$, *we have*

(28) $$\partial(U + V \circ A)(x) = \partial U(x) + A^* \partial V(A x).$$ ▲

Let us mention explicitely several useful consequences: U and V denote proper lower semicontinuous convex functions and K and M denote non-empty closed convex sets.

a) If $0 \in \operatorname{Int}(\operatorname{Im} A - \operatorname{Dom} V)$, then

(29) $$\partial(V \circ A)(x) = A^* \partial V(A x).$$

b) If $0 \in \operatorname{Int}(\operatorname{Dom} U - \operatorname{Dom} V)$, then

(30) $$\partial(U + V)(x) = \partial U(x) + \partial V(x).$$

c) If $0 \in \operatorname{Int}(K - \operatorname{Dom} V)$, then

(31) $$\partial(V|_K)(x) = \partial V(x) + N_K(x).$$

d) If $0 \in \operatorname{Int}(M - A \operatorname{Dom} U)$, then

(32) $$\partial(U|_{A^{-1}(M)})(x) = \partial U(x) + A^* N_M(A x).$$

The last formula implies that if $\bar{x} \in A^{-1}(M)$ minimizes a lower semicontinuous convex function U on $A^{-1}(M)$ and if $0 \in \operatorname{Int}(M - A \operatorname{Dom} U)$, then $\bar{x}$ is a solution to the inclusion

(33) $$0 \in \partial U(\bar{x}) + A^* N_M(A \bar{x})$$

and conversely.

Theorem 5. *Let us consider n convex lower semicontinuous functions* V_i *from* X *to* $]-\infty, +\infty]$. *Let V be defined by*

(34) $$V(x) \doteq \max_{i=1,\ldots,n} V_i(x)$$

and

(35) $$J(x) \doteq \{i = 1, \ldots, n \mid V_i(x) = V(x)\}.$$

Let us assume that the n functions V_i *are continuous at x. Then*

(36) $$\partial V(x) = \overline{\operatorname{co}}\left(\bigcup_{i=1}^{n} \partial V_i(x)\right).$$ ▲

Chapter 1. Set-Valued Maps

Introduction

We gather in this chapter the properties of set-valued maps which are needed for the study of differential inclusions.

The first issue we face is the problem of defining continuity of set-valued maps. In the case of single valued maps f from X to Y, continuous functions are characterized by two equivalent properties:

a) for any neighborhood $N(f(x_0))$ of $f(x_0)$, there exists a neighborhood $N(x_0)$ of x_0 such that $f(N(x_0)) \subset N(f(x_0))$;

b) for any generalized sequence of elements x_μ converging to x_0, the sequence $f(x_\mu)$ converges to $f(x_0)$.

These two properties can be adapted to the case of set-valued maps from X to Y: they become:

A) for any neighborhood $N(F(x_0))$ of $F(x_0)$, there exists a neighborhood $N(x_0)$ of x_0 such that $F(N(x_0)) \subset N(F(x_0))$;

B) for any generalized sequence of elements x_μ converging to x_0 and for any $y_0 \in F(x_0)$, there exists a sequence of elements $y_\mu \in F(x_\mu)$ that converges to y_0.

In the case of set-valued maps, these two properties are no longer equivalent. We call *upper semicontinuous* maps those that satisfy property A, *lower semicontinuous* maps those that satisfy property B and *continuous* maps the ones that satisfy both properties A and B.

We also introduce the concept of lipschitzean set-valued maps. After having defined these concepts and presented their properties, we list the examples of set-valued maps which will be used later in the book.

a) *Parametrized maps* of the form

$$F(x) \doteq f(x, U) \doteq \{f(x, u)\}_{u \in U}$$

provided by control theory, where f is a single valued map from $X \times U$ to Y. They naturally enjoy many properties which make them quite attractive. So, the purpose of some approximation procedures is to show that a given map

F, if it is not parametrizable, can still be written as

$$F(x) = \bigcap_{n \geq 0} f_n(x, U_n)$$

a decreasing intersection of parametrized maps.

We shall see that this is the case when F is upper semicontinuous with closed convex values (Approximation Theorem) and, above all, that this is also the case for the map that associates with any state x the set of trajectories of a differential inclusion issued from x.

b) *Marginal maps*, provided by optimization theory, which associate to a parameter $y \in Y$ the subset

$$M(y) \doteq \{\bar{x} \in G(y) \mid W(\bar{x}, y) = \inf_{x \in G(y)} W(x, y)\}$$

of solutions to a minimization problem.

c) *Feed-back maps*, which shall play an important role in Viability Theory, which associate to the state x of a system a subset of controls of the form

$$C(x) \doteq \{u \in U(x) \mid f(x, u) \in T(x)\}.$$

d) *closed convex cone valued maps*, examples of which are provided by tangent and normal cones to convex subsets.

e) *Set-valued maps with closed convex graph*, for which the Banach closed graph Theorem can be extended: they are lower semicontinuous on the interior of their domain (Robinson-Ursescu Theorem). This important theorem is at the origin of many important results on the calculus of tangent cones and subdifferentials of convex functions.

We proceed by studying more carefully upper semicontinuous maps with compact convex images to a vector space supplied with its weak topology. They enjoy a very handy characterization: if $\sigma(F(x), p) \doteq \sup_{y \in F(x)} \langle p, y \rangle$ denotes the support function of $F(x)$, they satisfy:

$$\forall p \in Y^*, \quad x \to \sigma(F(x), p) \quad \text{is upper semicontinuous.}$$

We call *upper hemicontinuous* the maps which satisfy this property. We then prove the *Convergence Theorem*, which shall play a crucial role for proving the existence of solutions to differential inclusions.

This theorem states that if F is upper semicontinuous from a Hilbert space X to the closed *convex* subsets of a Hilbert space Y, if a sequence of functions $x_n(\cdot)$ in $\mathscr{C}(0, T; X)$ converges strongly to $x(\cdot)$, if a sequence of functions $y_n(\cdot)$ in $L^1(0, T; X)$ converges *weakly* to $y(\cdot)$ and if for almost all t, $y_n(t)$ "almost belongs" to $F(x_n(t))$ in some precise sense, then $y(t)$ belongs to $F(x(t))$ for almost all t.

We shall then deal with the *selection* problem, i.e. the problem of finding single valued maps whose graph is contained in the graph of a given set-valued map. We shall first investigate the properties of some explicit selection schemes and, in doing so, we shall consider the parametrization problem i.e. the possibility of describing a set-valued map $F(x)$ as $\{f(x,u)|u\in U\}$. There U can be thought of as a parameter set. Next we shall consider a well known theorem of Michael yielding existence of a continuous selection for lower semicontinuous maps with convex images.

Upper semicontinuous maps with convex images in general do not have continuous selections. We shall present two ways to overcome this difficulty. First we shall show that this class of maps admits continuous single-valued *almost selections* in the sense that for every $\varepsilon>0$ there exists a continuous f_ε whose graph belongs to an ε-ball around the graph of the given map. This property is used to prove Kakutani's Fixed Point Theorem, the analogue for set-valued maps of Schauder's Theorem.

An alternative approach to the study of uppersemicontinuous convex valued maps is by inbedding the graph of such a map F into the graphs of a *decreasing* sequence of continuous convex valued maps. The approximating maps F_n can be chosen to satisfy

$$\forall x\in X, \qquad F_n(x) \doteq \sum_{i\in I} \psi_i(x)\, C_i$$

where the functions ψ_i are non negative and locally Lipschitzean and the sets C_i are closed and convex.

In general the existence of selections or of approximate selections is strictly connected with the convexity of the images of the given map. Other class of mappings enjoying one or the other such property and arising when considering differential equations or inclusions will be presented in Chapter 2. We shall conclude the present chapter by presenting some measurable selection theorems for maps with non necessarily convex images. These measurable selection theorems, that we shall repeatedly use in the book and that are needed in several branches of applied mathematics, have their origin in the pioneering works of Von Neumann and of Filippov.

1. Set-Valued Maps and Continuity Concepts

Set-Valued Maps

Let X and Y be two sets. A set valued map F from X to Y is a map that associates with any $x\in X$ a subset $F(x)$ of Y. The subsets $F(x)$ are called the *images* or the *values* of F.

The subset

$$\text{(1)} \qquad \operatorname{Dom}(F) \doteq \{x\in X \,|\, F(x) \neq \emptyset\}$$

is called the *domain* of F.

When $\mathrm{Dom}(F)=X$, we say that the map F is *strict*.

In order to avoid the trivial map $x\to\emptyset$, we say that a map F is *proper* if its domain is nonempty.

When $K\subset X$ is a nonempty subset and when F is a set valued map from K to Y with *nonempty* values, we can (and sometimes shall) "extend" it to the set valued map F_K defined on the whole set X by

$$F_K(x)\doteq\begin{cases}F(x) & \text{when } x\in K\\ \emptyset & \text{when } x\notin K\end{cases}\tag{2}$$

whose domain is K: $\mathrm{Dom}(F_K)=K$.

In this book, we shall most of the time *assume or check* that the images $F(x)$ are nonempty for all $x\in X$, i.e., assume that the map F is strict. Otherwise, we shall mention that F is a proper set-valued map, i.e., that its domain is nonempty and may be contained in X.

It is very convenient to characterize a set-valued map by its *graph*: the graph of F is the subset of pairs (x, y) where $y\in F(x)$:

$$\mathrm{Graph}(F)\doteq\{(x,y)\in X\times Y\,|\,y\in F(x)\}.\tag{3}$$

Conversely, let $\mathscr{F}$ be a non-empty subset of $X\times Y$; it defines a set-valued map in the following way: We set $F(x)\doteq\{y\in Y\,|\,(x,y)\in\mathscr{F}\}$.

Then $\mathscr{F}$ is the graph of the set-valued map F. The *range* $R(F)$ is, by definition, the subset

$$R(F)\doteq\bigcup_{x\in X}F(x).\tag{4}$$

The *inverse* F^{-1} of the set-valued map F from X to Y is the set-valued map from $R(F)$ to X defined by

$$x\in F^{-1}(y)\quad\text{if and only if }\ y\in F(x).\tag{5}$$

If F is associated to a non-empty subset $\mathscr{F}$ of $X\times Y$, the set-valued map F^{-1} is defined by

$$F^{-1}(y)=\{x\in X\,|\,(x,y)\in\mathscr{F}\}.$$

Note that $\mathrm{Dom}(F)$ is the projection of $\mathscr{F}$ onto X and that $R(F)=\mathrm{Dom}(F^{-1})$ is the projection of $\mathscr{F}$ onto Y.

A set-valued map F whose graph is a cone is called a *process* (or a positively homogeneous map), because $F(\lambda x)=\lambda F(x)$ for all $\lambda\geq 0$.

We shall say that a set-valued map is "compact", "bounded" if its range $R(F)$ is a compact (respectively bounded) subset of Y. We shall say that it is "*locally compact*" (locally bounded) if for every $x\in D(F)$, there exists a neighborhood of x on which F is compact (respectively bounded).

Continuity Concepts

In what follows X and Y denote Hausdorff topological spaces. Let F be a set-valued map with non-empty values.

Definition 1. We say that F is *upper semicontinuous* (*u.s.c.*) at $x^0 \in X$ if for any open N containing $F(x^0)$ there exists a neighborhood M of x^0 such that $F(M) \subset N$.

We say that F is upper semicontinuous if it is so at every $x^0 \in X$. ▲

Proposition 1. *Let F and G be two set-valued maps from X to Y and from Y to Z respectively. Define GF by*

$$GF(x) \doteq \bigcup_{y \in F(x)} G(y). \tag{6}$$

If F and G are upper semi-continuous, so is GF. ▲

Proof. Let $N \doteq N(GF(x_0))$ be an open neighborhood of $GF(x_0)$. Since G is u.s.c., the subset $M \doteq \{y \in Y$ such that $G(y) \subset N\}$ is open. Since F is also u.s.c., there exists a neighborhood $P \doteq P(x_0)$ of x_0 such that $F(P) \subset M$. Hence $GF(P) \subset N$.

Proposition 2. *The graph of an u.s.c. set-valued map with closed values from X to Y is closed.* ▲

Proof. Let us consider a sequence of elements (x_μ, y_μ) of the graph of F that converges to some $(x, y) \in X \times Y$. Since F is u.s.c., we can associate to any closed neighborhood $N(F(x))$ an index μ_0 such that $\forall \mu \geqslant \mu_0$, $y_\mu \in N(F(x))$. Hence, y belongs to every neighborhood of $F(x)$. □

There is a partial converse to this result.

Theorem 1. *Let F and G be two set-valued maps from X to Y such that, $\forall x \in X$, $F(x) \cap G(x) \neq \emptyset$. We suppose that*

$$\begin{cases} \text{i) } F \text{ is upper semicontinuous at } x_0, \\ \text{ii) } F(x_0) \text{ is compact}, \\ \text{iii) the graph of } G \text{ is closed.} \end{cases} \tag{7}$$

Then the set-valued map $F \cap G\colon x \to F(x) \cap G(x)$ is upper semi-continuous at x_0. ▲

Proof. Let $N \doteq N(F(x_0) \cap G(x_0))$ be an open neighborhood of $F(x_0) \cap G(x_0)$. We have to find a neighborhood $N(x_0)$ of x_0 such that, $\forall x \in N(x_0)$, $F(x) \cap G(x) \subset N$.

If $N \supset F(x_0)$, this follows from the upper semi-continuity of F. If $N \not\supset F(x_0)$, then we introduce the subset

$$K \doteq F(x_0) \cap \operatorname{Comp}(N)$$

that is compact (since $F(x_0)$ is compact). Let $P \doteq \operatorname{graph}(G)$, which is closed. For any $y \in K$, we have $y \notin G(x_0)$ and thus, $(x_0, y) \notin P$. Since P is closed, there exist open neighborhoods $N_y(x_0)$ and $N(y)$ such that $P\cap(N_y x_0)\times N(y))=\emptyset$. Therefore,

$$(8)\qquad \forall x\in N_y(x_0),\quad G(x)\cap N(y)=\emptyset.$$

Since K is compact, it can be covered by n neighborhoods $N(y_i)$. The union $M\doteq\bigcup_{i=1}^{n} N(y_i)$ is a neighborhood of K and $M\cup N$ is a neighborhood of $F(x_0)$. Since F is u.s.c. at x_0, there exists a neighborhood $N_0(x_0)$ of x_0 such that

$$(9)\qquad \forall x\in N_0(x_0),\quad F(x)\subset M\cup N.$$

We set $N(x_0)\doteq N_0(x_0)\cap\bigcap_{i=1}^{n} N_{y_i}(x_0)$. Hence, when $x\in N(x_0)$, properties (8) and (9) imply that

$$(10)\qquad \begin{cases}\text{i)}\ F(x)\subset M\cup N,\\ \text{ii)}\ G(x)\cap M=\emptyset.\end{cases}$$

Therefore, $F(x)\cap G(x)\subset N$ when $x\in N(x_0)$. □

Corollary 1. *Let G be a set-valued map from X to a compact space Y whose graph is closed. Then G is upper semicontinuous.* ▲

Proof. We take F to be defined by $F(x)\doteq Y$ for all $x\in X$ and we apply the above theorem. □

This corollary provides a very useful tool for proving that a given set valued map is upper semi-continuous. Remark, though, that the assumption of compactness of Y is essential. Consider the map F, from $\mathbb{R}$ into the subsets of $\mathbb{R}^2$, defined by $F(\xi)\doteq\{(x,y)\mid y=\xi x\}$. Then F has closed graph: assume that $\xi_n\to\xi^*$, $(x_n,y_n)\to(x^*,y^*)$, $(x_n,y_n)\in F(\xi_n)$. Then $y_n=\xi_n x_n$ and passing to the limit, $y^*=\xi^* x^*$, i.e., $(x^*,y^*)\in F(\xi^*)$. However, consider, for instance, $\xi^0=0$. For no $\varepsilon>0$ and for no $\xi\neq 0$ is $F(\xi)\subset F(0)+\varepsilon B$.

Proposition 3. *Let F be an upper semi-continuous map with compact values from a compact space X to Y. Then $F(X)$ is compact.* ▲

Proof. We shall prove that any open covering $\{U_\lambda\}_{\lambda\in\Lambda}$ of $F(X)$ contains a finite covering. Since each image $F(x)$ is compact, it can be covered by a finite number $n(x)$ of such U_λ. We write:

$$F(x)\subset U_x\doteq\bigcup_{1\le i\le n(x)} U_{\lambda_i}.$$

Since F is u.s.c., there exists an open neighborhood $N(x)$ of x such that $F(N(x))\subset U_x$. But X is compact: it is contained in the union of p such neighborhoods $N(x_j)$. Thus

$$F(X)\subset\bigcup_{1\le j\le p} F(N(x_j))\subset\bigcup_{1\le j\le p} U_{x_j}=\bigcup_{1\le j\le p}\ \bigcup_{1\le i\le n(x_j)} U_{\lambda_i}.$$

Hence, from the open cover $\{U_\lambda\}_{\lambda\in\Lambda}$ we have selected a finite open subcover $\{U_{\lambda_i} | 1 \le j \le p;\ 1 \le i \le n(x_j)\}$. This proves that $F(X)$ is compact. □

Proposition 4. *Let X be a topological space, Y a metric space, F from X into the subsets of Y, be u.s.c. Then inverse images of open sets are borellians.* ▲

Proof. Let V be open. In a metric space, every open set is an F_σ, i.e. $V = \bigcup_{n\ge 0} K_n$, where each K_n is closed. Then $F^{-1}(K_n)$ is closed: if a point x is such that $F(x)\cap K_n = \emptyset$, there is an open N containing $F(x)$ such that $N\cap K_n = \emptyset$. Images of points close to x are in N, and have empty intersection with K_n, i.e. the complement of $F^{-1}(K_n)$ is open. Then $F^{-1}(V) = \bigcup_{n\ge 0} F^{-1}(K_n)$ is an F_σ. □

In the case of single valued maps, as we have remarked, continuity can be characterized also in terms of generalized sequences or, equivalently, by requiring that inverse images of open sets be open. This second point of view leads to the following definition.

Definition 2. We say that F is *lower semicontinuous* (in short l.s.c.) *at* $x^0 \in X$ if for any $y^0 \in F(x^0)$ and any neighborhood $N(y^0)$ of y^0, there exists a neighborhood $N(x^0)$ of x^0 such that

$$\forall x \in N(x^0), \qquad F(x)\cap N(y^0) \neq \emptyset. \tag{11}$$

We say that F is *lower semicontinuous* if it is lower semi-continuous at every $x^0 \in X$.

The above definition could be phrased as follows: given any generalized sequence x_μ converging to x^0 and any $y^0 \in F(x^0)$, there exists a generalized sequence $y_\mu \in F(x_\mu)$ that converges to y^0. When X and Y are metric, this last characterization holds true with usual (i.e., countable) sequences.

Remark. The mapping F_+, from $\mathbb{R}$ into its subsets, defined by $F_+(0) = [-1, +1]$ and $F_+(x) = \{0\}$, $x \neq 0$, is u.s.c. while it is not l.s.c.

The mapping F_-, defined by $F_-(0) = \{0\}$, $F_-(x) = [-1, +1]$, $x \neq 0$, is l.s.c. while it is not u.s.c.

Definition 3. A set valued map F from X to Y is said to be continuous at $x_0 \in X$ if it is both u.s.c. and l.s.c. at x_0. It is said to be continuous if it is continuous at every point $x \in X$. ▲

An important class of continuous set-valued maps are the Lipschitzean set-valued maps. We denote by

$$B(K,\eta) \doteq \{y \in X \,|\, d(y,K) \le \eta\}$$

the ball of radius η around the subset K. When K is a subset of a normed space, we can write $B(K,\eta)=K+\eta B$.

Definition 4. Let F be a set valued map from a metric space X to a metric space Y. We say that F is *locally lipschitzean* if for any $x_0 \in X$, there exist a neighborhood $N(x_0)\subset X$ and a constant $L\geq 0$ (the lipschitz constant) such that

$$\forall x, x' \in N(x_0), \quad F(x)\subset B(F(x'), L d(x,x')). \tag{12}$$

It is *Lipschitzean* if there exists $L\geq 0$ such that

$$\forall x, x' \in X, \quad F(x)\subset B(F(x'), L d(x,x')). \tag{13}$$ ▲

Remark that the above definition is symmetric in x and x'.

Proposition 5. *Let X and Y be metric spaces, G from X into the subsets of Y be lower semicontinuous and $g\colon X\to Y$ be continuous. Let $x\to\varepsilon(x)$, from X into R_+, be lower semicontinuous. Then the map $x\to\Phi(x)$, defined by*

$$\Phi(x) \doteq \mathring{B}[g(x), \varepsilon(x)] \cap G(x)$$

is lower semicontinuous on its domain. ▲

Proof. Fix x^* in Dom Φ, y^* in $\Phi(x^*)$ and $\eta>0$.

For some $\sigma>0$, $d(y^*, g(x^*))=\varepsilon(x^*)-\sigma$. There exist δ_1 such that to any x with $d(x,x^*)<\delta_1$ we can associate y_x in $G(x)$ so that $d(y_x, y^*)<\min(\eta,\sigma/3)$, δ_2 such that $d(x,x^*)<\delta_2$ implies $\varepsilon(x)>\varepsilon(x^*)-\sigma/3$ and δ_3 such that $d(x,x^*)<\delta_3$ implies $d(g(x^*), g(x))<\sigma/3$. Then when $d(x,x^*)<\min(\delta_1,\delta_2,\delta_3)$, at once

$$\begin{aligned} d(y_x, g(x)) &\leq d(y_x, y^*)+d(y^*, g(x^*))+d(g(x^*), g(x)) \\ &<\sigma/3+\varepsilon(x^*)-\sigma+\sigma/3=\varepsilon(x^*)-\sigma/3<\varepsilon(x), \end{aligned}$$

i.e., $y_x \in \Phi(x)$ and:

$$d(y_x, y^*)<\eta.$$ □

Proposition 6. *Let F be a map from a metric space X to a normed space Y, Lipschitzean with Lipschitz constant L. Then the map $x\to\overline{\text{co}}(F(x))$ is Lipschitzean with the same constant L.*

Proof. Fix x and x' in X. For y in $\overline{\text{co}}(F(x))$ and $\varepsilon>0$, there exist y_i in $F(x)$ such that $\|y-\sum\lambda_i y_i\|<\varepsilon$. For each i there exists y'_i in $F(x')$, $\|y_i - y'_i\|\leq \mathfrak{d}(F(x), F(x'))+\varepsilon$. Hence

$$\|y-\lambda_i y'_i\| \leq \varepsilon + \|\sum \lambda_i (y_i - y'_i)\| \leq \varepsilon + \sum \lambda_i (\mathfrak{d}(F(x), F(x'))+\varepsilon) \leq L d(x,x')+2\varepsilon.$$

Being ε arbitrary $y\in B(\overline{\text{co}}(F(x)), Ld(x,x'))$. Interchanging x and x' we have that the map $x\to\overline{\text{co}}(F(x))$ is Lipschitzean.

Definition 5. Let X and Y be normed spaces. We say that F is *upper semicontinuous at* x^0 *in the* ε *sense* if, given $\varepsilon>0$, there exists $\delta>0$ such that $F(x_0+\delta B)\subset F(x^0)+\varepsilon B$.

We say that F is upper semi-continuous in the ε sense if it is so at every $x^0\in X$. ▲

Clearly, in a normed space, a map that is u.s.c. is also u.s.c. in the ε sense: in fact $F(x^0)+\varepsilon B$ and $x^0+\delta B$ are special neighborhoods.

The converse is not true, i.e., a map that is u.s.c. in the ε sense need not be u.s.c. Consider the map F, from $\mathbb{R}$ into the subsets of $\mathbb{R}^2$ defined by $F(\xi) \doteq \{(x, y)\,|\,x=\xi\}$. To show that it is u.s.c. in the ε sense it is enough to take, in the definition, $\delta=\varepsilon$. However F is not u.s.c. in the usual sense: for N take the set $\{(x, y)\,|\,|y|<1/|x|\}$. It is an open set containing $F(0)$, but for every $x\neq 0$, $F(x)\not\subset N$. □

In the case when the images $F(x)$ of F are compact, the two definitions coincide. In fact it is enough to show that, in this case, given any open N containing $F(x^0)$, there is a positive ε such that $F(x^0)+\varepsilon B\subset N$. To every y belonging to N we can associate a positive $\eta(y)$, the distance from y to the complement of N. This continuous and positive function has a positive minimum η on the compact $F(x^0)$. Setting $\varepsilon=\eta/2$, $F(x^0)+\varepsilon B$ is contained in N. □

Definition 6. Let X and Y be normed spaces. We say that F *is lower semicontinuous at* x^0 *in the* ε *sense*, if, given $\varepsilon>0$ there exists $\delta>0$ such that,

$$\forall x\in x_0+\delta B, \quad F(x_0)\subset F(x)+\varepsilon B. \tag{14}$$ ▲

Clearly, a map which is lower semicontinuous in the ε sense is also lower semicontinuous. If $F(x_0)$ is compact, F is lower semicontinuous at x_0 if and only if it is l.s.c. at x^0 in the ε sense:

Let y_i, $i=1,\dots,m$, be such that $\{y_i+\frac{1}{2}\varepsilon B\}$ covers $F(x^0)$, and let δ_i, $i=1,\dots,m$ be such that $d(x, x^0)<\delta_i$ implies $F(x)\cap(y_i+\frac{1}{2}\varepsilon B)\neq\emptyset$. Set $\delta=\inf\delta_i$. Then $d(x, x^0)<\delta$ implies $y_i\in F(x)+\frac{1}{2}\varepsilon B$, all i's, i.e., $y_i+\frac{1}{2}\varepsilon B\subset F(x)+\varepsilon B$, all i's, and a fortiori $F(x^0)\subset F(x)+\varepsilon B$. □

Remark. The following result shows how the definition of upper semi-continuity in the usual sense is restrictive: a u.s.c. map in the usual sense is "essentially" a u.s.c. map in the ε sense with compact values.

Theorem 2. *Let X and Y be normed with Y complete; let F be upper semicontinuous at x^0. Then there exists a compact $K\subset F(x_0)$ such that for every $\varepsilon>0$ there exists $\delta>0$: $F(x^0+\delta B)\subset F(x^0)\cup(K+\varepsilon B)$.* ▲

Proof. Let us set $S_\delta\doteq F(x^0+\delta B)\setminus F(x^0)$. To prove the theorem we have to provide a compact $K\subset F(x^0)$ such that, given $\varepsilon>0$, for all sufficiently small δ's, $S_\delta\subset K+\varepsilon B$.

We first claim that, for every sufficiently small δ, S_δ is bounded. Assume it is not so. Then there are two sequences $\{x_n\}$ and $\{y_n\}$, $x_n \to x^0$, $y_n \in F(x_n) \backslash F(x^0)$ and $\|y_n\| \uparrow \infty$. The image of the sequence $\{y_n\}$ is a closed set and $F(x^0)$ belongs to the open N, the complement of $\{y_n\}$. Then for no neighborhood V of x^0 can $F(V)$ be contained in N, a contradiction to the strict upper semi-continuity of F.

Now consider the Kuratowski's index of the (bounded) S_δ, $\alpha(\delta) = \alpha(S_\delta)$ [Kuratowski, 1958, Vol. I, p. 318]. The real function $\alpha(\cdot)$ is monotonically increasing, i.e., it decreases when δ decreases to zero. It cannot be that $\lim \alpha(\delta) = l > 0$. Assume it were so. Then we can define two sequences $\{x_n\}$ and $\{y_n\}$, such that $x_n \to x^0$, $y_n \in F(x_n) \backslash F(x^0)$ and $d(y_n, \{y_1, \ldots, y_{n-1}\}) \geq l/4$. In fact, assume we have defined $x_1, \ldots, x_m$ and $y_1, \ldots, y_m$ with the above properties, and define x_{m+1} and y_{m+1} as follows. Let $\delta^* = \frac{1}{2} \min\{d(x_n, x^0), n = 1, \ldots, m\}$ and consider S_{δ^*}. Since $\alpha(S_{\delta^*}) \geq l$, there is a y in S_{δ^*} such that $d(y^*, \{y_1, \ldots, y_m\}) \geq l/4$. If it were not so, $\{y_i + (l/4)B\}_{i=1}^m$ would cover S_{δ^*}, hence $\alpha(\delta^*)$ would not be larger that $l/2$, a contradiction. Since y belongs to some $F(x) \backslash F(x^0)$, set $y_{m+1} = y$ and $x_{m+1} = x$. Hence sequences $\{x_n\}$ and $\{y_n\}$ with the required properties can be defined. The sequence $\{y_n\}$ contains no convergent subsequences; then its image is a closed set. The set $F(x^0)$ is contained in N, the open complement of the image of $\{y_n\}$ and again there is no neighborhood V of x^0 such that $F(V) \subset N$.

There remains only the case $\lim \alpha(\delta) = 0$. Let $\{\delta_n\}$ be any sequence decreasing to zero and set K to be the non-empty and compact set $\bigcap_n \bar{S}_{\delta_n}$. Since the S_δ are a monotonic family, K is also $\bigcap_\delta \bar{S}_\delta$. We claim that K is the required set, i.e., that given any $\varepsilon > 0$, for every sufficiently small δ, $S_\delta \subset K + \varepsilon B$. Assume it is not so. Then for some positive ε there are two sequences, $x_n \to x^0$ and $y_n \in F(x_n) \backslash F(x^0)$, such that $d(y_n, K) \geq \varepsilon$. The index α of the image of the sequence $\{y_n\}$ has to be zero: in fact, disregarding a finite number of points (a finite set has no influence on the index itself), this set is contained in S_m with m arbitrarily large, i.e., with α arbitrarily small. Being then the image of the sequence $\{y_n\}$ totally bounded, we can extract a subsequence $\{y_{n_j}\}$ converging to some y^*. This y^* has to be in each $\bar{S}_{\delta_n}$, hence in their intersection, while, at the same time, $d(y^*, K) \geq \varepsilon$. This is a contradiction, and K is the set required by the theorem. □

2. Examples of Set-Valued Maps

We list below a series of examples of set-valued maps we shall use later in the book. We begin with "parametrized" maps.

Continuity Properties of Parametrized Maps

Control theory provides set-valued map of the following type. We consider three sets X, Y and U and a map f from $X \times U$ to Y. We associate with it the set-valued map F from X to Y defined by

(1) $$F(x) \doteq \{f(x,u)\}_{u\in U}.$$

We say that F is "parametrized" by elements of U.

Proposition 1. *Assume that X and Y are Hausdorff topological spaces.*

a) *If we suppose that*

(2) $$\forall u \in U, \quad x \to f(x,u) \quad \text{is continuous},$$

then F is lower semi-continuous.

b) *If we suppose that*

(3) $$\begin{cases} \text{i) } U \text{ is a compact topological space,} \\ \text{ii) } f \text{ is continuous from } X \times U \text{ to } Y, \end{cases}$$

then F is continuous. ▲

Proof. a) The first statement is obvious.

b) Let N be an open neighborhood of $F(x_0)$. For any $u \in U$, N is a neighborhood of $f(x_0, u)$. The continuity of f implies that there exist open neighborhoods $M_u(x_0)$ and $N(u)$ such that $f(x,v)$ belongs to N whenever x belongs to $M_u(x_0)$ and v to $N(u)$. We can cover the compact set U by n such open sets $N(u_i)$. Hence $M(x_0) \doteq \bigcap_{i=1}^{n} M_{u_i}(x_0)$ is still a neighborhood of x_0. If x belongs to this neighborhood and if v is chosen in U, hence in some set $N(u_i)$, then $f(x,v)$ belongs to N. Therefore, $F(x)$ is contained in N whenever x ranges over $M(x_0)$. □

When U is convex, and thus, connected, the images $F(x) \doteq f(x, U)$ are connected because f is continuous. This property of connectedness remains true for set valued maps which are decreasing limits of parametrized maps with convex control sets, despite the fact that connectedness is not stable by decreasing intersection.

We shall use later such set-valued maps (for instance, maps associating to each point the set of trajectories of differential equations issued from this point).

Theorem 1. *Assume that F maps a metric space X to the* closed *subsets of a Banach space Y and that there exists a sequence of* compact connected *subsets U_n and of continuous functions f_n from $X \times U_n$ to Y such that*

(4) $$\begin{aligned} &\text{i) } \forall x \in X,\ \forall n \in \mathbb{N}, \quad F(x) \subset \ldots f_{n+1}(x, U_{n+1}) \subset f_n(x, U_n) \ldots, \\ &\text{ii) } \forall x \in X,\ \forall \varepsilon > 0,\ \exists N \mid \forall n \geq N, \quad f_n(x, U_n) \subset F(x) + \varepsilon B. \end{aligned}$$

Then the set-valued map F is upper semi-continuous with compact connected *values.* ▲

It follows from the following lemma.

Lemma 1. *Let us consider a family of compact connected subsets* K_n *of a metric space satisfying*

$$(5)\quad \begin{cases} \text{i) } K \doteq \bigcap_{n>0} K_n \subset \ldots \subset K_{n+1} \subset K_n \subset \ldots \subset K_0, \\ \text{ii) } \forall \varepsilon > 0,\ \exists N \mid \forall n > N, \quad K_n \subset B(K, \varepsilon). \end{cases}$$

Then K *is also compact and connected.* ▲

Proof. Assume that K is not connected: then $K = K^1 \cup K^2$ where K^1 and K^2 are nonempty, disjoint and closed. Since the subsets K^i are compact and disjoint, there exists ε such that

$$(6)\qquad B(K^1, 2\varepsilon) \cap B(K^2, 2\varepsilon) = \emptyset.$$

Let n such that K_n is contained in $B(K, \varepsilon)$ (this is possible thanks to (5) ii)). We set

$$(7)\qquad K_n^i \doteq K_n \cap B(K^i, \varepsilon), \quad i = 1, 2.$$

Since K^i is contained in K_n^i, the subsets K_n^i are nonempty. They are obviously closed and property (6) implies that they are disjoint. Since K_n is contained in $B(K, \varepsilon)$, they cover it. This is a contradiction, since K_n is connected. □

Also, acyclity, a concept weaker than convexity but stronger than connectedness and simple connectedness, is stable by decreasing intersection. (For a precise definition of acyclicity, see Bourgin [1963].)

The following version of the Vietoris-Begle Theorem is useful for our purposes.

Theorem 2. *Let* Y *be a finite dimensional space. Let us consider a sequence of compact acyclic subsets* U_n *and of continuous functions* $f_n\colon X \times U_n \to Y$ *such that*

$$(8)\quad \begin{cases} \text{i) } \forall x \in X,\ y \in Y, \quad \{u \in U_n \mid f_n(x, u) = y\} \text{ is acyclic}, \\ \text{ii) } \forall x \in X,\ \forall n \geq 0, \quad f_{n+1}(x, U_{n+1}) \subset f_n(x, U_n). \end{cases}$$

Then the set-valued map F *from* X *to* Y *defined by*

$$(9)\qquad F(x) \doteq \bigcap_{n \geq 0} f_n(x, U_n)$$

is upper semi-continuous with nonempty compact acyclic images (which are then connected and simply connected). ▲

Proof. The Vietoris-Begle Theorem and assumption (8) imply that $f_n(x, U_n)$ is acyclic.

Since any decreasing intersection of acyclic subsets is still acyclic, $F(x)$ is then acyclic. What matters is the fact that acyclic subsets are connected and simply connected. □

Remark. We shall see that continuous maps with compact convex values are parametrized by elements of convex compact control sets (see Theorem 1.7.2 below).

Lower Semicontinuity of Feedback Control Maps

The following result will be used for proving the existence of feedback controls.

Theorem 3. *Let K be a topological space, X and Y be Banach spaces, $f: K \times Y \to X$ be a continuous map that is affine with respect to the second variable.*

Let U and T be lower semi-continuous convex valued maps from K to Y and X respectively. We suppose that U is locally bounded (U is bounded on a neighborhood of each point). We assume that there exists $\gamma > 0$ such that

$$\text{(10)} \quad \begin{cases} \forall x \in K, \ \forall y \in X, \ \|y\| \leq \gamma, \quad \exists u \in U(x) \text{ such that} \\ f(x,u) + y \in T(x). \end{cases}$$

Then the set-valued map C defined by

$$\text{(11)} \quad C(x) \doteq \{u \in U(x) \mid f(x,u) \in T(x)\}$$

is lower semicontinuous. ▲

Proof. Let $u \in C(x)$ and $\varepsilon > 0$. We have to prove that there exists a neighborhood $N(x)$ of $x \in K$ such that

$$\text{(12)} \quad \forall y \in N(x), \quad \exists u_y \in C(y) \cap (u + \varepsilon B).$$

Let $N_0(x)$ be the neighborhood on which U is bounded and $\delta = \sup\{\operatorname{diam}(U(y)) \mid y \in N_0(x)\}$, which is finite. We take $\varepsilon < 2\delta$ and $\alpha \doteq \gamma\varepsilon/(2\delta - \varepsilon)$. Since f is continuous, since U and T are lower semi-continuous, we can find a neighborhood $N(x) \subset N_0(x)$ and $\eta \leq \varepsilon/2$ such that, $\forall y \in N(x)$,

$$\text{(13)} \quad \begin{cases} \text{i) if } \|u - v\| \leq \eta, \text{ then } \|f(y,v) - f(x,u)\| \leq \alpha/2, \\ \text{ii) } u \in U(y) + \eta B, \\ \text{iii) } f(x,u) \in T(y) + (\alpha/2) B. \end{cases}$$

Hence,

$$\text{(14)} \quad \forall y \in N(x), \ \exists v_y \in U(y) \text{ such that } f(y, v_y) \in T(y) + \alpha B.$$

Assumption (10) can be written

$$\text{(15)} \quad \forall y \in K, \quad \gamma B \subset T(y) - f(y, U(y)).$$

Let us set $\theta = \dfrac{\gamma}{\alpha+\gamma} < 1$ and multiply the inclusion (14) by θ. We obtain, by noticing that $\theta\alpha = (1-\theta)\gamma$

$$\theta f(y, v_y) \in \theta T(y) + (1-\theta)\gamma B.$$

We multiply the inclusion (15) by $(1-\theta)$: we get

$$\theta f(y, v_y) \in \theta T(y) + (1-\theta) T(y) - (1-\theta) f(y, U(y)).$$

Since $T(y)$ is convex and since f is affine with respect to u, we have proved the existence of $w_y \in U(y)$ such that

$$f(y, \theta v_y + (1-\theta) w_y) \in T(y),$$

i.e., such that $u_y \doteq \theta v_y + (1-\theta) w_y$ belongs to $C(y)$. It remains to see that

$$\|u - u_y\| \leq \|u - v_y\| + (1-\theta)\|v_y - w_y\| \leq \frac{\varepsilon}{2} + \frac{\alpha}{\alpha+\gamma}\|v_y - w_y\| \leq \frac{\varepsilon}{2} + \frac{\alpha}{\alpha+\gamma}\delta = \varepsilon.$$

So, C is lower semi-continuous. □

Cone-Valued Maps

We recall that given a cone T, we denote by T^- its negative polar.

Proposition 2. *Let X be a finite dimensional space and $T(\cdot)$ be a set valued map associating with any $x \in K$ a closed convex cone $T(x)$. Let $N(\cdot)$ be the set valued map $x \to N(x) \doteq T(x)^-$. The following conditions are equivalent:*

(16) *The set-valued map $N(\cdot)$ has a closed graph*

and

(17) *The set-valued map $T(\cdot)$ is lower semicontinuous.* ▲

Proof. a) Let us assume that $T(\cdot)$ is lower semi-continuous. Let $(x_n, p_n) \in \operatorname{graph}(N(\cdot))$ be a sequence converging to (x, p). To prove that $p \in N(x)$, let us choose any $v \in T$ and check that $\langle p, v\rangle \leq 0$. Since $T(\cdot)$ is lower semicontinuous, $v = \lim_{n\to\infty} v_n$ where $v_n \in T(x_n)$. Since $\langle p_n, v_n\rangle \leq 0$ (for $p_n \in N(x_n)$), we deduce that $\langle p, v\rangle = \lim_{n\to\infty} \langle p_n, v_n\rangle \leq 0$. Hence (x, p) belongs to $\operatorname{graph}(N(\cdot))$ and thus, N has a closed graph.

b) Let us assume that the graph of $N(\cdot)$ is closed; let x_n converge to x and $v \in T(x)$. We shall prove that the projection $v_n \doteq \pi_{T(x_n)}(v)$ converges to v. Indeed, we know that $p_n \doteq \pi_{N(x_n)}(v) \in N(x_n)$ and that $\|p_n\| \leq \|v\|$. Since X is a finite dimensional space, some subsequence (again denoted by) p_n converges to $p \in X$. Since the graph of $N(\cdot)$ is closed, we deduce that $p \in N(x)$. Hence the subsequence $v_n = v - p_n$ converges to $v - p$. Since $\langle p_n, v_n\rangle = 0$, we deduce that $\langle p, v - p\rangle = 0$; Hence $\|p\|^2 = \langle p, v\rangle \leq 0$ and thus $p = 0$ and $v = \lim_{n\to\infty} v_n$. □

Upper Semicontinuity of the Epigraph

Proposition 3. *Let X and Y be metric spaces and W: $X \times Y \to R$ be a lower semicontinuous function. We assume that Y is compact. We denote by $V(x)$ the function $y \to V(x)(y) \doteq W(x, y)$ and by $Ep\,V(x)$ its epigraph. Then $x \to Ep\,V(x)$ is upper semicontinuous (in the ε-sense).* ▲

Proof. To say that $x \to Ep\,V(x)$ is upper semicontinuous at x_0 in the ε-sense means that for all $\varepsilon > 0$, there exists $\eta > 0$ such that, for all $x \in K \cap B[x_0, \eta]$, for all $y \in Y$ and $\lambda \geq W(x, y)$, there exist $y_0 \in Y \cap B[y, \varepsilon]$ and $\lambda_0 \leq \lambda + \varepsilon$. Hence it suffices to prove that there exists $y_0 \in Y \cap B[y, \varepsilon]$ satisfying $W(x_0, y_0) \leq W(x, y) + \varepsilon$.

Since W is lower semicontinuous, we can associate to each $y \in Y$ an $\eta_y \in]0, \varepsilon[$ such that

$$\forall x \in K \cap B[x_0, \eta_y],\ \forall z \in B[y, \eta_y], \qquad W(x_0, y) \leq W(x, z) + \varepsilon.$$

Since Y is compact, it is covered by n such balls $B[y_i, \eta_{y_i}]$. Set $\eta \doteq \min_{i=1,\ldots,n} \eta_{y_i} > 0$. Hence, for all $x \in K \cap B[x_0, \eta]$, for all $y \in Y$, there exists one of the points y_i such that $d(y, y_i) \leq \eta_{y_i}$ and $W(x_0, y_i) \leq W(x, y) + \varepsilon$. □

Continuity Properties of Marginal Functions and Maps

Let X and Y be two sets, G be a set-valued map from Y to X and W be a real-valued function defined on $X \times Y$. We consider the family of maximization problems

$$V(y) \doteq \sup_{x \in G(y)} W(x, y), \tag{18}$$

which depend upon the parameter y. The function V is called the *marginal function* and the set-valued map M associating to the parameter $y \in Y$ the set

$$M(y) \doteq \{x \in G(y) \mid V(y) = W(x, y)\} \tag{19}$$

of solutions to the maximization problem $V(y)$ is called the *marginal map*.

By *stability* of this family of optimization problems, we usually mean the study of various continuity properties of the marginal function V and the marginal map M.

We begin with

Theorem 4. *Suppose that*

$$\begin{cases} \text{i) } W \text{ is lower semicontinuous on } X \times Y, \\ \text{ii) } G \text{ is lower semicontinuous at } y_0. \end{cases} \tag{20}$$

Then V is lower semicontinuous at y_0. ▲

Proof. Fix $\varepsilon > 0$ and choose $x_0 \in G(y_0)$ such that $V(y_0) - \varepsilon/2 \leq W(x_0, y_0)$. Since W is lower semicontinuous, there exist neighborhoods $N(x_0)$ and $M_1(y_0)$

such that

$$\text{(21)} \qquad \forall x \in N(x_0),\ \forall y \in M_1(y_0), \quad W(x_0, y_0) - \varepsilon/2 \le W(x, y).$$

Since G is lower semicontinuous, there exists a neighborhood $M_2(y_0)$ such that

$$\text{(22)} \qquad \forall y \in M_2(y_0), \quad G(y) \cap N(x_0) \neq \emptyset.$$

So we take $M \doteq M_1(y_0) \cap M_2(y_0)$. For any $y \in M$, we know that there exists $x \in G(y)$ such that $W(x, y) \ge W(x_0, y_0) - \varepsilon/2 \ge V(y_0) - 2\varepsilon/2$. Hence V is lower semicontinuous at y_0. □

We single out a useful consequence

Corollary 1. *Let Y be a metric space. If G: $X \to Y$ is lower semicontinuous, then the function*

$$\text{(23)} \qquad (x, y) \in X \times Y \to d(G(x), y)$$

is upper semicontinuous. ▲

We consider now the case when W and G are upper semicontinuous.

Theorem 5. *Suppose that*

$$\text{(24)} \qquad \begin{cases} \text{i) } W \text{ is upper semicontinuous on } X \times Y, \\ \text{ii) } G(y_0) \text{ is compact and } G \text{ is upper semicontinuous at } y_0. \end{cases}$$

Then the marginal function V is upper semicontinuous at y_0. ▲

Proof. We have to prove that for any $\varepsilon > 0$, there exists a neighborhood $N(y_0)$ of y_0 such that

$$\text{(25)} \qquad \forall y \in N(y_0), \quad V(y) \le V(y_0) + \varepsilon.$$

Since W is upper semicontinuous, we can associate with any $x \in X$ open neighborhoods $N(x)$ of x and $N_x(y_0)$ of y_0 such that

$$\text{(26)} \qquad \forall x' \in N(x),\ \forall y \in N_x(y_0), \quad W(x', y) \le W(x, y_0) + \varepsilon.$$

Since $G(y_0)$ is compact, it can be covered by n neighborhoods $N(x_i)$. Therefore

$$\text{(27)} \qquad N \doteq \bigcup_{i=1}^{n} N(x_i) \text{ is a neighborhood of } G(y_0).$$

The set valued G being upper semicontinuous, there exists a neighborhood $N_0(y_0)$ such that:

$$\text{(28)} \qquad \forall y \in N_0(y_0), \quad G(y) \subset N.$$

We consider the following neighborhood of y_0:

$$N(y_0) \doteq N_0(y_0) \cap \bigcap_{i=1}^{n} N_{x_i}(y_0).$$

When y belongs to $N(y_0)$ and x belongs to $G(y)$, then x belongs to N, and thus, to some $N(x_i)$. Hence, since y belongs to $N_{x_i}(y_0)$, we deduce from (26) that

$$W(x, y) \le W(x_i, y_0) + \varepsilon \le V(y_0) + \varepsilon.$$

Hence, by taking the supremum when x ranges over $G(y)$ we obtain inequality (25). □

We point out the following often used consequence.

Corollary 2. *Let X be a Hausdorff topological space, Y be a compact space and W be an upper semicontinuous function from $X \times Y$ to R. Then the marginal function V defined on Y by*

(29) $$V(y) \doteq \sup_{y \in Y} W(x, y)$$

is upper semicontinuous. ▲

When the function W and the set-valued map G are continuous, so is the marginal function V, thanks to the above theorems. We now investigate the behavior of the sets of maximizers

(30) $$M(y) \doteq \{x \in G(y) \mid V(y) = W(x, y)\}.$$

Theorem 6. *Suppose that*

(31) $$\begin{cases} \text{i)} & W \text{ is continuous on } X \times Y, \\ \text{ii)} & G \text{ is continuous with compact values.} \end{cases}$$

Then the marginal function V is continuous and the marginal set-valued map M is upper semi-continuous. ▲

Proof. We note that $M(y) = G(y) \cap K(y)$ where

$$K(y) \doteq \{x \in X \mid V(y) = W(x, y)\}.$$

Since V and W are continuous functions, the graph of K is closed. The subsets $M(y)$ are obviously non-empty. Since G is upper semicontinuous, Theorem 1.1 implies that M is also upper semicontinuous. □

Finally, we consider the case when W and G are both Lipschitzean.

Theorem 7. *Suppose that*

(32) $$\begin{cases} \text{i)} & W \text{ is Lipschitzean on } X \times Y \text{ with Lipschitz constant } l, \\ \text{ii)} & G \text{ is Lipschitzean with Lipschitz constant } c. \end{cases}$$

Then the marginal function V is Lipschitz with Lipschitz constant $l(c+1)$. ▲

Proof. We take $y_1 \in Y$ and we choose $x_1 \in F^{-1}(y_1)$ satisfying $V(y_1) \leq W(x_1, y_1) + \varepsilon$. There exists $x_2 \in F^{-1}(y_2)$ such that

$$\begin{aligned}\|x_1 - x_2\| &\leq d(x_1, F^{-1}(y_2)) + \varepsilon \\ &\leq c\,\|y_1 - y_2\| + \varepsilon\end{aligned}$$

because G is Lipschitzean.

Since $V(y_2) \geq W(x_2, y_2)$, we deduce that

$$\begin{aligned}V(y_1) - V(y_2) &\leq W(x_1, y_1) - W(x_2, y_2) + \varepsilon \\ &\leq l(\|x_1 - x_2\| + \|y_1 - y_2\|) + \varepsilon \\ &\leq l(c+1)\,\|y_1 - y_2\| + (l+1)\,\varepsilon.\end{aligned}$$

Since ε is arbitrary, it follows that

$$V(y_1) - V(y_2) \leq l(c+1)\,\|y_1 - y_2\|. \qquad \square$$

3. Continuity Properties of Maps with Closed Convex Graph

Closed convex maps from Banach spaces to Banach spaces are lower semicontinuous on the interior of their domains: this important result is an extension of the closed graph theorem for continuous linear operators. Since it is customary to prove the closed graph theorem from the Banach open mapping principle, we shall state the Robinson-Ursescù theorem in the following way.

Theorem 1 (Robinson-Ursescù). *Let X and Y be Banach spaces, F be a proper closed convex map (i.e., map with closed convex graph) from X to Y the range of which has a nonempty interior.*

Let $y_0 \in \operatorname{Int}(R(F))$ and $x_0 \in F^{-1}(y_0)$ be chosen. There exists $\gamma > 0$ such that

$$\text{(1)} \quad \forall x \in \operatorname{Dom} F,\ \forall y \in y_0 + \gamma B, \quad d(x, F^{-1}(y)) \leq \frac{1}{\gamma}\, d(y, F(x))\,(1 + \|x - x_0\|).$$

In particular, F^{-1} is lower semicontinuous on $\operatorname{Int} R(F)$. *By taking $x = x_0$, we get*

$$\text{(2)} \qquad \forall y \in y_0 + \gamma B, \quad d(x_0, F^{-1}(y)) \leq \frac{1}{\gamma}\, d(y, F(x_0)). \qquad \blacktriangle$$

Corollary 1. *Let X and Y be Banach spaces, F be a proper closed convex map from X to Y the range of which has a nonempty interior. Assume furthermore that F^{-1} is locally bounded. Then F^{-1} is locally Lipschitz on* $\operatorname{Int} R(F)$. ▲

Let us apply Theorem 1 to the map F defined

$$\text{(3)} \qquad F(x) \doteq \begin{cases} Ax - M & \text{when } x \in L, \\ \emptyset & \text{when } x \notin L. \end{cases}$$

Corollary 2. *Let $L \subset X$ and $M \subset Y$ be nonempty closed convex subsets of Banach spaces X and Y and A belong to $\mathscr{L}(X, Y)$. We set*

(4) $$\forall y \in Y, \quad F^{-1}(y) \doteq \{x \in L \mid Ax \in M + y\}.$$

Let us assume that

(5) $$\mathrm{Int}(A(L) - M) \neq \emptyset.$$

Then for all $y_0 \in \mathrm{Int}(A(L) - M)$ and for all $x_0 \in F^{-1}(y_0)$, there exists $\gamma > 0$ such that

(6) $$\forall x \in L, \ \forall y \in y_0 + \gamma B, \quad d(x, F^{-1}(y)) \leq \frac{1}{\gamma} d_M(Ax - y)(1 + \|x - x_0\|). \quad \blacktriangle$$

Proof. The graph of the map F is obviously closed and convex, its range is equal to $A(L) - M$, its inverse is defined by (4) and $d(y, F(x)) = d_M(Ax - y)$. These remark made, Corollary 2 follows from *Theorem 1*. □

When L and M are cones, Corollary 2 implies the following statement.

Corollary 3. *Let $P \subset X$ and $Q \subset Y$ be closed convex cones of Banach spaces X and Y and A be a continuous linear operator from X to Y. We assume that*

(7) $$Y = A(P) - Q.$$

Let F^{-1} be the map defined from Y to X by

(8) $$\forall y \in Y, \quad F^{-1}(y) \doteq \{x \in P \mid Ax \in Q + y\}.$$

Then F^{-1} is Lipschitz: there exists $\gamma > 0$ such that

(9) $$\forall y_1, y_2 \in Y, \quad F^{-1}(y_2) \subset F^{-1}(y_1) + \frac{1}{\gamma}\|y_1 - y_2\| B. \quad \blacktriangle$$

By taking $P \doteq X$ and $Q \doteq \{0\}$, we obtain the Banach open mapping principle:

Corollary 4 (Banach's Open Mapping Principle). *Let A be a surjective continuous linear operator from a Banach space X to a Banach space Y. Then the inverse A^{-1} is a Lipschitz set-valued map from Y to X.* ▲

Proof of the Robinson-Ursescù Closed Graph Theorem. The proof of *Theorem 1* is rather involved. We begin by deducing it from a more concise statement. We set

(10) $$K \doteq \mathrm{Dom}\, F.$$

Proposition 1. *We posit the assumptions of* Theorem 1. *Let $y_0 \in \mathrm{Int}\, F(K)$ and $x_0 \in F^{-1}(y_0)$ be given and B denote the unit ball of X. Then,*

(11) $$y_0 \in \mathrm{Int}\, F(K \cap (x_0 + B)). \quad \blacktriangle$$

We shall decompose the proof of Proposition 1 in three lemmas. But, before, we deduce Theorem 1 from *Proposition 1.*

Proof of Theorem 1.1 from Proposition 1. Let $x \in K$ be fixed. By *Proposition* 1, there exists $\gamma > 0$ such that

$$y_0 + 2\gamma B \subset F(K \cap (x_0 + B)). \tag{12}$$

Let $y \in y_0 + \gamma B$ be given. If x belongs to $F^{-1}(y)$, then $d(x, F^{-1}(y)) = 0$ and the conclusion is satisfied. If not, for all $\varepsilon > 0$, there exists $z \in F(x)$ such that

$$\|y - z\| \leq d(y, F(x))(1 + \varepsilon).$$

Since $y + \gamma B \subset F(K \cap (x_0 + B))$, we obtain

$$\frac{\gamma(y - z)}{\|y - z\|} \in F(K \cap (x_0 + B)) - y. \tag{13}$$

Let us set $\lambda \doteq \dfrac{\|y - z\|}{\gamma + \|y - z\|}$ which belongs to $]0, 1[$. We can write inclusion (13) in the form

$$(1 - \lambda)(y - z) \in \lambda F(K \cap (x_0 + B)) - \lambda y. \tag{14}$$

Since $(1 - \lambda) z \in (1 - \lambda) F(x)$ and since the graph of F is convex, we obtain:

$$\begin{aligned} y &\in \lambda F(K \cap (x_0 + B)) + (1 - \lambda) F(x) \\ &\subset F(\lambda(K \cap (x_0 + B)) + (1 - \lambda) x). \end{aligned} \tag{15}$$

Then there exists $x_1 \in K \cap (x_0 + B)$ such that, by setting $x_y \doteq \lambda x_1 + (1 - \lambda) x$, we have $x_y \in F^{-1}(y)$. Furthermore, we observe that

$$\begin{aligned} \|x_y - x\| = \|\lambda x_1 + (1 - \lambda) x - x\| = \lambda \|x_1 - x\| &\leq \lambda(\|x - x_0\| + \|x_1 - x_0\|) \\ &\leq \lambda(\|x - x_0\| + 1) \end{aligned}$$

because $x_1 \in x_0 + B$.

Now, $\lambda \doteq \dfrac{\|y - z\|}{\gamma + \|y - z\|} \leq \dfrac{d(y, F(x))(1 + \varepsilon)}{\gamma}$ and thus,

$$d(x, F^{-1}(y)) \leq \|x - x_y\| \leq \frac{d(y, F(x))(1 + \varepsilon)}{\gamma}(\|x - x_0\| + 1). \tag{16}$$

By letting ε converge to 0, we obtain

$$d(x, F^{-1}(y)) \leq \frac{1}{\gamma} d(y, F(x))(\|x - x_0\| + 1). \tag{17}$$ □

The proof of *Proposition 1* is analogous to the proof of the open mapping Theorem. We use

Lemma 1. *Let T be a subset of a Banach space Y satisfying*

(18) $$\frac{1}{2}\sum_{k=0}^{\infty} 2^{-k} T \subset T.$$

If 0 belongs to the interior of the closure of T, it actually belongs to the interior of T. ▲

Proof. By assumption, there exists $\gamma > 0$ such that $2\gamma B \subset \bar{T}$; hence, for every $k > 1$, we have $2\,2^{-k}\gamma B \subset 2^{-k}\bar{T}$.

Let $y \in \gamma B$. Then there exists $v_0 \in T$ such that

$$2y - v_0 \in 2\,2^{-1}\gamma B \subset 2^{-1}\bar{T}$$

since $2y \in \bar{T}$.

Let us assume that we have constructed a sequence of elements $v_k \in T$ $(0 \le k \le n-1)$ such that

(19) $$2y - \sum_{k=0}^{n-1} 2^{-k} v_k \subset 2\,2^{-n}\gamma B.$$

Since $2\,2^{-n}\gamma B \subset 2^{-n}\bar{T}$, we can find $v_n \in T$ such that

$$2y - \sum_{k=0}^{n-1} 2^{-k} v_k - 2^{-n} v_n \in 2\,2^{-(n+1)}\gamma B \subset 2^{-(n+1)}\bar{T}.$$

In this way, we have constructed a sequence of elements $v_k \in T$ such that

$$y = \frac{1}{2}\sum_{k=0}^{\infty} 2^{-k} v_k \in \frac{1}{2}\sum_{k=0}^{\infty} 2^{-k} T.$$

By assumption, y belongs to T. So, we have proved that $\gamma B \subset T$. □

We apply this lemma to the subset

(20) $$T \doteq F(K \cap (x_0 + B)) - y_0.$$

Lemma 2 proves that 0 belongs to the interior of the closure of T and Lemma 3 states that this subset T satisfies property (18): hence we shall conclude that 0 belongs to the interior of T, i.e., the conclusion of Proposition 1.

Lemma 2. *We posit the assumptions of* Proposition 1. *Then* 0 *belongs to the interior of the closure of the subset $T \doteq F(K \cap (x_0 + B)) - y_0$.* ▲

Proof. We set $K_n \doteq K \cap (x_0 + nB)$. Hence $T = F(K_1) - y_0$.

We remark that $K = \bigcup_{n=1}^{\infty} K_n$ and thus, that $F(K) = \bigcup_{n=1}^{\infty} F(K_n)$. We note also that

$$\left(1 - \frac{1}{n}\right) x_0 + \frac{1}{n} K_n \subset K_1.$$

Since the graph of F is convex, we deduce that

(21) $$\left(1-\frac{1}{n}\right)F(x_0)+\frac{1}{n}F(K_n)\subset F(K_1).$$

Since $0\in\operatorname{Int}(F(K)-y_0)$, there exists $\gamma>0$ such that

$$\gamma B\subset F(K)-y_0=\bigcup_{n=1}^{\infty}(F(K_n)-y_0).$$

But $y_0\in F(x_0)$ and (21) implies that

$$\left(1-\frac{1}{n}\right)y_0+\frac{1}{n}F(K_n)\subset F(K_1),$$

i.e.,

$$F(K_n)-y_0\subset n(F(K_1)-y_0)=nT.$$

Therefore $\gamma B\subset\bigcup_{n=1}^{\infty}n\bar{T}$. Baire's *Theorem* (see Aubin [1979a], Theorem 5.3.1, p. 187) implies that the interior of some subset $n\bar{T}$ is nonempty. Then there exist $x_0\in\bar{T}$ and $\delta>0$ such that $x_0+\delta B\subset\bar{T}$. Since $-\frac{\gamma x_0}{\|x_0\|}$ belongs to γB, and thus, to the union of the $n\bar{T}$'s, there exists n such that $-\frac{\gamma x_0}{n\|x_0\|}$ belongs to $\bar{T}$.

Let $\lambda\doteq\gamma/(\gamma+n\|x_0\|)\in]0,1[$.

Hence $\lambda\delta B=\lambda x_0-(1-\lambda)\frac{\gamma x_0}{n\|x_0\|}+\lambda\delta B\subset\lambda\bar{T}+(1-\lambda)\bar{T}\subset\bar{T}$ because $\bar{T}$ is convex. We have shown that 0 belongs to the interior of $\bar{T}$. □

It remains to check that T satisfies property (18).

Lemma 3. *We posit the assumptions of* Proposition 1. *Then the subset* $T\doteq F(K\cap(x_0+B))-y_0$ *satisfies the property*

(18) $$\frac{1}{2}\sum_{k=0}^{\infty}2^{-k}T\subset T.$$ ▲

Proof. We take $y_0=0$ for the sake of simplicity. Let y_* belong to $\frac{1}{2}\sum_{k=0}^{\infty}2^{-k}T$. We set $\alpha_n\doteq1\Big/\sum_{k=0}^{n}2^{-k}$ so that $\alpha_n\to\frac{1}{2}$ and $y_n\doteq\alpha_n\sum_{k=0}^{n}2^{-k}v_k$ where $v_k\in T$ so that y_n converges to y_*. By definition of T, we can find $u_k\in K\cap(x_0+B)$ such that $v_k\in F(u_k)$. Since the graph of F is convex, we deduce that

(22) $$y_n\in\alpha_n\sum_{k=0}^{n}2^{-k}F(u_k)\subset F\left(\alpha_n\sum_{k=0}^{n}2^{-k}u_k\right).$$

Let us consider the sequence of elements $x_n\doteq\alpha_n\sum_{k=0}^{n}2^{-k}u_k$.

Since $u_k \in x_0 + B$, we deduce that it is a Cauchy sequence, which converges to some x_* for X is complete. Since $K \cap (x_0 + B)$ is convex, x_n belongs to $K \cap (x_0 + B)$. This set being also closed, x_* belongs to $K \cap (x_0 + B)$. Inclusion (22) states that the sequence of elements (x_n, y_n) belongs to the graph of F. Since it is closed, it follows that $(x_*, y_*) \in \operatorname{graph} F$, i.e., that $y_* \in F(x_*) \subset F(K \cap (x_0 + B)) \doteq T$. □

4. Upper Hemicontinuous Maps and the Convergence Theorem

The idea of upper semicontinuity is to require that, given x^0, for any neighborhood N belonging to a certain class and containing $F(x^0)$, there should exist a neighborhood M of x^0 such that $F(M) \subset N$. The larger the class of M allowed, the more restrictive the definition becomes and vice-versa. We shall now consider upper hemicontinuity, that is upper semicontinuity with respect to the rather narrow class of neighborhoods which define the weak topology on a locally convex space.

Let Y and Y^* be two vector spaces.

We say that a bilinear form $(p, x) \to \langle p, x \rangle$ on $Y^* \times Y$ is a *duality pairing* if and only if

(1) $$\begin{cases} \text{i) } \langle p, y \rangle = 0 & \textit{for all } y \in Y \textit{ implies that } p = 0, \\ \text{ii) } \langle p, y \rangle = 0 & \textit{for all } p \in Y^* \textit{ implies that } y = 0. \end{cases}$$

When Y is a locally convex Hausdorff space and $Y^* \doteq \mathscr{L}(Y, R)$ its dual, then the bilinear form defined by $\langle p, y \rangle \doteq p(y)$ is a duality pairing.

In this section, we supply Y with the weak topology $\sigma(Y, Y^*)$. For this topology, a neighborhood of $F(x^0)$ is a set containing a weakly open subset containing $F(x^0)$, and a map F is upper semicontinuous at x^0 whenever the definition holds with respect to this class of neighborhoods.

We recall that the *support function* $\sigma_K \colon Y^* \to]-\infty, +\infty]$ is defined as

(2) $$\sigma_K(p) \doteq \sup_{y \in K} \langle p, y \rangle. \qquad ▲$$

We also set for convenience $\sigma_K(p) \doteq \sigma(K, p)$. Support functions are known to characterize any nonempty closed convex set $K \subset Y$:

(3) $$K = \{ y \in Y \mid \forall p \in Y^*, \ \langle p, y \rangle \leq \sigma_K(p) \}.$$

Assume we are given a set-valued map F from X to Y. For every $p \in Y^*$ we can consider the function with values in $]-\infty, +\infty]$ given by

$$x \to \sigma(F(x), p).$$

Definition 1. We say that F is *upper hemicontinuous* (in short, u.h.c.) at $x^0 \in X$ if, for every $p \in Y^*$, the function $x \to \sigma(F(x), p)$ is upper semicontinuous at x^0. F is called *upper hemicontinuous* if it is u.h.c. at every $x^0 \in X$. ▲

Let (P, E) be any finite set of pairs (p_i, ε_i), $i=1, \ldots, m$, where p_i is a continuous linear functional and ε_i a positive real.

We set

(4) $$N(P, E) \doteq \{y \mid \langle p_i, y\rangle \leq \varepsilon_i,\ i=1, \ldots, m\}.$$

The collection $N(P, E)$ is a system of neighborhoods of the origin for the weak topology.

When K is a closed convex subset of Y, we set

(5) $$N(K; (P, E)) \doteq \{y \mid \langle p_i, y\rangle \leq \sigma(K, p_i) + \varepsilon_i,\ i=1, \ldots, m\}.$$

Clearly $N(K; (P, E))$ is an open set containing $K + N(P, E)$, and thus, K.

According to the preceding definition, a map F is u.h.c. at x^0 if for every $p \in Y^*$ and every $\varepsilon > 0$ there is $M \doteq N(p, \varepsilon)$ such that $\sigma(F(M), p) \leq \sigma(F(x^0), p) + \varepsilon$. Equivalently, we can say that F is u.h.c. at x^0 if for every finite set of pairs (p_i, ε_i), $i=1, \ldots, m$, there exists M such that $\sigma(F(M), p_i) \leq \sigma(F(x^0), p_i) + \varepsilon_i$, $i=1, \ldots, m$. Hence:

Proposition 1. *Any upper semicontinuous map from X to Y supplied with the weak topology is upper hemicontinuous.* ▲

The Convergence Theorem

The following property of upper hemicontinuous maps with closed values shall play a crucial role in this book.

Theorem 1 (Convergence Theorem). *Let F be a proper hemicontinuous map from a Hausdorff locally convex space X to the closed convex subsets of a Banach space Y. Let I be an interval of R and $x_k(\cdot)$ and $y_k(\cdot)$ be measurable functions from I to X and Y respectively satisfying*

(6) *for almost all $t \in I$, for every neighborhood $\mathcal{N}$ of 0 in $X \times Y$ there exists $k_0 \doteq k_0(t, \mathcal{N})$ such that*
$$\forall k \geq k_0, \quad (x_k(t), y_k(t)) \in \operatorname{graph}(F) + \mathcal{N}.$$

If

(7) i) *$x_k(\cdot)$ converges almost everywhere to a function $x(\cdot)$ from I to X,*
ii) *$y_k(\cdot)$ belongs to $L^1(I, Y)$ and converges weakly to $y(\cdot)$ in $L^1(I, Y)$,*

then

(8) *for almost all $t \in I$,*
$$(x(t), y(t)) \in \operatorname{graph}(F), \quad \text{i.e.,} \quad y(t) \in F(x(t)).$$ ▲

Proof. a) We recall that for convex subsets of a Banach space (here $L^1(I, Y)$), the strong closure coincides with the weak closure (for the weakened topology $\sigma(L^1, L^\infty)$). We apply this result: for any h, $y(\cdot)$ belongs to the weak closure of the convex hull $\mathrm{co}\{y_k(\cdot)\}_{k\ge h}$ of the subset $\{y_k(\cdot)\}_{k\ge h}$. It coincides with the strong closure of $\mathrm{co}\{y_k(\cdot)\}_{k\ge h}$. Hence, we can choose

$$v_h(\cdot) \doteq \sum_{k=h}^{\infty} a_h^k y_k(\cdot) \in \mathrm{co}\{y_k(\cdot)\}_{k\ge h}$$

$\left(\text{where } a_h^k \ge 0,\ a_h^k = 0, \text{ except for a finite number of } k\text{'s and } \sum_{k=h}^{\infty} a_h^k = 1\right)$ such that $v_h(\cdot)$ converges strongly to $y(\cdot)$ in $L^1(I; Y)$. Then there exists a subsequence (again denoted) $v_h(t)$ that converges to $y(t)$ for almost all $t \in I$.

b) Let $t \in I$ such that $x_h(t)$ converges to $x(t)$ in X and $v_h(t)$ converges to $y(t)$ in Y. Let us fix λ, $p \in Y^*$ such that $\sigma(F(x(t)), p) < +\infty$ and $\lambda > \sigma(F(x), p)$. Since F is upper hemicontinuous, there exists a neighborhood $\mathcal{M}_0$ of 0 in X such that when $u \in x(t) + \mathcal{M}_0$, we have $\sigma(F(u), p) \le \lambda$. Let k_1 be an index such that $x_k(t) \in x(t) + \frac{1}{2}\mathcal{M}_0$ when $k \ge k_1$. By assumption (6), there exist $k_0 \ge k_1$ and (u_k, w_k) in graph(F) such that $u_k \in x_k(t) + \frac{1}{2}\mathcal{M}_0$ and $\|y_k(t) - w_k\| \le \eta$ for all $k \ge k_0$. Therefore, $u_k \in x(t) + \mathcal{M}_0$ and we obtain

$$\begin{aligned}\langle p, y_k(t)\rangle &\le \langle p, w_k\rangle + \eta\,\|p\| \\ &\le \sigma(F(u_k), p) + \eta\,\|p\| \le \lambda + \eta\,\|p\|.\end{aligned}$$

Let $h \ge k_0$ be fixed. By multiplying the above inequalities by $a_h^k \ge 0$ and summing them, we deduce that

$$\langle p, v_h(t)\rangle \le \lambda + \eta\,\|p\|$$

and thus, by letting $h \to \infty$, that

$$\langle p, y(t)\rangle \le \lambda + \eta\,\|p\|.$$

This implies that $\sigma(F(x(t), p)) > -\infty$. By letting λ converge to $\sigma(F(x(t), p))$ and η to 0, we deduce that

$$\langle p, y(t)\rangle \le \sigma(F(x(t)), p).$$

Since $F(x(t))$ is closed and convex, we deduce that $y(t)$ belongs to $F(x(t))$. □

Let us mention the following corollary.

Corollary 1. *Let F be a proper hemicontinuous map from X to the closed convex subsets of Y. Let $p, q \ge 1$ and $\mathcal{F}$ be the map from $L^p(I, X)$ to $L^q(I, Y)$ defined by*

(9) $$y(\cdot) \in \mathcal{F}(x(\cdot)) \Leftrightarrow \text{for almost all } t \in I,\ y(t) \in F(x(t)).$$

Then the graph of $\mathcal{F}$ is closed in $L^p(I, X) \times L^q(I, Y)$ when $L^p(I, X)$ is supplied with the strong topology and $L^q(I, Y)$ is supplied with the weak topology. ▲

Upper Hemicontinuity and Upper Semicontinuity

Consider the map F from R into the closed subsets of R given by

(10) $$F(t) \doteq \{(x_1, x_2) \mid x_2 \geq (1+t)\, x_1^2\}.$$

Then F is upper hemicontinuous at $t=0$. In fact fix $p=(p_1, p_2)$, $p \neq (0,0)$. Either $p_2 \geq 0$, and there is nothing to prove since $\sigma(F(0), p) = +\infty$, or $p_2 < 0$. In this second case an easy calculation shows that $(F(t), p) = \frac{-1}{4} p_1^2 (p_2(1+t))^{-1}$ is continuous at $t=0$. However, it is easy to see that F is not upper semicontinuous at $t=0$.

Theorem 2. *Let F be upper hemicontinuous at x_0. If $F(x_0)$ is convex and weakly compact, it is also upper semicontinuous at x_0.* ▲

It follows from the following *Lemma:*

Lemma 1. *Let M be a weakly open set containing $F(x^0)$. Then there is (P, E), a finite set of pairs (p_i, ε_i), such that $N(F(x_0), (P, E))$ is contained in M.* ▲

Proof. The proof proceeds into two steps.

a) The result holds if Y is a finite dimensional space, isomorphic to R^n with the usual topology. Consider the bounded open set $F(x^0)+B$ and its intersection with M. From now on we shall consider M to be this intersection. Set K to be the complement of M with respect to the closure of $F(x^0)+2B$: K is a compact set.

Fix any $p \in Y^*$, $\|p\|=1$, and any ε, $0<\varepsilon<1$. The set

(11) $$K(p, \varepsilon) \doteq \{y \mid \langle p, y\rangle \leq \sigma(F(x^0), p)+\varepsilon\} \cap K$$

is (compact and) nonempty: consider a point $y_m \in F(x^0)$ such that $\langle p, y_m\rangle = \inf\{\langle p, y\rangle \mid y \in F(x^0)\}$ and add a vector $(-2z)$, where z is such that $\langle p, z\rangle = \|z\| = 1$: then the distance of $y_m - 2z$ from $F(x^0)$ is exactly 2, hence it belongs to K. Assume that the Lemma is false: then the family $\{K(p, \varepsilon)\}$ has the finite intersection property, i.e., no matter which finite set of pairs (P, E) we choose, we have

(12) $$\bigcap_{(p_i, \varepsilon_i) \in (P, E)} K(p_i, \varepsilon_i) \neq \emptyset.$$

Since K is compact, the intersection over all the $K(p, \varepsilon)$ must be nonempty: let ξ be in this intersection. By a basic separation argument there are $\bar{p}, \bar{\varepsilon}$ such that $\langle \bar{p}, \xi\rangle > \sigma(F(x^0), \bar{p}) + \bar{\varepsilon}$. Hence $\xi \notin K(\bar{p}, \bar{\varepsilon})$, a contradiction. This proves the Lemma.

b) We show now that the result is true in general. For every $y \in F(x^0)$ let $(P, E) \doteq (P, E)_y$ be a finite set of pairs such that $y + N(P, E) \subset M$, and let $\{y_j + N(P, E)_j\}$ be a finite subcover of $F(x^0)$. Consider the finite set of all functionals $\{p_{ij}\}$ for all points y_j and write $Y=U+V$, U a finite dimensional space and V the intersection of all the null spaces of these functionals. Since U is a finite dimensional space, the results is true on U. Set Π to be the projector onto U. For each j, $\Pi(y_j+N(P,E)_j)$ is open and the union over j covers $\Pi(F(x^0))$.

Consider, on U, the bounded closed convex set $\Pi F(x^0)$, its open neighborhood $\bigcup_j \Pi(y_j+N(P,E)_j)$ and let $\sigma^U(\Pi F(x^0),p)$ be its support function. Then there is a finite set of pairs (P^U,E) with $p_i^U\in U^*$, such that the set

$$\{y\in U\,|\,\langle p_i^U, y\rangle \le \sigma^U(\Pi F(x^0), p_i^U)+\varepsilon_i\}$$

is contained in $\bigcup_j (y_j+N(P,E_j))_j$.

Extend each p_i^U to a $p_i\in Y^*$ by $\langle p_i, y\rangle = \langle p_i, u+v\rangle \doteq \langle p_i^U, u\rangle$. Then it is easy to verify that

$$\sigma(F(x^0),p_i)=\sigma^U(\Pi F(x^0),p_i^U).$$

We claim that the set

$$W\doteq\{y\,|\,\langle p_i,y\rangle\le\sigma(F(x^0),p_i)+\varepsilon_i\}$$

is contained in M. Let $\bar{y}$ belong to it and consider $\Pi\bar{y}$. For every i,

$$\langle p_i^U,\Pi\bar{y}\rangle=\langle p_i,\bar{y}\rangle\le\sigma(F(x^0),p_i)+\varepsilon_i=\sigma^U(\Pi F(x^0),p_i^U)+\varepsilon_i.$$

By the above there exists some j^* such that

$$\Pi\bar{y}\in\Pi(y_{j^*}+N(P,E)_{j^*}),\qquad \text{i.e. } \Pi\bar{y}=y_{j^*}+\Pi v,$$

with $v\in N(P,E)_{j^*}$. Fix (p_{ij^*},ε_i) in $(P,E)_{j^*}$. Note that, since V is contained in the null space of p_{ij^*}, for every vector ξ, $\langle p_{ij^*},\xi\rangle=\langle p_{ij^*},\Pi\xi\rangle$. Hence

$$\begin{aligned}\langle p_{ij^*},\bar{y}\rangle&=\langle p_{ij^*},\Pi\bar{y}\rangle=\langle p_{ij^*},\Pi y_{j^*}+\Pi v\rangle\\&\le\langle p_{ij^*},y_{j^*}\rangle+\varepsilon_i.\end{aligned}$$

The above means that $\bar{y}\in y_{j^*}+N(P,E)_{j^*}\subset M$. □

Corollary 2. *Let Y be a finite dimensional space and let $F(x^0)$ be convex and compact. If F is upper hemicontinuous at x^0, then it is upper semicontinuous at x_0.* ▲

The following are some further properties of upper hemicontinuous maps. We recall that Y is supplied with the weak topology.

Proposition 2. *The graph of an upper hemicontinuous set-valued map with closed convex values is closed in $X\times Y$.* ▲

Proof. We shall show that the complement of the graph is open. Let (x^0,y^0) be in the complement of graph(F). Since y^0 and $F(x^0)$ are disjoint closed convex sets and one of the two is compact, there is a functional that strictly separates the two, i.e., there are $p \in Y$ and $\varepsilon > 0$:

$$\langle p, y^0 \rangle \geq \sigma(F(x^0), p) + \varepsilon.$$

By the upper hemicontinuity of F, there is a neighborhood M of x^0 such that when $x \in M$, then $\sigma(F(x), p) \leq \sigma(F(x^0), p) + \varepsilon/2$. On the other hand, the set

$$N \doteq \{y \mid |\langle p, y - y^0 \rangle| \leq \varepsilon/2\}$$

is a neighborhood of y^0. For $(x, y) \in M \times N$,

$$\langle p, y \rangle \geq \langle p, y^0 \rangle - \varepsilon/2 \geq \sigma(F(x^0), p) + \varepsilon/2 > \sigma(F(x), p),$$

i.e., (x, y) is in the complement of graph(F). □

Proposition 3. *If X is compact and if F is upper hemicontinuous with bounded values, then $F(X)$ is bounded in Y.* ▲

Proof. Since the images of F are bounded, the functions $x \to \sigma(F(x), p)$ are finite. Since X is compact and F is upper hemicontinuous, then

$$\forall p \in Y, \quad \phi(p) \doteq \sup_{x \in X} \sigma(F(x), p) < +\infty.$$

Hence ϕ is a *finite* positively homogeneous lower semicontinuous convex function: it is the support function of a *bounded* closed convex subset K of Y. Hence $F(X) \subset K$. □

Remark. We recall that when Y is a reflexive Banach space, any bounded closed convex subset is weakly *compact.*

Proposition 4. *A finite sum and a finite product of upper hemicontinuous set-valued maps is upper hemicontinuous.* ▲

Proof. Let us consider n upper hemicontinuous set-valued maps F_j from X to Y. Since $\sigma(F_1(x) +, \ldots, + F_n(x), p) = \sum_{j=1}^{n} \sigma(F_j(x), p)$. we deduce that $\sum_{j=1}^{n} F_j$: $x \to \sum_{j=1}^{n} F_j(x)$ is upper hemicontinuous.

We consider now n upper hemicontinuous set-valued maps F_j from X_i to Hausdorff locally convex vector spaces Y_j. Set

$$Y \doteq \prod_{j=1}^{n} Y_j \quad \text{and} \quad F \doteq \prod_{j=1}^{n} F_j \colon x \to \prod_{j=1}^{n} F_j(x).$$

Since $\sigma\left(\prod_{j=1}^{n} F_j(x), p\right) = \sum_{j=1}^{n} \sigma(F_j(x), p_j)$, we deduce that $\prod_{j=1}^{n} F_j$ is upper hemicontinuous. □

5. Hausdorff Topology

Let Y be a Hausdorff uniform space whose topology is defined by a family of semi-distances d_K depending upon a parameter K. Let $\mathscr{F}(Y)$ be the family of nonempty closed subsets M of Y. We set $d_K(y, N) \doteq \inf_{x \in N} d_K(x, y)$ and we make the convention $d_K(X, \emptyset) \doteq \infty$. Also set

(1) $$\delta_K(M, N) \doteq \sup_{y \in M} \inf_{x \in N} d_K(x, y) = \sup_{y \in M} d_K(y, N) \in [0, \infty].$$

We associate with the semi-distance d_K the function $\mathfrak{d}_K$ defined by

(2) $$\mathfrak{d}_K(M, N) \doteq \max(\delta_K(M, N), \delta_K(N, M)).$$

Proposition 1. *The functions $\mathfrak{d}_K$ are semi-distances on $\mathscr{F}(Y)$, which define a topology for which $\mathscr{F}(Y)$ is Hausdorff.* ▲

Proof. By definition, $\mathfrak{d}_K$ is symmetric. If M, N, P are three closed subsets, then $\delta_K(M, N) \le \delta_K(M, P) + \delta_K(P, N)$. Indeed, if $y \in M$, $x \in N$, $z \in P$, we know that

$$d_K(y, N) \le d_K(z, N) + d_K(y, z) \le \delta_K(P, N) + d_K(y, z).$$

Taking the infimum when z ranges over P yields

$$d_K(y, N) \le \delta_K(P, N) + d_K(y, P).$$

Taking the supremum when y ranges over N yields

(3) $$\delta_K(M, N) \le \delta_K(P, N) + \delta_K(M, P).$$

Therefore $\mathfrak{d}_K$ is a semi-distance.

To prove that $\mathscr{F}(Y)$ is a Hausdorff space, we have to prove that if $\mathfrak{d}_K(M, N) = 0$ for all parameters K, then $M = N$.

Let $y \in M$. Since $\delta_K(M, N) = 0$ for all K, then $d_K(y, N) = 0$ for all K. This implies that y belongs to the closure of N, which is equal to N. Hence $M \subset N$. In the same way, $N \subset M$. Hence $N = M$. □

Remark. Often one uses triangle inequalities mixing d_K, δ_K and $\mathfrak{d}_K$. Clearly while a chain like $d_K(x, A) \le d_K(x, B) + \delta_K(B, C) + \mathfrak{d}_K(C, A)$ is true, a chain like $\mathfrak{d}_K(A, B) \le \delta_K(y, A) + \delta_K(y, B)$ is false. ▲

Let us set

(4) $$B_K(N, \varepsilon) \doteq \{y \in Y \mid d_K(y, N) \le \varepsilon\}.$$

We see that

(5) $$M \subset B_K(N, \varepsilon) \Rightarrow \delta_K(M, N) \le \varepsilon.$$

Conversely, we note that

(6) $$\delta_K(M, N) \le \varepsilon \Rightarrow M \subset B_K(N, 2\varepsilon).$$

We can use these remarks for characterizing upper and lower semicontinuous set-valued maps.

Proposition 2. *Let F be a set-valued map with closed values from X to the uniform space Y. If F is upper semicontinuous at x_0, then*

$$(7)\qquad \begin{cases} \forall K,\ \forall \varepsilon>0, & \exists \text{ a neighborhood } N(x_0) \text{ of } x_0 \text{ such that} \\ \forall x\in N(x_0), & \delta_K(F(x), F(x_0))\leq\varepsilon. \end{cases}$$

The converse is true when $F(x_0)$ is compact. ▲

Proof. a) If F is upper semicontinuous at x_0, then for any K and for any $\varepsilon>0$, there exists a neighborhood $N(x_0)$ of x_0 such that

$$\forall x\in N(x_0), \quad F(x)\subset B_K(F(x_0), \varepsilon).$$

Hence $\forall x\in N(x_0)$, $\delta_K(F(x), F(x_0))\leq\varepsilon$.

b) Conversely, let $N(x_0)$ be a neighborhood of x_0 such that

$$\forall x\in N(x_0), \quad \delta_K(F(x), F(x_0))\leq\frac{\varepsilon}{2}.$$

Then, $\forall x\in N(x_0)$, $F(x)\subset B_K(F(x_0), \varepsilon)$. Since $F(x_0)$ is compact, any neighborhood N of $F(x_0)$ contains a "ball" $B_K(F(x_0), \varepsilon)$. □

Proposition 3. *Let F be a set-valued map with closed values from X to the uniform space Y. If*

$$(8)\qquad \begin{cases} \forall K,\ \forall \varepsilon>0, & \exists \text{ a neighborhood } N(x_0) \text{ of } x_0 \text{ such that} \\ \forall x\in N(x_0), & \delta_K(F(x), F(x_0))\leq\varepsilon. \end{cases}$$

then F is lower semicontinuous at x_0. The converse is true when we assume that $F(x_0)$ is compact. ▲

Proof. a) Let $N(x_0)$ be a neighborhood of x_0 such that

$$\forall x\in N(x_0), \quad \delta_K(F(x_0), F(x))\leq\frac{\varepsilon}{3}.$$

Then $\forall x\in N(x_0)$, $F(x_0)\subset B_K\left(F(x), \frac{2\varepsilon}{3}\right)$. Hence, for any $y_0\in F(x_0)$ and any $x\in N(x_0)$, $y_0\in B_K(F(x), \varepsilon)$ and thus, there exists $y\in F(x)$ such that $d_K(y_0, y)\leq\varepsilon$. Hence F is lower semicontinuous at x_0.

b) Conversely, suppose that F is lower semicontinuous at x_0 and that $F(x_0)$ is compact. We can cover $F(x_0)$ by n open balls $B_K^0(y_i, \varepsilon)$. Since F is lower semicontinuous, there exist n neighborhoods $N_{y_i}(x_0)$ such that:

$$x\in N_{y_i}(x_0) \quad \text{implies that } F(x)\cap B_K^0(y_i, \varepsilon)\neq\emptyset.$$

Let $N(x_0) \doteq \bigcap_{i=1}^{n} N_{y_i}(x_0)$. Then any $y \in F(x_0)$ belongs to a ball $B_K^0(y_i, \varepsilon)$. Furthermore, we know that for any $x \in N(x_0)$, $F(x) \cap B_K^0(y_i, \varepsilon) \neq \emptyset$. Thus

$$d(y, F(x)) \leq d(y, y_i) + d(y_i, F(x)) < 2\varepsilon.$$

Hence

$$\delta_K(F(x_0), F(x)) = \sup_{y \in F(x_0)} d_K(y, F(x)) \leq 2\varepsilon \quad \text{when } x \in N(x_0). \qquad \square$$

Corollary 1. *A set-valued map with nonempty compact values from X to the uniform space Y is continuous if and only if F is a continuous map from X to the uniform space $\mathscr{F}(Y)$ endowed with the Hausdorff topology.* ▲

Remark. The most important case is the one where Y is a Hausdorff topological vector space whose topology is defined by semi-norms p_K.

When Y is a Banach space, the space $\mathscr{F}(Y)$ is a metric space for the Hausdorff distance

$$\mathfrak{d}(M, N) \doteq \max(\delta(M, N), \mathfrak{d}(N, M)) \quad \text{where}$$
$$\delta(M, N) \doteq \sup_{y \in M} d(y, N).$$

When Y is supplied with the weak topology, the semi-norms p_K are defined by

(9) $$p_K(x) \doteq \sup_{p \in K} \langle p, x \rangle$$

where K is the symmetric convex hull of a finite subset of points of Y^*.

When M and N are closed and convex, we can write

(10) $$\delta_K(M, N) = \sup_{p \in K}(\sigma(M, p) - \sigma(N, p)).$$

Indeed,

$$\delta_K(M, N) = \sup_{y \in M} \inf_{x \in N} \sup_{p \in K} \langle p, y - x \rangle.$$

Since N is convex, since K is convex and compact, we can apply the minimax theorem:

$$\inf_{x \in N} \sup_{p \in K} \langle p, y - x \rangle = \sup_{p \in K} \inf_{x \in N} \langle p, y - x \rangle.$$

Therefore

$$\begin{aligned} \delta_K(M, N) &= \sup_{p \in K} \sup_{y \in M} \inf_{x \in N} \langle p, y - x \rangle \\ &= \sup_{p \in K} (\sup_{y \in M} \langle p, y \rangle - \sup_{x \in N} \langle p, x \rangle) \\ &= \sup_{p \in K} (\sigma(M, p) - \sigma(N, p)). \end{aligned}$$

6. The Selection Problem

Given a family of sets $\{F_\alpha\colon \alpha\in A\}$, a *selection* is a map $\alpha\to f_\alpha$ in F_α. The existence of at least one such map when the index set A is finite is obvious. For an arbitrary set A, it is an axiom: the *Axiom of Choice*. In what follows to the family F_α we shall give the usual pattern of a set-valued map and impose a structure on the map F and on the spaces it acts upon and we shall investigate two main aspects of the selection problem: first we shall study the properties of some explicit selection schemes, i.e., of some selections arising from some prescribed rule. Then we shall prove the existence of selections satisfying some regularity conditions, like continuity or measurability.

Continuous maps need not have, in general, continuous selections.

Example 1. A continuous map from the interval $]-1,+1[$ into the subsets of R^2 with no continuous selection. ▲

Define Φ as

$$\Phi(t) \doteq \begin{cases} \left\{(v_1,v_2) \mid v_1=\cos\theta,\ v_2=t\sin\theta \text{ and} \right. \\ \left. \dfrac{1}{t}\le\theta\le\dfrac{1}{t}+2\pi-|t|\right\} & \text{whenever } t\neq 0 \\ \{(v_1,v_2) \mid -1\le v_1\le 1,\ v_2=0\} & \text{for } t=0 \end{cases}$$

For $t\neq 0$, $\Phi(t)$ is a subset of an ellipse in R^2, whose small axis shrinks to zero as $t\to 0$, so that the ellipse collapses to a segment, $\Phi(0)$. The subset of the ellipse given by $\Phi(t)$ is obtained by removing from it a section, from the angle $\dfrac{1}{t}-|t|$ to the angle $\dfrac{1}{t}$. As t gets smaller, the arclength of this hole decreases while the initial angle increases as $\dfrac{1}{t}$, i.e., it spins around the

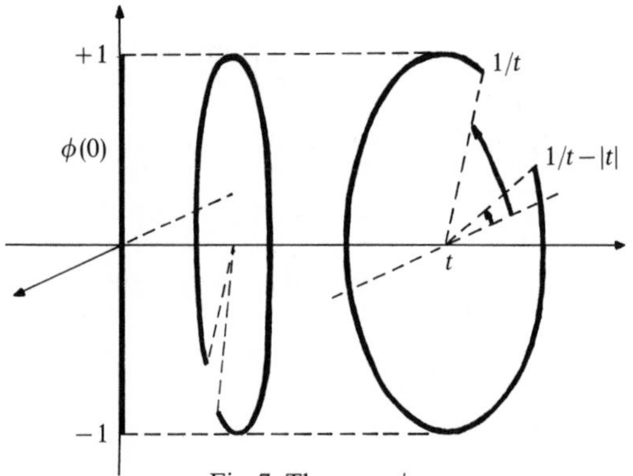

Fig. 7. The map ϕ

origin with increasing angular speed. However the map is Hausdorff continuous at the origin while any continuous selection $f(t)$ defined on $]-1,0[$ or on $]0,+1[$ (for instance, $v_1=\cos 1/t$, $v_2=t\sin 1/t$) could not be continuously extended to the whole $]-1,+1[$. In fact the hole in the ellipse would force this selection to rotate around the origin with an angle $\rho(t)$ between $\frac{1}{t}$ and $\frac{1}{t}+2\pi-|t|$ and $\lim_{t\to 0} f(t)$ cannot exist. □

Example 2. A continuous map from the unit ball of R^2 into its compact subsets that has neither a continuous selection nor fixed points. ▲

Let B be the closed unit ball in R^2, Q the rectangle $0\leq\rho\leq 1$, $0\leq\theta<2\pi$. Introduce the map π defined by $\pi(\rho,\theta)\doteq(\rho\cos\theta,\ \rho\sin\theta)$; it is a bijection from Q onto $B\backslash\{0\}$. On Q define the set-valued map Φ to the subsets of R by

$$\Phi(\rho,\theta)\doteq\{\omega\,|-\theta-2\pi(1-\rho)\leq\omega\leq-\theta+2\pi(1-\rho)\}.$$

In particular, $\Phi(1,\theta)=\{-\theta\}$.

Set $\tau\colon R\to[0,2\pi[$ to be $\tau(x)\doteq x \bmod (2\pi)$ and define ψ, from Q to the subsets of $[0,2\pi[$ to be

$$\psi(\rho,\theta)\doteq\tau(\Phi(\rho,\theta)).$$

Finally set F, from B into its subsets by

$$F(x,y)\doteq\begin{cases}\pi(1,\psi(\pi^{-1}(x,y))), & (x,y)\neq(0,0),\\ \{(\xi,\eta)\,|\,(\xi^2+\eta^2)=1\}, & x=y=0.\end{cases}$$

The map F is continuous with compact values, subsets of $F(0,0)$.

Assume F admits a continuous selection f. Then f maps B into itself and, by Brouwer's Theorem, it has a fixed point x_*. Hence $x_*=f(x_*)\in F(x_*)$, and $\|x_*\|=1$. However, then, $F(x_*)=\{-x_*\}$, a contradiction. □

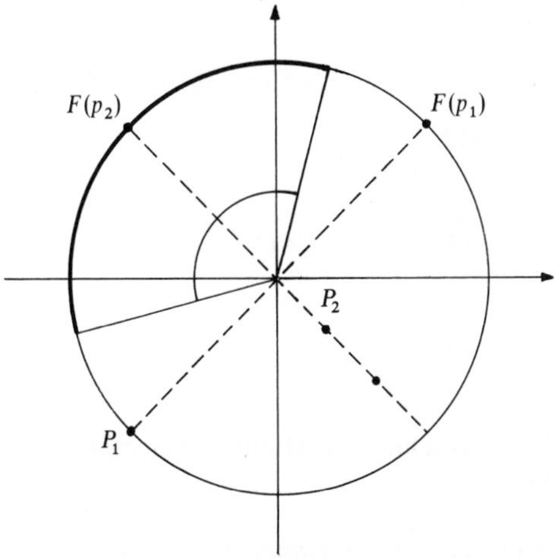

Fig. 8. The map F

7. The Minimal Selection

In this section, we show that a natural selection process of a closed convex valued map, the minimal selection, is continuous whenever the map is such. Unfortunately, there are lipschitzean maps whose minimal selections are not lipschitzean.

Theorem 1. *Let X be a metric space, Y a Hilbert space and F, from X into the closed convex subsets of Y, be continuous. Then the mapping $x \to m(F(x))$ is a continuous selection from F.* ▲

Proof. Fix $\varepsilon > 0$ and $x \in X$. By the continuity of F, whenever x is sufficiently close to x^*, then $\|m(F(x)\|^2 \le \|m(F(x^*))\|^2 + \varepsilon$. This proves the claim when $\|m(F(x^*))\| = 0$ since then $m(F(x)) \in \sqrt{\varepsilon}\, B$. Assume then that $\|m(F(x^*))\| > 0$. For every x sufficiently close to x^* there is some $y_x \in F(x^*)$ such that $\|m(F(x)) - y_x\| \le \varepsilon / \|m(F(x^*))\|$. Hence

$$\begin{aligned} &\langle m(F(x^*)), m(F(x))\rangle \\ (1) \qquad &\quad = \langle m(F(x^*)), y_x\rangle + \langle m(F(x^*)), m(F(x)) - y_x\rangle \\ &\quad \ge \langle m(F(x^*)), y_x\rangle - \varepsilon. \end{aligned}$$

The variational characterization of the projector of best approximation yields, for every $y \in F(x^*)$ (in particular, for $y = y_x$) that

$$\langle m(F(x^*)), y\rangle \ge \langle m(F(x^*)), m(F(x^*))\rangle$$

so that (1) becomes

$$(2) \qquad \langle m(F(x^*)), m(F(x)) - m(F(x^*))\rangle \ge -\varepsilon.$$

On the other hand

$$\begin{aligned} \|m(F(x))\|^2 = \|m(F(x^*))\|^2 + \|m(F(x^*)) - m(F(x))\|^2 \\ + 2\langle m(F(x^*)), m(F(x)) - m(F(x^*))\rangle. \end{aligned}$$

This equality and (2) yield

$$\begin{aligned} &\|m(F(x^*)) - m(F(x))\|^2 - 2\varepsilon \\ &\quad \le \|m(F(x^*)) - m(F(x))\|^2 + 2\langle m(F(x^*)), m(F(x)) - m(F(x^*))\rangle \\ &\quad \le \varepsilon, \text{ i.e.} \end{aligned}$$

$$\|m(F(x^*)) - m(F(x))\|^2 \le 3\varepsilon. \qquad \square$$

Counterexample: Non-Lipschitzean Minimal Selection of a Lipschitzean Map

We wish to define a closed convex valued F, from a closed subset $K \subset R^2$ into R^2 that is lipschitzean but is such that $x \to m(F(x))$ is not lipschitzean.

We begin by defining a map Φ, from R into the subsets of R^2, by setting for each δ,

$$\Phi(\delta) \doteq \{(x, y) \mid x = \sin\delta - t\cos\delta,\ y = \cos\delta + t\sin\delta,\ t\in[0, 2\sin\delta] \text{ or } t\in[\sin\delta, 0]\}.$$

For each δ, $\Phi(\delta)$ is a segment of length $|2\sin\delta|$, tangent to the unit circle at the point $(\sin\delta, \cos\delta)$. In particular, $\Phi(0) = \{(1, 0)\}$.

The Hausdorff distance between images of two different values of δ can be estimated as follows.

a) $\mathfrak{d}(\Phi(\delta), \Phi(-\delta)) \leq 2\sin^2\delta$.

b) Whenever δ_1 and δ_2 are of the same sign and $|\delta_1| > |\delta_2|$ (we can assume that both are positive)

$$\mathfrak{d}(\Phi(\delta_1), \Phi(\delta_2)) \leq 6(|\delta_1| - |\delta_2|).$$

In fact, call AB and $A'B'$ the end points of $\Phi(\delta_1)$ and $\Phi(\delta_2)$, tangent to the unit circle in B and B' respectively. It is easy to see that $\mathfrak{d}(B', B) = 2\sin\dfrac{|\delta_1| - |\delta_2|}{2}$. Also

$$A = (\sin\delta - 2\sin\delta\cos\delta, \cos\delta + 2\sin\delta\sin\delta)$$
$$A' = (\sin\delta' - 2\sin\delta'\cos\delta', \cos\delta' + 2\sin\delta'\sin\delta')$$

so that

$$\|A - A'\|^2 = [\sin\delta - \sin\delta' - (\sin 2\delta - \sin 2\delta')]^2 + [\cos\delta - \cos\delta' + 2\sin^2\delta - 2\sin^2\delta']^2.$$

Without asking for the best possible estimate the mean value theorem yields

$$\|A - A'\|^2 \leq (\delta_1 - \delta_2)^2 c^2$$

where c is at most 6. Hence in this case

$$\mathfrak{d}(\Phi(\delta_1), \Phi(\delta_2)) \leq c(\delta_1 - \delta_2).$$

Set $\delta(x) = \arcsin\sqrt{|x|}$. Define K as

$$K \doteq \left\{(x, y) \mid y = \delta(x),\ x\in\left[-\frac{\pi}{2}, \frac{\pi}{2}\right]\right\}$$

and F, from K into the subsets of R^2, by

$$F(x, \delta(x)) \doteq \Phi(\operatorname{sign}(x)\delta(x)).$$

We claim that: a) F is lipschitzean on K; b) $m(F(x))$ is not lipschitzean on K.

a) Let $P_1 = (x_1, \delta(x_1))$, $P_2 = (x_2, \delta(x_2))$, $|x_1| \geq |x_2|$, and set $P_2^* = (|x_2| \operatorname{sign} x_2, \delta(x_2))$. Then $\mathfrak{d}(\Phi(P_1), \Phi(P_2)) \leq \mathfrak{d}(\Phi(P_1), \Phi(P_2^*)) + \mathfrak{d}(\Phi(P_2^*), \Phi(P_2))$ and

$$\mathfrak{d}(\Phi(P_1), \Phi(P_2^*)) \leq c(|\delta_1| - |\delta_2|) \leq c\|P_1 - P_2^*\| \leq c\|P_1 - P_2\|$$

$$\mathfrak{d}(\Phi(P_2^*), \Phi(P_2)) = \max\{0, 2\sin^2|\delta(x_2)|\}$$
$$\leq |x_2| \leq \|P_2 - P_2^*\| \leq \|P_1 - P_2\|.$$

Hence

$$\mathfrak{d}(\Phi(P_1), \Phi(P_2)) \leq (c+1)\,\|P_1 - P_2\|.$$

b) For $P_2 = -P_1$ we have that

$$\|m(F(P_1)) - m(F(-P_1))\| = 2\sin\delta(x_1) = 2\sqrt{|x_1|} = 2\sqrt{2\frac{|x_1|}{2}} = 2\sqrt{\|P_2 - P_1\|/2}$$

and the map $x \to 2\sqrt{|x|/2}$ is not lipschitzean.

An Application: The Parametrization Problem

To simplify our description of a map we are interested in knowing whether a given set-valued map $F(x)$ can be represented as the image of a fixed set $\mathscr{U}$ by a function, i.e., whether, given F, we can find a suitable set $\mathscr{U}$ and a single valued function f of the two variables x and u such that, for every x,

$$F(x) = f(x, \mathscr{U}).$$

If this is the case we say that F is parametrized by $\mathscr{U}$. In general we call the problem of representing F in the above form a parametrization problem.

Since working with a map F that admits a parametrization is definitely simpler, one wonders how general this parametrization property can be. Remark that, in case when f is continuous, for every fixed u^0 the map $x \to f(x, u^0)$ would be a continuous selection from F. Hence, for instance, the maps described in Section 6 *cannot be parametrized* as above. From this remark it follows that the parametrization problem is more demanding than the selection problem and we should expect cases when there are continuous selections from F while F cannot be parametrized as $f(x, \mathscr{U})$.

The possibility of parametrizing F depends largely on having a selection procedure that on one hand is general enough to yield a selection through any previously chosen point of $F(x)$ while, on the other, is "canonical" i.e. has a form in which the chosen point enters only as a parameter. The preceding minimal selection has exactly these properties.

Lemma 1. *Set K to be a compact subset of a Hilbert space Y and $\mathscr{C}(K)$ to be the set of compact convex subsets of K.*

Let us denote by $\pi(H, x)$ the projection of x onto the compact convex subset H.

We supply $\mathscr{C}(K)$ with the Hausdorff distance (see Section 5). Then the map $(H, x) \to \pi(H, x)$ is continuous from $\mathscr{C}(K) \times K$ to K. ▲

Proof. To check the continuity of π, remark that the real valued map $(H, x) \to d(x, H)$ is continuous on the product of $\mathscr{C}(K)$ (supplied with the

Hausdorff metric) and Y. Hence the map B

$$B\colon\ (H,x)\to\{y\in K \mid \|y-x\|\leq d(x,H)\}$$

is a continuous map with values in the non-empty compact subsets of K. Certainly so is the map $(H,x)\to H$.

By the very definition of π,

$$\pi(H,x)=H\cap B(H,x).$$

Since both maps at the right hand side have closed graph, π has a closed graph. Being single valued, it is continuous. □

Theorem 2. *Let X be a metric space, F, from X into the nonempty compact convex subsets of R^n, be continuous. Then there exists a continuous map f, from the product of X and B, the unit ball of R^n, into Y such that*

$$\forall x\in X,\qquad F(x)=f(x,B).$$ ▲

Proof. Set $p(x)\doteq\max\{1,\|F(x)\|\}$: p is continuous and so is the map

$$(x,u)\to\pi(F(x),p(x)\,u).$$

Set f to be $f(x,u)\doteq\pi(F(x),p(x)\,u)$.

Then f is continuous and $f(x,B)\in F(x)$. Moreover $p(x)B\supset F(x)$, hence $f(x,B)=F(x)$.

8. The Chebishev Selection

Another explicit selection process is to pass through the Chebishev center. In order to introduce it we need the following property of Hilbert spaces.

Proposition 1. *Let X be a Hilbert space, $a,b,x\in X$. Let $\delta\doteq\|b-a\|$ and $\rho\doteq\max\{\|x-a\|,\|x-b\|\}$. Then for any t, $0\leq t\leq 1$, we have that*

$$\|x-[t\,a+(1-t)\,b]\|^2\leq\rho^2-t(1-t)\,\delta^2.$$

Proof. Assume that $\rho=\|x-a\|$. Then since

$$\|x-b\|^2=\rho^2+\delta^2-2\langle x-a,b-a\rangle,$$

we must have $\delta^2-2\langle x-a,b-a\rangle\leq 0$. Equivalently, for any t, $0\leq t\leq 1$,

$$-2(1-t)\,\langle x-a,b-a\rangle\leq-(1-t)\,\delta^2.$$

On the other hand,

$$\|x-(t\,a+(1-t)\,b)\|^2=\rho^2+(1-t)^2\,\delta^2-2\langle x-a,t\,a+(1-t)\,b-a\rangle$$
$$=\rho^2+(1-t)^2\,\delta^2-2(1-t)\,\langle x-a,b-a\rangle,$$

and, by the preceding remark, the right-hand side is bounded above by

$$\rho^2+(1-t)^2\,\delta^2-(1-t)\,\delta^2=\rho^2-t(1-t)\,\delta^2.$$

Hence we have shown that

$$\|x-[t\,a+(1-t)\,b]\|^2\le\rho^2-t(1-t)\,\delta^2.$$ □

Definition 1. Let K be a bounded subset of a Hilbert space X. The *Chebishev radius* r_c of K is defined to be

$$r_c \doteq \inf\{\rho\,|\,\exists k\,|\,K\subset k+\rho B\}.$$

For each $\rho>r_c$, set $C_\rho\doteq\{k\,|\,K\subset k+\rho B\}$. ▲

Proposition 2. *The set* $\bigcap_{\rho>r_c} C_\rho$ *consists of a single point,* $c(K)$, *called* the Chebishev center *of* K. *It has the property that*

$$c(K)+r_c B\supset K.$$

Moreover, K *convex implies that* $c(K)\in\overline{K}$. ▲

Proof. By the very definition, for each $\rho>r_c$, C_ρ is a bounded, closed and non-empty subset of X.

By **Proposition 1**, it is also convex, since $\sqrt{\rho^2-t(1-t)\delta^2}\le\rho$. Hence it is weakly compact and consequently, $\bigcap_{\rho>r_c} C_\rho\neq\emptyset$.

Moreover, as one sees by contradiction, for every c in $\bigcap_{\rho>r_c} C_\rho$, $c+r_c B\supset K$. We claim that this intersection contains only one point. Assume it contains two, a and b, and consider their average $\frac{1}{2}(a+b)$. Every x in K is contained in $a+r_c B$ and $b+r_c B$, hence in

$$(a+b)/2+\sqrt{r_c^2-\tfrac{1}{4}\,\|a-b\|^2}\,B.$$

The radius of this last ball is strictly smaller than r_c, a contradiction.

Finally, assume that $c(K)\notin\overline{K}$. Then by displacing $c(K)$ slightly in the direction orthogonal to any hyperplane strictly separating $c(K)$ from the closed and convex $\overline{K}$, it would possible to decrease all the distances from $c(K)$ to the points of K by a factor not depending on the given point, a contradiction to the definition of $c(K)$. □

Theorem 1. *Let* F *be a continuous map from* K *into the bounded closed convex subsets of a Hilbert space* X. *Then the single-valued map* c *defined by*

$$c(x)\doteq c(F(x))$$

is a continuous selection. ▲

Proof. a) To begin with, the map $x \to r_c(F(x))$ is continuous. In fact, fix $x^0 \in K$ and $\eta > 0$. There exists a point a such that $F(x^0) \subset a + (r_c(F(x^0)) + \eta/2)B$; since F is u.s.c. (in the ε sense) there exists $\delta^1 > 0$ such that $d(x, x^0) < \delta^1$ implies $F(x) \subset F(x^0) + \eta/2\,B$, so that $F(x) \subset a + (r_c(F(x^0)) + \eta)B$. Therefore $r_c(x) \le r_c(x^0) + \eta$ when x is sufficiently close to x^0. On the other hand, since F is also lower semi-continuous (in the ε sense), there exists a $\delta_2 > 0$ such that $d(x, x^0) < \delta_2$ implies $F(x) + \eta/2B \supset F(x^0)$. Hence for some a, $a + r_c(F(x)) + \eta B \supset F(x) + \eta/2B \supset F(x^0)$, i.e., $r_c(F(x^0)) \le r_c(F(x)) + \eta$.

b) Assume $c(x)$ is not continuous at x^0. There exist $\eta > 0$ and points x_n arbitrarily close to x^0 such that $|c(x_n) - c(x^0)| > \eta$. Let $\varepsilon > 0$ be so small that

$$\sqrt{(r_c(F(x^0)) + \varepsilon)^2 - \tfrac{1}{4}\eta^2} < r_c(F(x^0)) - 2\varepsilon$$

and let n be so large that $F(x_n) \subset F(x^0) + \varepsilon B$ and that $|r(F(x_n)) - r(F(x_0))| < \varepsilon$.

Then on one hand

$$F(x_n) \subset c(x_n) + r_c(F(x_n))\,B \subset c(x_n) + (r_c(F(x^0)) + \varepsilon)\,B$$

while on the other

$$F(x_n) \subset F(x^0) + \varepsilon B \subset c(x^0) + (r_c\,F(x^0) + \varepsilon)\,B.$$

The distance of any point of $F(x_n)$ from the average $\tfrac{1}{2}(c(x_n) + c(x^0))$ is at most

$$\sqrt{(r_c(F(x^0)) + \varepsilon)^2 - \tfrac{1}{4}\eta^2}$$

which is strictly smaller than $r_c(F(x^0)) - 2\varepsilon$ i.e., strictly smaller than $r_c(F(x_n))$, contradicting the definition of the Chebishev radius. □

We note that although F continuous implies that $c(F(x))$ is continuous, the center does not inherit all the regularity properties of a map F. The following is an example to show that F lipschitzean does not imply that $c(F(x))$ is lipschitzean.

Proposition 3. *There exists a lipschitzean map F with compact convex values such that $x \to c(F(x))$ is not lipschitzean.* ▲

Proof. We shall define a map F from a subset $D(F) \subset \mathbb{R}^2$ into the compact convex subsets of $\mathbb{R}^2$. Let $I = [-\tfrac{1}{4}, \tfrac{1}{4}]$. For $k \in I$, consider the closed circle of radius 1 centered at $(k, 0)$ and call $\Phi(k)$ its intersection with the strip $|x| \le |k|$: it is a closed and convex subset of $\mathbb{R}^2$. In particular, $\Phi(0)$ is the segment $x = 0$, $|y| \le 1$.

We first claim that when a and b have the same sign (we take $|b| > |a|$) $\mathfrak{d}(\Phi(a), \Phi(b)) \le |b - a|$. In fact, on the one hand, for every point $(x, y) \in \Phi(b)$, there is a point $(x', y) \in \Phi(a)$ such that $|x' - x| \le |b - a|$, i.e., a horizontal displacement of points of $\Phi(a)$ by at most $|b| - |a|$ covers $\Phi(b)$. On the other hand, let $(x, y) \in \Phi(a)$ and consider the point (x, y') of $\Phi(b)$ nearest to (x, y). Their distance is at most

$$\sqrt{1 - 4|a|^2} - \sqrt{1 - |b + a|^2},$$

i.e., at most

$$(|b|^2+2|a|\,|b|-3|a|^2)(\sqrt{1-4|a|^2}+\sqrt{1-|b+a|^2})^{-1}.$$

The term inside the first parentheses, being $(|b|-|a|)\,(|b|+3|a|)$ is bounded by $(|b|-|a|)$ while the term inside the second is at least $\sqrt{\frac{3}{2}}$. Hence $\Phi(a) \subset \Phi(b)+(|b|-|a|)B$. Combining it with the preceding inequality we obtain that $\mathfrak{d}(\Phi(b), \Phi(a)) \leq |b-a|$.

Let us define next the mappings $f\colon I \to \mathbb{R}$ and $g\colon I \to \mathbb{R}$ given by $g(x) = \frac{1}{2}\operatorname{sign}(x)\sqrt{x^2+2|x|}$ and $f(x)=\sqrt{\frac{3}{2}}-|g(x)|$. Define the map F, with domain the graph of $f(x)$ and range the compact convex subsets of $\mathbb{R}^2$, by setting

$$F((x, f(x))) = \Phi(g(x)).$$

We claim that F is lipschitzean with constant $\sqrt{2}$, and that its center is not lipschitzean.

We consider two points $P_1=(a, f(a))$ and $P_2=(b, f(b))$ on the graph of f, and we evaluate $\mathfrak{d}(F(P_1), F(P_2))$.

i) Let a and b be of the same sign. Then

$$\begin{aligned}\mathfrak{d}(F(P_1), F(P_2)) &= \mathfrak{d}(\Phi(g(a)), \Phi(g(b))) \leq |g(a)-g(b)| \\ &= |f(a)-f(b)| \leq d(P_1, P_2).\end{aligned}$$

ii) Assume $b=-a$. Then

$$D \doteq \mathfrak{d}(\Phi(g(a)), \Phi(g(-a))) = \mathfrak{d}(\Phi(g(a)), \Phi(-g(a))) = \sqrt{1+4|g(a)|^2}-1.$$

By the very definition of g,

$$\sqrt{1+4|g(a)|^2}-1 = 2|a| = |b-a|.$$

iii) Let a and b be of different signs; we assume that $|a|<|b|$. Set $P^* \doteq (-a, f(-a))$. Combining points i) and ii) we have that

$$\mathfrak{d}(F(P_1), F(P_2)) \leq \mathfrak{d}(F(P_1), F(P^*)) + \mathfrak{d}(F(P^*), F(P_2)) \leq |P_1-P^*|+|P^*-P_2|.$$

On the other hand, since

$$\langle P_1-P^*, P_2-P^*\rangle \leq 0,$$

$|P_1-P_2| \geq \sqrt{|P_1-P^*|^2+|P^*-P_2|^2}$. Combining this inequality with

$$|P_1-P^*|+|P^*-P_2| \leq \sqrt{2}\sqrt{|P_1-P^*|^2+|P^*-P_2|^2}$$

one obtains

$$\mathfrak{d}(F(P_1), F(P_2)) \leq \sqrt{2}\,|P_1-P_2|$$

proving that F is lipschitzean with constant $\sqrt{2}$.

We show next that the center of F is not lipschitzean. Take two symetric points on the graph of f, $P_1 \doteq (a, f(a))$ and $P_2 \doteq (-a, f(-a))$.

The Chebishev center of $F(P_1)$ is in $(g(a), 0)$ while that of $F(P_2)$ is in $(-g(a), 0)$. Their distance is $2|g(a)| = \sqrt{a^2+|a|}$, while the distance of P_1 and P_2 is $2|a|$. The ratio $\sqrt{a^2+|a|}/2|a|$ gets unbounded as $a \to 0$. Hence no lipschitz constant for the center can exist.

9. The Barycentric Selection

The aim of this section is to present a selection procedure of a map F with *compact convex* images, that is lipschitzean whenever F is lipschitzean. This stronger regularity of the selection is obtained at the expenses of stronger requirements on the space where F takes values and on the global boundedness of F.

Moreover the lipschitz constant K of the selection, besides being different from the lipschitz constant of F, depends on the dimension n of the space $\mathbb{R}^n$ containing the images of F: the same map F, when the images are considered as subsets of spaces of different dimensions, has selections with different lipschitz constants.

The following is the main result of this section.

Theorem 1. *Let X be a metric space, F, from X into the compact convex subsets of $\mathbb{R}^n$, be lipschitzean. Assume moreover that for some M, $F(x) \subset MB$, for every x in X. Then there exist a constant K and a single valued map f from X into $\mathbb{R}^n$, lipschitzean with constant K, a selection from F.* ▲

The proof will be obtained by considering the barycentric selection of F. We define it in the following way.

Let $A \subset \mathbb{R}^n$ be a compact convex body, i.e. a compact convex set with nonempty interior, and let m_n be the n-dimensional Lebesgue measure. Since $m_n(A)$ is positive we can define the barycenter of A as

(1) $$b(A) = \frac{1}{m_n(A)} \int_A x \, dm_n.$$

Proposition 1. *The barycenter of A, $b(A)$, belongs to A.* ▲

Proof. Assume the contrary: $d(b(A), A)$ is positive. Set a to be $\pi_A(b(A))$, b to be $b(A)$ and $p \doteq b - a$.

By the characterization of the best approximation (*Corollary* 0.6.1) we have that for all x in A, $\langle x-a, p\rangle \leq 0$. However from

$$p = b - a = \frac{1}{m_n(A)} \int_A (x-a)\, dm_n$$

we have

$$\begin{aligned} \|p\|^2 &= \left\langle \frac{1}{m_n(A)} \int_A (x-a)\, dm_n, p \right\rangle \\ &= \frac{1}{m_n(A)} \int_A \langle x-a, p\rangle \, dm_n \leq 0, \end{aligned}$$

a contradiction; hence $b(A)$ belongs to A. □

In order to take advantage of the barycenter to obtain a selection procedure we have to fix a measure (say m_n) and make sure that the sets we are considering have positive measures. Since this is not the case, in general, for the images of a set valued map, a possible trick consists in taking the barycenter of a ball around our sets.

Proposition 2. *Let $A\subset\mathbb{R}^n$ be compact and convex and consider $A^1 \doteq A+B$. Then $b(A^1)$ belongs to A.* ▲

Proof. As above assume it is not so. Set a to be $\pi_A(b(A^1))$, the point of A nearest to $b=b(A^1)$, set $p\doteq b-a$ and $\hat{p}=p/\|p\|$. Then

$$\|p\|^2=\frac{1}{m_n(A^1)}\int_{A^1}\langle x-a,p\rangle\, dm_n \tag{2}$$

and as, before, to reach a contradiction it is enough to show that the right hand side is non positive.

It is convenient to consider Sp, the linear transformation mapping x into its symmetric with respect to the hyperplane orthogonal to p through a:

$\mathrm{Sp}(x)=a+(x-a)-2\langle x-a,\hat{p}\rangle\,\hat{p}$.

Set $A^1_+\doteq\{x\in A^1|\langle x-a,p\rangle>0\}$, $A^1_-\doteq\{x\in A^1|\langle x-a,p\rangle\le 0\}$.

We remark that $\mathrm{Sp}(A^1_+)\subset A^1$. In fact fix x in A^1_+ and consider $\mathrm{Sp}(x)$:

Set x' to be the projection of $\pi_A(x)$ on the line through x and $\mathrm{Sp}(x)$. By the Pythagorean theorem to show that

$\|x-\pi_A(x)\|\ge\|\mathrm{Sp}(x)-\pi_A(x)\|$ it is enough to show that

$\|x-x'\|\ge\|\mathrm{Sp}(x)-x'\|$. We have that

$\|x-x'\|=\langle x-x',\hat{p}\rangle=\langle x-a,\hat{p}\rangle-\langle x'-a,\hat{p}\rangle$

and

$$\begin{aligned}\|\mathrm{Sp}(x)-x'\|&=-\langle \mathrm{Sp}(x)-x',\hat{p}\rangle=-\langle \mathrm{Sp}(x)-a,\hat{p}\rangle+\langle x'-a,\hat{p}\rangle\\&=\langle x-a,\hat{p}\rangle+\langle x'-a,\hat{p}\rangle.\end{aligned}$$

Since, again by the characterization of the best approximation, x' belongs to A^1_-,

$d(\mathrm{Sp}(x),A)\le\|\mathrm{Sp}(x)-\pi_A(x)\|\le\|x-\pi_A(x)\|=d(x,A)\le 1$,

Then $\mathrm{Sp}(x)$ belongs to A^1.

Write A^1 as $(A^1_+\cup\mathrm{Sp}(A^1_+))\cup(A^1\backslash(A^1_+\cup\mathrm{Sp}(A^1_+)))$

and consider the integral in (2) separately on these two subsets. Remark that the first is invariant with respect to the transformation Sp, that the determinant of the Jacobian of the transformation Sp is one and that the map $x\to\langle x-a,\hat{p}\rangle$ is antisymmetric with respect to Sp. The change of variables formula hence yields

$$(3) \qquad \int_{\mathrm{Sp}(A'_+ \cup \mathrm{Sp}(A'_+))} \langle x-a, p\rangle = \int_{(A'_+ \cup \mathrm{Sp}(A'_+))} \langle \mathrm{Sp}(x)-a, p\rangle = -\int_{\mathrm{Sp}(A'_+ \cup \mathrm{Sp}(A'_+))} \langle x-a, p\rangle.$$

Hence this integral is zero.

Since $A^1 \backslash (A^1_+ \cup \mathrm{Sp}(A^1_+))$ is contained in A^1_-,

$$\int_{A^1} \langle x-a, p\rangle \leq 0$$

the desired contradiction. □

Proof of Theorem 1. The above proposition shows that the single valued map $b^1 \doteq x \to b(F(x)+B)$ is a selection from F. We have to prove that it is a lipschitzean selection.

Fix x and x'. Call $\Phi \doteq F(x)+B$, $\Phi' \doteq F(x')+B$. Then

$$\begin{aligned}
&\frac{1}{m_n(\Phi)}\int_{\Phi} x\, dm_n - \frac{1}{m_n(\Phi')}\int_{\Phi'} x\, dm_n \\
&\leq \left|\left(\frac{1}{m_n(\Phi)} - \frac{1}{m_n(\Phi')}\right)\int_{\Phi\cap\Phi'} x\, dm_n\right| \\
(4) \qquad &\quad + \left|\frac{1}{m_n(\Phi)}\int_{\Phi\backslash\Phi'} x\, dm_n - \frac{1}{m_n(\Phi')}\int_{\Phi'\backslash\Phi} x\, dm_n\right| \\
&\leq |m_n(\Phi)-m_n(\Phi')|\,(M+1)/(m_n(B))^2 \\
&\quad + \{m_n(\Phi\backslash\Phi')+m_n(\Phi'\backslash\Phi)\}\,(M+1)/m_n(B).
\end{aligned}$$

We wish to express the above estimate in terms of $\mathfrak{d}(\Phi, \Phi')$. For this purpose, we begin to compare $m_n(\Phi+\delta B)$, $\delta>0$, and $m_n(\Phi)$. Since the unit ball of $\mathbb{R}^n$ is contained in the unit cube $\{|x_i|\leq 1,\ i=1,\ldots,n\}$, we can as well estimate

$$(5) \qquad m_n\{\varphi+\sum \delta_i e_i \,|\, \varphi\in\Phi,\ |\delta_i|\leq\delta\}$$

where $\{e_i\}$ is an orthonormal basis.

From elementary calculus we have that when S is a convex set and v a unit vector, the measure of $\{S+\delta_x v \,|\, |\delta_x|\leq\delta\}$ is $m_n(S)+|\delta|\, m_{n-1}(P_v(S))$ where P_v is the projection of S into the hyperplane normal to v through the origin ($P_v(S)$ is the "shadow" of S).

Denote by

$$\Phi_v \doteq \left\{\varphi+\sum_{i=1}^{v}\delta_i e_i \,|\, \varphi\in\Phi,\ \delta_i\leq\delta\right\}$$

and by P_i the projection along the direction e_i.

Recursively we obtain

$$m_n(\Phi_n) \leq m_n(\Phi) + \delta\sum_{j=0}^{n-1} m_{n-1}(P_{n-j}(\Phi_{n-j})).$$

Since Φ is contained in $(M+1)B$, each element of each $P_j(\Phi_j)$ has a distance from the origin of at most $(M+1)+\delta\sqrt{n}$, so that, setting B_{n-1} the unit ball in $\mathbb{R}^{n-1}$,

$$\begin{aligned} m_n(\Phi+\delta B) &\leq m_n(\Phi_n) \\ &\leq m_n(\Phi)+\delta n\, m_{n-1}((M+1+\delta\sqrt{n})\, B_{n-1}). \\ &\leq m_n(\Phi)+\delta K_1 \end{aligned}$$

for some constant K_1.

Set δ to be $\mathfrak{d}(\Phi, \Phi')$. Then $\Phi'\subset\Phi+\delta B$ and $\Phi\subset\Phi'+\delta B$, hence

$$m_n(\Phi\backslash\Phi')\leq m_n(\Phi'+\delta B)-m_n(\Phi'), \quad \text{and} \quad m_n(\Phi'\backslash\Phi)\leq m_n(\Phi+\delta b)-m_n(\Phi).$$

Analogously, $|m_n(\Phi)-m_n(\Phi')|\leq K_1\delta$. Hence by (4) we obtain

$$|b(F(x)+B)-b(F(x')+B)|\leq K_2\,\mathfrak{d}(F(x)+B, F(x')+B)$$

for a suitable K_2. Finally, let L be the lipschitz constant of F and set K to be LK_2. We have

$$\begin{aligned} |b^1(x)-b^1(x')| &\leq K_2\,\mathfrak{d}(F(x)+B, F(x')+B) \\ &\leq K_2\,\mathfrak{d}(F(x), F(x'))\leq K\,d(x, x'), \end{aligned}$$

i.e. $f=b^1$ is the required lipschitzean selection. □

10. Selection Theorems for Locally Selectionable Maps

The concept of locally selectionable maps has been introduced because these set valued maps do possess continuous selections when they are defined on paracompact topological spaces.

Definition 1. We say that a set valued map F from X to Y is "locally (continuously) selectionable" at $x_0\in X$ if for all $y_0\in F(x_0)$, there exist an open neighborhood $N(x_0)$ of x_0 and a continuous map $f\colon N(x_0)\to Y$ such that

(1) $\quad f(x_0)=y_0 \quad$ and for all $x\in N(x_0)$, $f(x)\in F(x)$

F is said to be "locally selectionable" if it is locally selectionable at every $x_0\in X$. ▲

We begin by pointing out:

Proposition 1. *Any locally selectionable set valued map is lower semicontinuous.* ▲

Proof. Let $v_0\in F(x_0)$ and $N(v_0)$ be a neighborhood of v_0. Let $f\colon N_1(x_0)\to Y$ be a local selection of F at x_0 passing through v_0. Since f is continuous, there exists a neighborhood $N_2(x_0)$ such that $f(x) \in N(v_0)$ for all $x \in N_2(x_0)$. Hence $F(x) \cap N(v_0)$ is not empty for all $x \in N_2(x_0)$ because it contains $f(x)$. □

Proposition 2. *Let X be paracompact and F be a locally selectionable map with convex values from X to a topological vector space Y. Then F has a continuous selection.* ▲

Proof. We associate with any $y \in X$ a point $z \in F(y)$ and a local continuous selection f_y: $N(y) \to Y$ (satisfying $f_y(y) = z$ and $f_y(x) \in F(x)$ when $x \in N(y)$).

Since X is paracompact, there exists a continuous partition of unity $a_y(\cdot)$ associated with the covering of X by the open neighborhoods $N(y)$. We define the map f by

$$f(x) = \sum_{y \in X} a_y(x)\, f_y(x) = \sum_{y \in I(x)} a_y(x)\, f_y(x)$$

where $I(x)$ is the non-empty *finite* set of points $y \in X$ such that $a_y(x) > 0$. Hence f is continuous, as a sum of products of continuous functions. If $y \in I(x)$, then $x \in N(y)$ because the support of a_y is contained in $N(y)$. Therefore $f_y(x) \in F(x)$ and consequently, since $F(x)$ is convex, $f(x) = \sum_{y \in I(x)} a_y(x) f_y(x) \in F(x)$. Hence f is a continuous selection of F.

Examples of Locally Selectionable Maps

We note that if the graph of F is open, then $F^{-1}(y)$ is open for all $y \in Y$.

Proposition 3. *If $F^{-1}(y)$ is open for all $y \in Y$ (and, in particular, if the graph of F is open), then F is locally selectionable.* ▲

Proof. Indeed, if $x_0 \in X$ and $y_0 \in F(x_0)$, we take $N(x_0) \doteq F^{-1}(y_0)$, which is open and contains x_0, and we define f by $f(x) = y_0$, which is a local continuous selection of F. □

Proposition 4. *Let F be locally selectionable at x_0; let the graph of G be open and $F(x_0) \cap G(x_0) \neq \emptyset$, then $F \cap G$ is locally selectionable at x_0.* ▲

Proof. Let $y_0 \in F(x_0) \cap G(x_0)$. We consider a local continuous selection f: $N(x_0) \to Y$ such that $f(x_0) = y_0$ and $f(x) \in F(x)$. Since the graph of G is open, there exists a neighborhood $N_1(x_0) \times N(y_0)$ contained in graph(G).

Since f is continuous, there exists a neighborhood $N_2(x_0) \subset N(x_0)$ such that $f(x) \in N(y_0)$ for all $x \in N_2(x_0)$. Then f: $N_1(x_0) \cap N_2(x_0) \to Y$ is a local selection of $F \cap G$. □

We shall use this corollary

Corollary 1. *Let X be a metric space, X_0 an open subset of X and F a locally selectionable map from X_0 into a topological vector space Y, such that $F(x)$ is a convex cone. There exists a continuous function $f: X \to Y$ satisfying*

$$\forall x \in X_0, \quad f(x) \in F(x) \quad and \quad \forall x \notin X_0, \quad f(x)=0. \qquad \blacktriangle$$

Proof. We set $X_1 \doteq X \backslash X_0$ and we define $H: X_0 \to Y$ by

$$H(x) \doteq F(x) \cap d_{X_1}(x)\, \mathring{B}.$$

We have that $H(x) \neq \emptyset$ because $F(x)$ is a cone. Since X_0 is paracompact (for metrizable), since F is locally selectionable (by assumption) and since the graph of $d_{X_1}(x)\mathring{B}$ is open, *Proposition 4* implies the existence of a continuous selection $h: X_0 \to Y$ of $H(x)$.

We define f by

$$f(x)=\begin{cases} h(x) & \text{if } x \in X_0, \\ 0 & \text{if } x \in X_1. \end{cases}$$

It is clear that f is a selection of F.

It remains to check that f is continuous, i.e., that f is continuous on X_1. Let $x \in X_1$, $\varepsilon > 0$ and $y \in x + \varepsilon B$. Then $f(x)=0$ and either $f(y)$ is equal to 0 (when $y \in X_1$) or $\|f(y)\| \leq d_{X_1}(y) \leq \varepsilon$ (when $y \in X_0$). Hence $\|f(x)-f(y)\| = \|f(y)\| \leq \varepsilon$ when $y \in x + \varepsilon B$.

11. Michael's Selection Theorem

The most famous continuous selection theorem is the following result by Michael.

Theorem 1. *Let X be a metric space, Y a Banach space. Let F from X into the closed convex subsets of Y be lower semi-continuous. Then there exists f: $X \to Y$, a continuous selection from F.* ▲

Proof. a) Let us begin by proving the following claim: *given any convex (not necessarily closed) valued lower semi-continuous map Φ and every $\varepsilon > 0$, there exists a continuous $\varphi: X \to Y$ such that for ξ in X, $d(\varphi(\xi), \Phi(\xi)) \leq \varepsilon$.*

In fact, for every $x \in X$, let $y_x \in \Phi(x)$ and let $\delta_x > 0$ be such that $(y_x + \varepsilon \mathring{B}) \cap \Phi(x') \neq \emptyset$ for x' in $B(x, \delta_x)$. Since X is metric, it is paracompact. Hence there exists a locally finite refinement $\{\mathscr{U}_x\}_x \in X$ of $\{B(x, \delta_x)\}_x$. Let $\{\pi_x(\cdot)\}_x$ be a partition of unity subordinate to it. The mapping $\varphi: X \to Y$ given by

$$\varphi(\xi) = \sum \pi_x(\xi)\, y_x$$

is continuous since it is locally a finite sum of continuous functions. Fix ξ. Whenever $\pi_x(\xi) > 0$, $\xi \in \mathscr{U}_x \subset B(x, \delta_x)$, hence $y_x \in \Phi(\xi) + \varepsilon \mathring{B}$. Since this latter set is convex, any convex combination of such y's (in particular, $\varphi(\xi)$) belongs to it.

b) Next we claim that we can define a sequence $\{f_n\}$ of continuous mappings from X into Y with the following properties

i) for each $\xi \in X$, $d(f_n(\xi), F(\xi)) \leq \frac{1}{2^n}$, $\quad n = 1, 2, \ldots,$

ii) for each $\xi \in X$, $\|f_n(\xi) - f_{n-1}(\xi)\| \leq \frac{1}{2^{n-2}}$, $n = 2, \ldots$.

For $n=1$ it is enough to take in the claim of part a), $\Phi = F$ and $\varepsilon = 1/2$.

Assume we have defined mappings f_n satisfying i) up to $n = v$. We shall define f_{v+1} satisfying i) and ii) as follows.

Consider the set $\Phi(\xi) \doteq \left(f_v(\xi) + \frac{1}{2^v} \mathring{B}\right) \cap F(\xi)$. By i) it is not empty, and it is a convex set. The map $\xi \to \Phi(\xi)$ is lower semicontinuous by *Proposition 1.5* and, by the claim of a), there exists a continuous φ such that $d(\varphi(x), \Phi(x)) \leq \frac{1}{2^{v+1}}$.

Set $f_{v+1}(\xi) \doteq \varphi(\xi)$. A fortiori $d(f_{v+1}(\xi), F(\xi)) \leq \frac{1}{2^{v+1}}$, proving i). Also $f_{v+1}(\xi) \in \Phi(\xi) + \frac{1}{2^{v+1}} \mathring{B} \subset f_v(\xi) + \left(\frac{1}{2^v} + \frac{1}{2^{v+1}}\right) \mathring{B}$ i.e.,

$$\|f_{v+1}(\xi) - f_v(\xi)\| \leq \frac{1}{2^{v-1}}$$

proving ii).

c) Since the series $\sum \frac{1}{2^n}$ converges, $\{f_n(\cdot)\}$ is a Cauchy sequence, uniformly converging to a continuous $f(\cdot)$. Since the values of F are closed, by i) of part b), f is a selection from F. □

Corollary 1. *Let F be a lower semi-continuous map with closed convex values from a paracompact space X to a Banach space Y. Let G: $X \to Y$ a set valued map with open graph. If $F(x) \cap G(x) \neq \emptyset$ for all $x \in X$, then there exists a continuous selection of $F \cap G$.* ▲

Proof. Michael's selection theorem implies that F is locally selectionable. Indeed, for any $y_0 \in F(x_0)$, the set valued map F_0 defined by

$$F_0(x_0) = \{y_0\}, \quad F_0(x) = F(x) \quad \forall x \neq x_0$$

is also l.s.c. with convex values. By Michael's theorem, there exists a continuous selection f_0 of F_0, which is a continuous selection of F passing through y_0. Hence the conclusion follows from Proposition 10.4.

12. The Approximate Selection Theorem and Kakutani's Fixed Point Theorem

There are examples of upper semi-continuous maps from a topological space X to the closed convex subsets that do not possess continuous selections. The easiest example is the set valued map from $\mathbb{R}$ to $\mathbb{R}$ defined by

$$F(x) \doteq \begin{cases} \{-1\} & \text{for } x<0, \\ [-1, +1] & \text{for } x=0, \\ \{+1\} & \text{for } x>0. \end{cases}$$

For most purposes, however, a weaker property, the existence of approximate selections, is enough.

We recall the following definition

Definition 1. We say that a map φ is locally compact (resp. locally bounded) if for each point in $\operatorname{Dom}\varphi$, there exists a neighborhood which is mapped in a compact subset (resp. bounded subset).

Theorem 1 (The Approximate Selection Theorem). *Let M be a metric space, Y a Banach space, F a map from M into the convex subsets of Y, upper semicontinuous. Then for every $\varepsilon>0$ there exists a locally Lipschitzean map f_ε from M to Y such that its range is contained in the convex hull of the range of F and*

$$\operatorname{Graph}(f_\varepsilon) \subset \operatorname{Graph}(F) + \varepsilon B.$$

If the minimal selection is locally compact (resp. locally bounded), the family $\{f_\varepsilon\}$ is locally equicompact (resp. equibounded) in the sense that for every

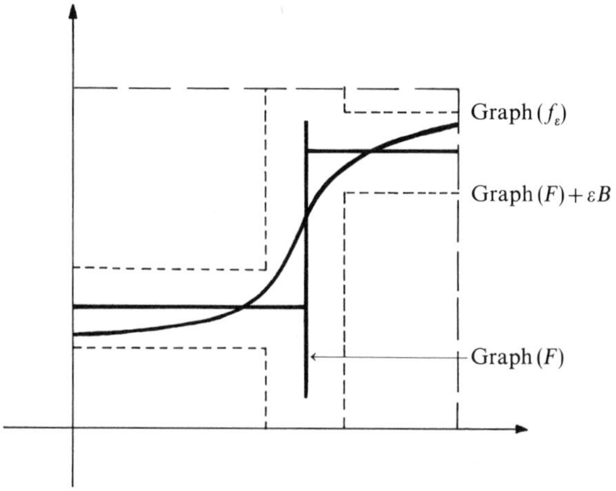

Fig. 9. Approximate selection of a set-valued map

$x \in M$ there exist a compact (resp. bounded) subset $K \subset Y$, a neighborhood $N(x)$ of x and $\varepsilon^0 > 0$, such that

$$\text{for every } \varepsilon < \varepsilon^0, \quad f_\varepsilon(N(x)) \subset K. \qquad \blacktriangle$$

Proof. a) Fix $\varepsilon > 0$. For every $x \in M$, there exists $\delta(x)$ such that for any $x^* \in B(x, \delta(x))$, we have $F(x^*) \subset F(x) + \frac{\varepsilon}{2} B$. We can take $\delta(x) < \varepsilon/2$.

The family of balls $\{B(x, \delta(x)/4\}_{x \in M}$ covers the paracompact space M. Let $\{\mathcal{U}_i\}_{i \in I}$ be a locally finite refinement and $\{a_i(\cdot)\}$ a locally Lipschitzean partition of unity subordinate to it. Choose for each i an $\bar{x}_i \in \mathcal{U}_i$ and define f_ε to be

$$f_\varepsilon(x) \doteq \sum a_i(x)\, m(F(\bar{x}_i)).$$

It is obvious that f_ε is well defined and locally Lipschitzean and that its range is in the convex hull of the range of F.

Fix $x \in M$. Then $a_i(x)$ is strictly positive only for a finite subset $I(x) \subset I$. For $i \in I(x)$, set x_i to be such that $\mathcal{U}_i \subset B(x_i, \delta(x_i)/4)$.

Set $\delta_i \doteq \delta(x_i)$ and let $j \in I(x)$ be such that $\delta_j \doteq \max_{i \in I(x)} \delta_i$. Then $x_i \in B(x_j, \delta_j/2)$ and thus $\mathcal{U}_i \subset B(x_j, \delta_j)$. Therefore, for any $i \in I(x)$,

$$m(F(x_i)) \in F(\mathcal{U}_i) \subset F(B(x_j, \delta_j)) \subset F(x_j) + \frac{\varepsilon}{2} B.$$

Since the latter set is convex, we deduce that $f_\varepsilon(x) \in F(x_j) + \frac{\varepsilon}{2} B$. So, there exists $y_j \in F(x_j)$ such that $\| f_\varepsilon(x) - y_j \| \le \frac{\varepsilon}{2}$. Then

$$d((x, f_\varepsilon(x)), (x_j, y_j)) \le d(x, x_j) + \| f_\varepsilon(x) - y_j \| \le \varepsilon,$$
$$\text{i.e. } (x, f_\varepsilon(x)) \in \text{graph}(F) + \varepsilon B.$$

b) Let us prove that the family $\{f_\varepsilon\}$ is equicompact (equibounded). Fix $x \in M$. There exist $\eta > 0$ and a convex compact (resp. bounded) subset K such that for any x^* in $B(x, \eta)$ we have $m(F(x^*)) \in K$.

Set $\varepsilon^0 \doteq \eta/4$ and $N(x)$ to be $B(x, \eta/2)$. For any $\varepsilon < \varepsilon^0$, the above construction shows that we can associate with any x^* in $N(x)$ a point $\bar{x} \in B(x^*, \varepsilon)$ and $f_\varepsilon(x^*)$, which is a convex combination of images through $m(f)$ of points in $B(\bar{x}, \varepsilon) \subset B(x, \eta)$. These images are in K and so is their convex combination. □

Corollary 1 (Kakutani's Fixed Point Theorem). *Let K be a compact convex subset of a Banach space X and let F be an upper semicontinuous map from K into its compact convex subsets. Then F has a fixed point, i.e. there exists $\bar{x}$ belonging to $F(\bar{x})$.* ▲

Proof. Let $\{f_n\}$ be a sequence of continuous maps from K, such that $\text{graph}\, f_n \subset \text{graph}\, F + \varepsilon_n B$. By Schauder's Theorem there are $x_n = f_n(x_n)$ and a suitable subsequence converges to some $\bar{x}$. Hence $d((\bar{x}, \bar{x}), \text{Graph}(F))$

$= \lim d((x_n, f_n(x_n)), \operatorname{Graph}(F)) = 0$. Being Graph$(F)$ closed, $(\bar{x}, \bar{x})$ belongs to Graph(F), i.e. $\bar{x} \in F(\bar{x})$. □

As we shall see another class of mappings shares this property of being approximated by single valued maps in the sense presented before: the maps associated with the set of solutions to a differential equations without uniqueness, or more generally, to a differential inclusion with upper semi-continuous convex values. Remark that these maps do not have convex values. Still they have the approximate selection property since what counts is the possibility of approximating the right hand side of the differential equation or inclusion by Lipschitzean maps, such that the flow defined by the approximating equations is single valued. Since in general the approximations will not map the set K exactly into itself, the following slight generalization of Schauder's Theorem will be handy.

Theorem 2. *Let X be a Banach space, K a compact convex subset of X, f a continuous map from K into K. Then there is $\bar{x}$ in K such that*

$$d(f(\bar{x}), \bar{x}) = d(f(\bar{x}), K).$$

In particular, when $f(K)$ is contained in $K + \varepsilon B$, there is a $\bar{x}$ such that $d(\bar{x}, f(\bar{x})) \leq \varepsilon$. ▲

Proof. Consider the map $x \to \pi_K(f(x))$. It is upper semicontinuous with compact convex values, hence, by Kakutani's theorem, there exists $\bar{x} \in \pi_K(f(\bar{x}))$. Then $d(f(\bar{x}), \bar{x}) = d(f(\bar{x})), \pi_K(f(\bar{x}))) = d(f(\bar{x}), K)$. □

13. σ-Selectionable Maps

Definition 1. We say that a map F from X into the subsets of Y is *σ-selectionable* if there exists a decreasing sequence of compact valued maps F_n with closed graph satisfying.

i) $\forall n \geq 0$, F_n has a continuous selection,
ii) $\forall x \in x$, $F(x) = \bigcap_n F_n(x)$. ▲

We shall show that upper semicontinuous maps with compact convex values are σ-selectionable.

In a later section we shall show that the same is true for maps associated with the set of solutions of a differential inclusions with compact convex valued upper semicontinuous right hand side.

Theorem 1. *Let X be a metric space, Y be a Banach space and F be a strict upper semicontinuous map from X to the closed convex subsets of Y. It is σ-*

selectionable. Actually, there exists a sequence of upper semicontinuous maps F_n from X to $\overline{\text{co}}(F(X))$, which approximate F in the sense that for all $x \in X$,

$$(1) \quad \begin{array}{l} \text{i) } \forall n \geq 0, \quad F(x) \subset \ldots \subset F_{n+1}(x) \subset F_n(x) \subset \ldots \subset F_0(x), \\ \text{ii) } \forall \varepsilon > 0, \ \exists N(\varepsilon, x) \mid \forall n \geq N(\varepsilon, x), \ F_n(x) \subset F(x) + \varepsilon B. \end{array}$$

The maps F_n can be written in the following form:

$$(2) \qquad \forall x \in X, \quad F_n(x) \doteq \sum_{i \in J_n} \Psi_i^n(x)\, C_i^n$$

where the subsets C_i^n are closed and convex and where the functions Ψ_i^n form a locally Lipschitzean locally finite partition of unity. ▲

Remark. a) Conditions (1) imply that

$$(3) \qquad F(x) = \bigcap_{n \geq 0} F_n(x).$$

b) If F maps X to a compact subset of Y, then the maps F_n are *compact convex* valued because the subsets $C_i^n \subset \overline{\text{co}}\, F(X)$ are compact.

c) When X is compact, the subsets I_n which appear in formula (1) are finite

Proof. Let K denote the closed convex hull of $F(X)$.

We fix $\rho > 0$. Let us then cover X with the open balls $\{B(x, \rho)\}_{x \in X}$. We shall now give a construction of F_0.

By *Proposition* 0.1.1, we can associate to the preceding covering of X by open balls, a locally finite open covering $(\Omega_i^{(0)})_{i \in I_{(0)}}$ such that:

for any $i \in I_{(0)}$ there exists $x_i^{(0)} \in X$ with $\Omega_i^{(0)} \subset \mathring{B}(x_i^{(0)}, \rho)$.

We then define for any $i \in I_{(0)}$, $C_i^{(0)} = \overline{\text{co}}\, F(\mathring{B}(x_i^{(0)}, 2\rho))$ which is a non empty closed convex subset of $K \doteq \overline{\text{co}}\, F(X)$.

Now, thanks to *Theorem* 0.1.1, we can associate a locally Lipschitzean partition of unity $(\Psi_i^{(0)})_{i \in I_{(0)}}$ to the open covering $(\Omega_i^{(0)})_{i \in I_{(0)}}$. We then define for all $x \in X$:

$$(4) \qquad F_0(x) \doteq \sum_{i \in I_{(0)}} \Psi_i^{(0)}(x)\, C_i^{(0)}.$$

The so defined set valued map F_0 is obviously a nonempty convex-valued map from X into K, well defined and upper semicontinuous on X since $(\Psi_i^{(0)})_{i \in I_{(0)}}$ is a locally finite, locally Lipschitzean partition of unity in X.

In order to define F_1 we do the same as before with the open covering $\{\mathring{B}(x, \rho/3)\}_{x \in X}$.

Thus we define a locally finite open covering $(\Omega_i^{(1)})_{i \in I_{(1)}}$ of X associated to $\{\mathring{B}(x_i^{(1)}, \rho/3)\}_{i \in I_{(1)}}$ and an associated locally Lipschitzean partition of unity $(\Psi_i^{(1)})_{i \in I_{(1)}}$.

As before we set for all $i \in I_{(1)}$:

$$(5) \qquad C_i^{(1)} \doteq \overline{\text{co}}\, F(\mathring{B}(x_i^{(1)}, 2\rho/3)) \subset K$$

and then we can define for any $x \in X$:

(6) $$F_1(x) \doteq \sum_{i \in I_{(1)}} \Psi_i^{(1)}(x)\, C_i^{(1)}.$$

The set valued map F_1 enjoys the same properties as F_0.

We shall now prove that for any $x \in X$, $F_1(x) \subset F_0(x)$. Let us fix $x \in X$. We define:

$$I_{(0)}^x = \{i \in I_{(0)} \mid x \in \mathring{B}(x_i^{(0)}, \rho)\}$$

and

$$I_{(1)}^x = \{i \in I_{(1)} \mid x \in \mathring{B}(x_i^{(1)}, \rho/3)\}.$$

Let $i_{(0)} \in I_{(0)}^x$ and $i_{(1)} \in I_{(1)}^x$ be given. Then if $y \in \mathring{B}(x_{i_{(1)}}^{(1)}, 2\rho/3)$ we get

$$d(y, x_{i_{(1)}}^{(1)}) \leq \frac{2\rho}{3} \quad \text{with } d(x, x_{i_{(0)}}^{(0)}) \leq \rho$$

and $d(x, x_{i_{(1)}}^{(1)}) \leq \rho/3$. Thus we have

$$d(y, x_{i_{(0)}}^{(0)}) \leq \frac{2\rho}{3} + \frac{\rho}{3} + \rho = 2\rho.$$

Then $\mathring{B}(x_{i_{(1)}}^{(1)}, 2\rho/3) \subset \mathring{B}(x_{i_{(0)}}^{(0)}, 2\rho)$ for all $i_{(0)} \in I_{(0)}^x$ and all $i_{(1)} \in I_{(1)}^x$. This leads to $C_{i_{(1)}}^{(1)} \subset C_{i_{(0)}}^{(0)}$ for such indexes.

Then for all $i_{(1)} \in I_{(1)}^x$ we have:

$$C_{i_{(1)}}^{(1)} \subset \sum_{i \in I_{(0)}^x} \Psi_i^{(0)}(x)\, C_i^{(0)} = \sum_{i \in I_{(0)}} \psi_i^{(0)}(x)\, C_i^{(0)} = F_0(x),$$

this being true by convexity arguments and since $(\Psi_i^{(0)})_{i \in I_{(0)}}$ is a locally finite partition of unity associated to $(\Omega_i^{(0)})_{i \in I_{(0)}}$ and thus also to $\{\mathring{B}(x_i^{(0)}, \rho)\}_{i \in I_{(0)}}$ which in particular says that $\Psi_i^{(0)}(x) = 0$ if $i \notin I_{(0)}^x$.

And then for the same reasons we get:

$$F_1(x) = \sum_{i \in I_{(1)}} \Psi_i^{(1)}(x)\, C_i^{(1)} = \sum_{i \in I_{(1)}^x} \Psi_i^{(1)}(x)\, C_i^{(1)} \subset F_0(x).$$

In fact when x is given, only a finite number of $i \in I_{(0)}$ and of $i \in I_{(1)}$ has to be considered since $(\Psi_i^{(0)})_{i \in I_{(0)}}$ and $(\Psi_i^{(1)})_{i \in I_{(1)}}$ are locally finite.

Hence we have effectively proved that $F_1(x) \subset F_0(x)$ for all $x \in X$. We shall now prove for example that for any $x \in X$, $F(x) \subset F_0(x)$. Indeed from the construction of $C_i^{(0)}$ we have $F(x) \subset C_i^{(0)}$ for all $i \in I_{(0)}^x$.

Then always for convexity reasons and since $(\Psi_i^{(0)})_{i \in I_{(0)}}$ is a locally finite partition of unity associated to $(\Omega_i^{(0)})_{i \in I_{(0)}}$, hence to $\{\mathring{B}(x_i^{(0)}, \rho)\}_{i \in I_{(0)}}$, we get:

$$F(x) \subset \sum_{i \in I_{(0)}^x} \Psi_i^{(0)}(x)\, C_i^{(0)} = \sum_{i \in I_{(0)}} \Psi_i^{(0)}(x)\, C_i^{(0)} = F_0(x).$$

Now let us define $\rho_n = (\frac{1}{3})^n \rho$ for any $n \in \mathbb{N}$. Then as for F_0 associated to $\rho_0 = \rho$ and for F_1 associated to $\rho_1 = \rho/3$ we can build by induction a sequence of set valued maps F_n: $X \to K$, $n \in \mathbb{N}$, each of them being upper semicontinuous nonempty convex compact valued and verifying properties (1) i) of Theorem 1.

To end the proof we have to show that (1) is verified.

Let $x \in X$ be given. Since F is upper semicontinuous, for any $\varepsilon > 0$, there exists $\eta_{\varepsilon,x} > 0$ such that $d(y,x) \leq \eta_{\varepsilon,x}$ implies $F(y) \subset F(x) + \varepsilon B$.

Then there obviously exists $N_{\varepsilon,x}$ such that for $n \geq N_{\varepsilon,x}$ we have $\rho_n \leq \frac{\eta_{\varepsilon,x}}{3}$.

Let us define as before $I^x_{(n)} \doteq \{i \in I_{(n)} | x \in \mathring{B}(x_i^{(n)}, \rho_n)\}$. For the same reasons as for F_0 and F_1 we can write:

$$\text{(7)} \quad \begin{aligned} &\text{i)}\ F_n(x) \doteq \sum_{i \in I^x_{(n)}} \Psi_i^{(n)}(x)\, C_i^{(n)}, \\ &\quad \text{where} \\ &\text{ii)}\ C_i^{(n)} \doteq \overline{\text{co}}\, F(\mathring{B}(x_i^{(n)}, 2\rho_n)) \subset K. \end{aligned}$$

Then for all $y \in \mathring{B}(x_i^{(n)}, 2\rho_n)$ with $i \in I^x_{(n)}$, we have

$$\begin{aligned} d(y,x) &\leq d(y, x_i^{(n)}) + d(x_i^{(n)}, x) \\ &\leq 2\rho_n + \rho_n = 3\rho_n < \eta_{\varepsilon,x} \qquad \text{if we take } n \geq N_{\varepsilon,x}. \end{aligned}$$

Thus for all $n \geq N_{\varepsilon,x}$ we have $F(y) \subset F(x) + \varepsilon B$ for all $y \in \mathring{B}(x_i^{(n)}, 2\rho_n)$ with $i \in I^x_{(n)}$.

But since $F(x) + \varepsilon B$ is closed convex we get:

$$C_i^{(n)} \subset F(x) + \varepsilon B \qquad \text{for all } i \in I^x_{(n)}.$$

And then always by convexity we get:

$$\text{(8)} \qquad F_n(x) \doteq \sum_{i \in I^x_{(n)}} \Psi_i^{(n)}(x)\, C_i^{(n)} \subset F(x) + \varepsilon B$$

for all $n \geq N_{\varepsilon,x}$.

Thus (2) ii) is verified and the proof of *Theorem 1* is complete. □

Theorem 2. *Let K be a convex compact subset of a Banach space X and F be a σ-selectionable map from K to K. Then F has a fixed point.* ▲

Proof. Let f_n be a continuous selection from F_n.

By *Theorem 12.2*, there exist x_n such that

$$d(x_n, f_n(x_n)) = d(f_n(x_n), K).$$

For any $\nu \leq n$ hence we have the chain of inequalities

$$\begin{aligned} d(x_n, F_\nu(x_n)) &\leq d(x_n, F_n(x_n)) \leq d(x_n, f_n(x_n)) = d(f_n(x_n), K) \\ &\leq \delta(F_n(x_n), K) \leq \delta(F_\nu(x_n), K). \end{aligned}$$

The compactness of K allows us to assume that x_n converges to x_*, and by the upper semicontinuity of F_ν we can write

$$d(x_*, F_\nu(x_*)) \leq \liminf d(x_n, F_\nu(x_n)) \leq \liminf \delta(F_\nu(x_n), K) \leq \delta(F_\nu(x_*), K).$$

Since $F_\nu(x_*)$ decreases to $F(x_*)$ contained in K, the last term converges to zero, and

$$d(x_*, F(x_*)) = \lim d(x_*, F_\nu(x_*)) = 0. \qquad □$$

14. Measurable Selections

The following construction is due to Kuratowski and Ryll-Nardzewski.

Theorem 1. *Let F, from a closed I of $\mathbb{R}$ into the closed subsets of a complete separable metric space Y, be such that inverse images of open sets are borellians. Then there exists a borel measurable map $s: I \to Y$, a selection from F.* ▲

Proof. Let us divide the proof into three steps.

a) We begin by remarking that we can change the metric of Y into an equivalent metric, preserving completeness and separability, so that Y becomes a bounded (say, with diameter M) complete metric space. This technical trick is used only to give a faster construction of the first step in the following induction argument.

b) Let C be a countable dense subset of Y. Set $\varepsilon_0 = M$, $\varepsilon_i = M/2^i$. We claim that we can define a sequence of mappings s_m such that:

$$\text{(1)} \qquad \begin{aligned} &\text{i) } s_m \text{ is Borel measurable with values in } C,\\ &\text{ii) } s_m(x) \in \mathring{B}(F(x), \varepsilon_m),\\ &\text{iii) } s_m(x) \in \mathring{B}(s_{m-1}(x), \varepsilon_{m-1}), \qquad m > 0. \end{aligned}$$

In fact, arrange the points of C into a sequence $\{c_j\}_0^\infty$, and define s_0 to be $\equiv c_0$. Then i) and ii) are clearly satisfied. Assume we have defined functions s_m satisfying i) and ii) up to $m = p-1$, and define s_p, satisfying i), ii) and iii), as follows.

Set

$$A_j \doteq F^{-1}(\mathring{B}(c_j, \varepsilon_p)) \cap s_{p-1}^{-1}(\mathring{B}(c_j, \varepsilon_{p-1}))$$

and

$$E_0 = A_0, \qquad E_j \doteq A_j \setminus (E_0 \cup \ldots \cup E_{j-1}).$$

We claim that:

$$\bigcup_{j=0}^{\infty} E_j = I.$$

In fact let $x \in I$ and consider $s_{p-1}(x)$ and $F(x)$. By ii) $s_{p-1}(x) \in \mathring{B}(F(x), \varepsilon_{p-1})$; by the density of C there is a c_j such that at once $s_{p-1}(x) \in B(c_j, \varepsilon_{p-1})$ and $F(x) \cap \mathring{B}(c_j, \varepsilon_{p-1}) \neq \emptyset$, i.e. $x \in A_j$. Finally, either x is in E_j or it is in some E_i with $i < j$. In either case it belongs to $\bigcup E_j$.

Define s_p to be c_j on E_j. Then the domain of s_p is I and ii) and iii) of the induction process certainly hold. Each E_j is a borellian.

c) Condition iii) implies that $\{s_m(x)\}_m$ is a Cauchy sequence, which converges to $s(x)$. Clearly, $s(x)$ belongs to $F(x)$. It remains to show that s is measurable.

This is equivalent to show that counter images of closed sets are borellians. Let K be closed and set $K_m = \mathring{B}(K, \varepsilon_m)$. Each $s_m^{-1}(K_m)$ is a borellian. We shall complete the proof by showing that $s^{-1}(K) = \bigcap s_m^{-1}(K_m)$. In fact on one hand, when $x \in s^{-1}(K)$, $s(x) \in K$ and since $d(s_m(x), s(x)) \le \varepsilon_m$, $s_m(x) \in \mathring{K}_m$ for every m. On the other hand, when $x \in s_m^{-1}(K_m)$ for all m, $s_m(x) \in \mathring{B}(K, \varepsilon_m)$ and since $s_m(x)$ converges to $s(x)$ and K is closed, $s(x) \in K$. □

Corollary 1. *Let $f: X \times U \to X$ be continuous where U is a compact separable metric space and assume that there exist an interval I and an absolutely continuous $x: I \to \mathbb{R}^n$, such that*

$$\text{a.e. in } I, \quad x'(t) \in f(x(t), U).$$

Then there exists a Lebesgue measurable $u: I \to U$ such that a.e. in I,

$$x'(t) = f(x(t), u(t)). \qquad \blacktriangle$$

Proof. Consider the map Φ defined in I by

$$\Phi(t) \doteq \{v \in U / f(x(t), v) = x'(t)\}.$$

We have to show the existence of a measurable selection from Φ.

In case x' were continuous, Φ would be upper semicontinuous: Indeed, the continuity of f, of x and of x' implies that Φ has a closed graph and, since it takes values in the compact U, Φ is u.s.c.. Upper semicontinuous maps to satisfy the assumptions of *Theorem 1*. Hence a selection would exist.

In general, by Lusin's Theorem, write I as $I^0 \cup I^1 \cup \ldots \cup I_n \cup \ldots$ where I^0 has Lebesgue measure zero, I^i is closed, $i \ge 1$, and $x'|_{I^i}$ is continuous, $i \ge 1$.

Theorem 1 applied on each I^i, $i \ge 1$, yields a borel measurable selection s^i from $\Phi|_{I^i}$, $i \ge 1$. Define arbitrarily s^0 and I^0 and set $u: I \to U$ to be:

$$u|_{I^i} \doteq s^i \qquad i \ge 0.$$

Inverse images of open sets are countable unions of borel subsets of I^i (hence of borellians in $\mathbb{R}$) and of a subset of a set of measure zero, hence they are measurable. □

Theorem 2. *Let F, from I into the closed subsets of $\mathbb{R}^n$, be continuous. Let v, from I into $\mathbb{R}^n$, be Lebesgue measurable. Then there exists a Lebesgue measurable function f such that*

$$(2) \qquad \begin{array}{ll} \text{i)}\ f(t) \in F(t), & \text{a.e. on} I, \\ \text{ii)}\ \|v(t) - f(t)\| = d(v(t), F(t)), & \text{a.e. on } I. \end{array} \qquad \blacktriangle$$

Proof. Consider v. By Lusin's Theorem, I can be written as the union of N, a set of Lebesgue measure zero, and of a countable collection of disjoint closed subsets J_n such that the restriction of v to each J_n is continuous. On each J_n consider the continuous map $t \to F(t) - v(t)$. We have that $t \to \|m(F(t) - v(t))\|$ is continuous, hence bounded by some β_n on J_n. The map $t \to \|m(F(t) - v(t))\|$ has a closed graph: its restriction to any J_n has images contained in the compact set $\beta_n B$, hence it is u.s.c. Such a map is borellian. Hence the preceding theorem applies to the restriction of $m(F(t)-v(t))$ to J_n: there exists $s_n(\cdot)$, a selection from $m(F-v)$, such that inverse images of open sets are borel subsets of J_n. Since J_n is closed, its borellians are borellians in I. Define s to be s_n on J_n, and arbitrarily on N. Then inverse images through s of open sets are a countable union of borellians and of a subset of a set of Lebesgue measure zero, i.e. they are Lebesgue measurable, and so is s.

Finally set $f(t)$ to be $v(t)+s(t)$. We claim that f is the required function.

Whenever $t\notin N$, $f(t)=v(t)+s(t)\in v(t)+m(F(t)-v(t))\subset v(t)+F(t)-v(t)=F(t)$ and i) holds. Moreover $\|f(t)-v(t)\|=\|s(t)\|=\|m(F(t)-v(t))\|=d(v(t),F(t))$, concluding the proof. □

Chapter 2. Existence of Solutions to Differential Inclusions

Introduction

In what follows we shall deal with the existence and properties of solutions to differential inclusions of the form

(1) $$x'(t) \in F(x(t))$$

or

(2) $$x'(t) \in F(t, x(t)).$$

For the corresponding problems for an ordinary differential equation of the form $x' = f(x)$ or $x' = f(t, x)$, it is clear what is meant by a solution, at least as long as we make the assumption of continuity of f. A solution has to be a continuous everywhere differentiable function. Hence the continuity of f allows us, as well, to define a solution as a continuously differentiable function on some interval. Only, in the time dependent case, when we decide to take into consideration functions that are measurable with respect to the time variable, we are forced to accept functions that are differentiable except on a set of measure zero, and we take solutions in the Carathéodory sense.

For differential inclusions the problem is not so easy. For instance, let F be constant, equal to $\{-1, +1\}$, and let us consider the set of solutions through 0 at $t=0$. There are only two C^1 solutions, namely $x_1(t) = t$ and $x_2(t) = -t$, and we feel that we should accept more functions as solutions, allowing the derivative not to exist, for instance on a finite number of points, or on a countable set, or on a set of measure zero. The choice of one or the other of these classes might depend on non-mathematical considerations, or on the simple need to be able to prove an existence theorem: the larger the class, the easier the existence result. However, some limits are imposed, just as for the single-valued case. In order to obtain the equivalence between a (single-valued) differential equation and the corresponding integral equation:

$$x(t) = x^0 + \int_0^t f(s, x(s))\, ds$$

we had to assume, by definition, that the solutions $x(\cdot)$, besides continuous, had to be absolutely continuous. We make the same assumption for a differential inclusion: a solution $x(\cdot)$ has to be, at least, an absolutely continuous function, i.e., roughly speaking, the primitive of its derivative. As for the acceptance of non-continuously differentiable solutions, this definition relies on convenience, and on physical consideration; it might be a matter of opinion. We feel uncomfortable in face of continuous functions, like the classic example of Vitali, whose derivative exists almost everywhere and is zero whenever it exists, and still is such that its value is zero at time zero, and one at time one. The absolute continuity gets rid of this behavior.

Since the absolutely continuous functions are the weakest acceptable kind of solution, it will be of interest to inquire about the existence of solutions in some stronger sense, i.e., more regular solutions.

For a differential inclusion, the conditions to be imposed on the set-valued mapping F in order to have solutions are of two kinds: regularity conditions on the map (i.e. the various kinds of continuity or semicontinuity) and conditions of topological or geometric type (compactness, convexity) on the images of points. Various combinations are possible: we would not expect to obtain solutions under weak assumptions of both types, while it should be quite easy to prove existence under strong assumptions of both types. In general, the intermediate cases will be more interesting.

In this chapter, we choose to study the following compromises

a) F is upper semicontinuous (or lower semicontinuous),
b) the values of F are compact and convex,

and

a) F is continuous (or lower semicontinuous, absolutely continuous), but not upper semicontinuous only,
b) the values of F are compact, but not necessarily convex.

The case of Lipschitzean maps F bridges those two classes of differential inclusions, because the Relaxation Theorem states that, in this case, the set of trajectories of the differential inclusion

(1) $$x'(t) \in F(x(t)), \quad x(0) = x_0$$

is dense in the set of trajectories of the differential inclusion

(3) $$x'(t) \in \overline{\text{co}}\, F(x(t)).$$

We shall study in the next chapter another compromise, where:

$$-F \text{ is maximal monotone.}$$

In this case, the values of F are closed and convex.

The simplest approach to the existence problem for a differential inclusion would be to reduce it to the corresponding problem for an ordinary differential equation. To begin with, we would like to know whether there exists a differential equation in some sense concealed into the differential inclusion, i.e., whether there exists a continuous function $f(\cdot)$ such that for every x in some domain, $f(x) \in F(x)$. If it were so, every solution of

$$x'(t) = f(x(t))$$

would be a solution of

$$x'(t) \in F(x(t)).$$

This is the purpose of the first section, in which we use the selection theorems presented in the preceding chapter. The minimal selection Theorem provides "slow solutions" to the above differential inclusion, i.e., solutions to the differential equation

(4) $$x'(t) = m(F(x(t))).$$

Apart from this case, we will always find a difficulty which does not appear for ordinary differential equations, where the convergence of the approximate solutions implies the convergence of the derivatives of the approximate solutions. Since this is not the case anymore, the *convergence of the derivatives will be the main issue in each existence proof.* Several techniques can be thought of: getting strong convergence from weak convergence via the Convergence Theorem (and this requires the convexity of the images), using the Cauchy criterion (as in the case of maximal monotone or Lipschitzean case), or using a (strong) compactness theorem (an analog of the Ascoli-Arzelà theorem) as in the theorem devised by Filippov for the continuous case.

In the case of upper semicontinuous maps with closed convex images, we prove an analog of Peano's Theorem for ordinary differential equations, yielding local existence of solutions to the initial value problem. This theorem is so basic that we supply it with several proofs.

We consider then the map $\xi \to \mathcal{T}_T(\xi)$ from a subset Ω of R^n into the subsets of $\mathscr{C}(I; R^n)$, that associates to a point ξ the set of solutions to the differential inclusion

(1) $$x'(t) \in F(t, x(t)), \quad x(0) = \xi$$

and the map $\xi \to \mathscr{A}_T(\xi)$, from Ω into its subsets, that associates to ξ its "attainable set" i.e. the set of final values $x(T)$ when $x(\cdot)$ ranges over $\mathcal{T}_T(\xi)$. We prove that these maps are upper semicontinuous. Moreover the map $\xi \to \mathcal{T}_T'(\xi)$ associating to ξ the set of derivatives of functions in $\mathcal{T}_T(\xi)$ is upper semicontinuous from Ω to the weak compact subsets of $L^\infty(0, T; R^n)$. Although the maps $\xi \to \mathcal{T}_T(\xi)$ and $\xi \to \mathscr{A}_T(\xi)$ are not, in general, convex valued, we shall verify that they enjoy essentially all the properties, encountered in Chapter 1, related to convex valued upper semicontinuous maps: the existence of approximate selections, the fixed point property, and the approximation property. We shall prove that $\mathcal{T}_T(\xi)$ is a compact connected set, as it is the case for an ordinary differential equation. As an application of connectedness we shall present the Hukuhara property. We shall show that the map $\xi \to \mathscr{A}_T(\xi)$ has acyclic values, from which it follows again that the map $\xi \to \mathscr{A}_T(\xi)$ has the fixed point property and has connected values.

We then tackle the case when F has no longer convex values. We still use adaptations of Euler's construction of approximate solutions, but, this time, by involving a careful (and astute) choice of the derivatives of these polygonal lines.

Indeed, we no longer can benefit from the Convergence Theorem, in which convexity of the images played a crucial role. We need a stronger topology, which yields at least pointwise convergence of the derivatives of the approximate solutions.

In the case of continuous maps, Filippov proposed a construction allowing the derivatives of the approximate solutions to remain in a compact subset of the space of bounded functions. When F is absolutely continuous, another construction shows that the derivatives of the approximate solutions remain in a compact subset of the space of continuous function: we then obtain the existence of continuously differentiable solutions to the differential inclusion.

As in the case of ordinary differential inclusions, we shall derive from Gronwall inequality further informations when the set-valued map F is Lipschitzean: in this case, for instance, the map $\mathscr{T}_T(\cdot)$ is *Lipschitzean*, and not only upper semicontinuous. We then prove the Relaxation Theorem we have already presented.

The last two sections are devoted to considering the problem on the existence of solutions to (1) as a fixed point problem. They are based on the selection approach devised by Antosiewicz-Cellina, that is the analogue to Michael's Selection Theorem, presented in Chapter 1 in the framework of maps having convex images. In fact, even though Examples 1 and 2 of Section 1.6 show that, in general, maps whose images are not convex do not possess continuous selections, we shall prove that this is not the case for the maps that are naturally associated with differential inclusions. Since the existence of continuous selections is intrinsically connected to lower semicontinuity, we shall conclude the chapter by proving an existence theorem for (1) that requires only the lower semicontinuity of the map F.

1. Convex Valued Differential Inclusions

We consider in this section differential inclusions

$$(1) \qquad x'(t) \in F(t, x(t)), \quad x(0) = x_0$$

where the images $F(t, x)$ of the set-valued map F are *convex*.

Case of Lower Semicontinuous Maps with Closed Convex Images

If F is a lower semicontinuous map with closed convex values, Michael's Selection Theorem guarantees the existence of continuous selections.

Theorem 1. *Let F be a lower semicontinuous set-valued mapping, from some open region $\Omega \subset R \times R^n$ into the non-empty, closed and convex subsets of R^n. Let $(0, x_0) \in \Omega$. Then there exists some interval $I = (\omega_-, \omega_+)$, $\omega_- < 0 < \omega_+$ and at least one continuously differentiable function $x: I \to R^n$, a solution to the Cauchy problem for the differential inclusion*

(1) $$x'(t) \in F(t, x(t)), \quad x(0) = x_0$$

Moreover either $\omega_+ = +\infty$ or the solution $x(t)$ tends to the boundary of Ω as $t \to \omega_+$, and analogously for ω_-. ▲

Proof. The set Ω, an open subset of $R \times R^n$, is a metric space, hence paracompact. Michael's theorem applies and there exists a continuous selection $f: \Omega \to R^n$.

The Cauchy problem

(2) $$x'(t) = f(t, x(t)), \quad x(t_0) = x_0$$

have at least one solution defined on a maximal interval of existence. The behavior at the end points of this interval is a classical result. (See Hartman [1973]). ▲

Case of Continuous Maps with Closed Convex Images

We can also use the minimal selection theorem when F is continuous. We recall that if K is a closed convex subset,

(3) $m(K) \doteq \pi_K(0)$ is the element of K with the smallest norm.

Definition 1. We say that a solution to the differential inclusion (1) is a *slow solution* if, for almost all $t \in [0, T]$, $x'(t) = m(F(t, x(t)))$. ▲

Theorem 2. *Let F be a continuous map with closed convex values defined on an open subset $\Omega \subset R \times R^n$ that contains $(0, x_0)$. Then there exists some interval (ω_-, ω_+), $\omega_- < 0 < \omega_+$, on which a slow solution issued from x_0 does exist. Moreover, $\omega_+ = +\infty$ or the slow solution tends to the boundary of Ω when t tends to ω_+.* ▲

Remark. We shall see in the next chapter that when $-F$ is a maximal monotone map (which has closed convex values), slow solutions still exists (and are the unique solutions to the Cauchy problem).

Case of Upper Semicontinuous Maps with Compact Convex Images

We begin by proving the analog of Peano's theorem stating the existence of a local solution to a differential inclusion for upper semicontinuous maps with compact convex values.

Definition 2. We say that a map ϕ is *locally compact* (resp. *locally bounded*) if for each point in $\mathrm{Dom}(\phi)$ there exists a neighborhood which is mapped into a compact subset (resp. bounded subset). ▲

Theorem 3. *Let X be a Hilbert space, $\Omega \subset R \times X$ be an open subset containing $(0, x_0)$. Let F be an upper semicontinuous map from Ω into the non-empty closed convex subsets of X. We assume that $(t,x) \to m(F(t,x))$ is locally compact. Then there exist $T>0$ and an absolutely continuous function $x(\cdot)$ defined on $[0,T]$, a solution to the differential inclusion*

$$x'(t) \in F(t, x(t)), \qquad x(0) = x_0. \tag{1}$$ ▲

Since $(t,x) \to m(F(t,x))$ is locally compact, there exist a compact convex subset K and scalars $a>0$, $b>0$ such that

$$Q \doteq \{(t,x) \in \Omega \,|\, |t| < a, \ \|x_0 - x\| < b\} \subset \Omega$$

and $m(F(t,x)) \in K$ for all (t,x) in Q. We set $T \doteq \min(a, b/\|K\|)$ and we shall establish the existence of a solution on $[0,T]$.

We shall provide four proofs of this theorem, the last one, using the fixed-point approach, being postponed to Section 5. The first one uses the approximate selections of F, the second the integral representation whereas the third follows Euler's approach by constructing polygonal lines.

First Proof. Let f_n be a sequence of continuous single-valued maps approaching F in the sense of the Approximate Selection Theorem 1.12.1. Let $x_n\colon [0,T] \to \Omega$ be solutions to

$$x_n'(t) = f_n(t, x_n(t)), \qquad x_n(0) = x^0.$$

In particular each $\|x_n'\|$ is bounded by $\|K\|$ and each x_n takes values in the compact set $x^0 + TK$. By Theorem 0.3.4 there exists a subsequence x_{n_k} such that x_{n_k} converges uniformly to $x(\cdot)$ on I and x_{n_k}' converges weakly to $x'(\cdot)$ in $L^1(I)$. Moreover

$$d(((t, x_{n_k}(t)), x_{n_k}'(t)), \operatorname{graph} F) = d(((t, x_{n_k}(t)), f_{n_k}(t, x_{n_k}(t))), \operatorname{graph} F) \to 0.$$

Set in the Convergence Theorem 1.4.1 $x_k(t)$ to be $(t, x_{n_k}(t))$ and $y_k(t)$ to be $x_{n_k}'(t)$. Then $((t, x(t)), x'(t)) \in \operatorname{graph}(F)$ i.e.

$$x'(t) \in F(t, x(t)).$$ □

Second Proof. This proof below makes use of the following Integral Representation Lemma, instead of the Convergence Theorem. We define simply the integral of a map as follows:

Definition 3. Let F be upper semicontinuous on I. Set

$$\int_I F(s)\,ds \doteq \left\{\int_I f(s)\,ds \,\middle|\, f \text{ an integrable selection from } F\right\}.$$ ▲

Lemma 1 (Integral Representation). *Let F be an upper semicontinuous map from $I \times X$ into the compact convex subsets of X. Then the continuous function $x(\cdot)$ is a solution on I to the inclusion*

$$x'(t) \in F(t, x(t)) \tag{1}$$

if and only if for every pair (t_1, t_2),

$$x(t_2) \in x(t_1) + \int_{t_1}^{t_2} F(s, x(s))\, ds. \tag{4}$$ ▲

Proof. i) When x is a solution to (1) on I, its derivative is a measurable selection from $F(s, x(s))$; hence (4) holds.

ii) To prove the converse, remark first that for a closed convex set A,

$$\int_{t_1}^{t_2} A\, ds = (t_2 - t_1) A. \tag{5}$$

It is obvious, in fact, that $(t_2 - t_1) A \subset \int_{t_1}^{t_2} A\, ds$. Let z be in $\int_{t_1}^{t_2} A\, ds$, i.e. $z = \int_{t_1}^{t_2} \zeta(s)\, ds$, $\zeta(\cdot)$ measurable with values in A. By the mean-value Theorem 0.5.3, $z = (t_2 - t_1)\xi$, $\xi \in \overline{\mathrm{co}}\{\zeta(s) : t_1 \leq s \leq t_2\}$, i.e. $z \in (t_2 - t_1) A$, proving (5).

Assume that (4) holds. Then $\|x(t_2) - x(t_1)\| \leq \|F\|\, |t_2 - t_1|$, i.e. x is Lipschitzean, hence differentiable a.e. Let t be a point where $x'(t)$ exists. Fix $\varepsilon > 0$. Let $\delta > 0$ be such that $|t - t'| \leq \delta$ implies $F(t', x(t')) \subset F(t, x(t)) + \varepsilon B$. Then

$$\begin{aligned} x(t_1) - x(t) &\in \int_t^{t_1} F(s, x(s))\, ds \\ &\subset \int_t^{t_1} (F(t, x(t)) + \varepsilon B)\, ds = (t_1 - t)(F(t, x(t)) + \varepsilon B), \end{aligned}$$

i.e.

$$x'(t) \in F(t, x(t)) + \varepsilon B.$$

Being ε arbitrary and $F(t, x)$ closed, $x'(t) \in F(t, x(t))$. □

Let x_n and f_n be as in the first proof. Let $x_{n_k} \to x$. Fix t_1 and t_2 in I. Set for simplicity $f_k(s) \doteq f_{n_k}(s, x_{n_k}(s))$, $\Phi(s) \doteq F(s, x(s))$.

$$\begin{aligned} & d\left(x(t_2) - x(t_1), \int_{t_1}^{t_2} \Phi(s)\, ds\right) \\ & = d(x(t_2) - x(t_1), x_{n_k}(t_2) - x_{n_k}(t_1)) + d\left(x_{n_k}(t_2) - x_{n_k}(t_1), \int_{t_1}^{t_2} \Phi(s)\, ds\right) \end{aligned} \tag{6}$$

$$\begin{aligned} d\left(x_{n_k}(t_2) - x_{n_k}(t_1), \int_{t_1}^{t_2} \Phi(s)\, ds\right) &= d\left(\int_{t_1}^{t_2} f_k(s), \int_{t_1}^{t_2} \Phi(s)\, ds\right) \\ &= \inf\left\{\left|\int_{t_1}^{t_2} f_k(s)\, ds - \int_{t_1}^{t_2} \varphi(s)\, ds\right| \mid \varphi(\cdot) \in \Phi(\cdot)\right\} \\ &\leq \inf\left\{\int_{t_1}^{t_2} |f_k(s) - \varphi(s)|\, ds \mid \varphi(\cdot) \in \Phi(\cdot)\right\}. \end{aligned}$$

Since: $x_{n_k}(s) \to x(s)$, Φ is upper semicontinuous, $d(((s,x), f_{n_k}(s,x)), \operatorname{graph}\Phi) \to 0$, it follows that $d(f_k(s), \Phi(s))$ converges *pointwise* to zero. By Corollary 1.14.1 for each k there exists φ_k, a measurable selection from Φ, such that $|f_k(s) - \varphi_k(s)| = d(f_k(s), \Phi(s))$ a.e. Hence

$$\inf\{\int |f_k - \varphi| \mid \varphi(\cdot) \in \Phi(\cdot)\} \leq \int |f_k - \varphi_k| = \int d(f_k, \Phi).$$

By the Dominated Convergence Theorem the last integral converges to zero. Hence from (6) we have

$$d\left(x(t_2) - x(t_1), \int_{t_1}^{t_2} \Phi(s)\, ds\right) = 0. \qquad \square$$

Third Proof (Polygonal Approximations). We define the sequence x^j by induction. We choose $v^0 = m(F(0, x_0))$ and set $x^1 \doteq x_0 + \frac{T}{k} v_0 \in x_0 + \frac{T}{k} K$. Then $\|x^1 - x_0\| \leq \frac{T}{k} \|K\| \leq b$ and $\left(\frac{T}{k}, x^1\right) \in Q$.

If $\left(\frac{(j-1)T}{k}, x^{j-1}\right) \in Q$ is constructed (for $j \leq k$), we set $v^{j-1} \doteq m\left(F\left(\frac{(j-1)T}{k}, x^{j-1}\right)\right)$ and $x^j \doteq x^{j-1} + \frac{T}{k} v^{j-1}$, which belongs to $x^{j-1} + \frac{T}{k} K$. Hence $x^j \in x_0 + \frac{jT}{k} K$. Since $\|x^j - x^0\| \leq \frac{jT}{k} \|K\| \leq b$, when $j \leq k$, we have constructed a sequence $x^j (0 \leq j \leq k)$ such that

i) $\left(\frac{jT}{k}, x^j\right) \in Q \subset \Omega$,

ii) $\frac{k}{T}(x^{j+1} - x^j) = m\left(F\left(\frac{jT}{k}, x^j\right)\right) \in K$.

We interpolate this sequence at the nodal points $\frac{jT}{k}$ by the piecewise linear function defined on each interval $\left[\frac{jT}{k}, \frac{(j+1)T}{k}\right]$ by

$$x_k(t) = x^j + \left(t - \frac{iT}{k}\right) v^j.$$

Assumptions of the Compactness Theorem are satisfied. When $t \in \left[\frac{jT}{k}, \frac{(j+1)T}{k}\right]$

$$\begin{aligned} ((t, x_k(t)), x_k'(t)) &= \left(\frac{jT}{k}, x^j, v^j\right) + \left(t - \frac{jT}{k}, \left(t - \frac{jT}{k}\right) v^j, 0\right) \\ &\in \operatorname{graph}(F) + \left(\frac{T}{k} B \times \frac{\|K\| T}{k} B \times \{0\}\right). \end{aligned} \tag{7}$$

Since (subsequences of) $t \to (t, x_k(t))$ and $t \to x'_k(t)$ converge pointwise to $t \to (t, x(t))$ and weakly in $L^1(0, T; X)$ to $x'(\cdot)$ respectively, the Convergence Theorem yields the existence of a solution on $[0, T]$. □

The interval on which the solution is defined depends upon the size of Ω and upon the neighborhood which is mapped in a compact set. In the case where $\Omega = R \times X$ and when $m(F(t,x))$ remains in a compact set K, we can take $a = \infty$ and $b = \infty$; therefore we can take T arbitrary in the proof of the preceding theorem, and consequently, obtain global results.

Theorem 4. *Let F be an upper semicontinuous map from $[0, \infty[\times X$ to X with non-empty closed convex values. We suppose that $m(F(t,x))$ remains in a compact subset of X. For any $x_0 \in X$, there exists an absolutely continuous function defined on $[0, \infty[$, a solution to*

$$x'(t) \in F(t, x(t)), \quad x(0) = x_0. \tag{1}$$ ▲

Application: Regularization of Differential Equations with Discontinuous Right-Hand Side

In order to provide existence for solutions of differential equations

$$x'(t) = f(x(t)) \tag{8}$$

when $f: R^n \to R^n$ is not continuous, the easiest way is to consider the smallest upper semicontinuous convex valued map F whose graph contains the graph of f. When f is locally bounded, this set-valued map F is defined by

$$F(x) \doteq \bigcap_{\varepsilon > 0} \overline{\text{co}}\, f(x + \varepsilon B). \tag{9}$$

It is clear that

$$\begin{cases} \text{i) } \forall x, f(x) \in F(x), \\ \text{ii) the map } x \to F(x) \text{ is upper semicontinuous with convex values,} \\ \text{iii) whenever } f \text{ is continuous at } x, F(x) = \{f(x)\}. \end{cases} \tag{10}$$

Certainly, any solution to the differential equation (8) is a solution to the differential inclusion

$$x'(t) \in F(x(t)). \tag{11}$$

We stress the point that whenever f is continuous at $x(t)$, then a solution to the differential inclusion (11) satisfies the equation $x'(t) = f(x(t))$.

In order to obtain this result, we do not need property (10) i) at points when f is not continuous. We can look for "smaller" set-valued maps ϕ which still satisfy properties (10) ii) and iii). Hence differential inclusions

$$x'(t) \in \phi(x(t)) \tag{12}$$

yield trajectories $x(\cdot)$ satisfying the equation $x'(t)=f(x(t))$ whenever f is continuous at $x(t)$.

We describe one such map ϕ.

Proposition 1. *Let f be a single-valued map from an open subset $\Omega \subset R^n$ to R^n which is locally bounded. We set*

$$\phi(x) \doteq \bigcap_{\varepsilon>0} \bigcap_{\text{meas}(N)=0} \overline{\text{co}}\, f(((x+\varepsilon B)\cap\Omega)\backslash N) \tag{13}$$

Then

(14) i) *the map $x \to \phi(x)$ is upper semicontinuous with nonempty convex values,*

ii) *whenever f is continuous at x, $\phi(x)=\{f(x)\}$.*

Assume moreover that f is measurable on Ω. Then

(15) iii) *$f(x)$ belongs to $F(x)$ at almost every x in Ω.* ▲

Proof. a) For ε sufficiently small, $x+\varepsilon B$ is contained in Ω and f is bounded on it. Since the sets $\bigcap_{\text{meas}(N)=0} \overline{\text{co}}\, f(((x+\varepsilon B)\cap\Omega)\backslash N)$ decrease with respect to $\varepsilon>0$, and are compact, it is enough for proving that $\phi(x)$ is non-empty, to show that each of the sets above is non-empty. Since each

$$\overline{\text{co}}\, f(((x+\varepsilon B)\cap\Omega)\backslash N) = \overline{\text{co}}\, f((x+\varepsilon B)\backslash N)$$

is compact, it is enough to show that for any finite family $\{N_i\}_1^m$, $\text{meas}(N_i)=0$,

$$\bigcap_{i=1}^{m} \overline{\text{co}}\, f((x+\varepsilon B)\backslash N_i)$$

is non-empty. Indeed, since $\text{meas}\left((x+\varepsilon B)\backslash \bigcup_{i=1}^{m} N_i\right)>0$ then there exists $\bar{x}\in(x+\varepsilon B)\backslash N_i$ for all $i=1,\ldots,m$.

b) Let us show that ϕ has a closed graph. Let $x_n \to x_*$, x_n and x_* in Ω and let $y_n \to y_*$, y_n in $\phi(x_n)$. Let $k(n)$ be such that $x_n \in x_* + \varepsilon_{k(n)} B$; hence $y_n \in \bigcap_{\text{meas}(N)=0} \overline{\text{co}}\, f((x_*+2\varepsilon_{k(n)}B)\backslash N)$, a closed set. Therefore

$$y_* \in \bigcap_{\varepsilon_{k(n)}} \overline{\text{co}}\, f((x_*+2\varepsilon_{k(n)}B)\backslash N).$$

c) Let f be continuous at $x\in\Omega$. For arbitrary $\delta>0$ and for ε sufficiently small, we obtain

$$\phi(x) \subset f(x)+\delta B.$$

Therefore, $\phi(x)=\{f(x)\}$.

d) By Lusin's Theorem, Ω can be written as

$$\Omega = \left(\bigcup_{n\in N} E_n\right) \cup M,$$

where meas$(M)=0$ and where the restrictions $f|_{E_n}$ are continuous. By the Vitali-Lebesgue Theorem, each E_n can be thought to coincide with the set of its points of density $\left(\text{i.e., points } x \text{ such that } \lim_{\varepsilon\to 0} \frac{\text{meas}(E_n\cap(x+\varepsilon B))}{\text{meas}(x+\varepsilon B)}=1\right)$. Let us fix x in some E_n. For every $\varepsilon>0$, we know that meas$(E_n\cap(x+\varepsilon B))$ is positive. So, by the same reasoning as is point a), we deduce that the set

$$\bigcap_{\text{meas}(N)=0} \overline{\text{co}}\, f(((x+\varepsilon B)\cap E_n)\backslash N)$$

is nonempty, and that for arbitrary $\delta>0$ and sufficiently small $\varepsilon>0$, it is contained in $f(x)+\delta B$. By letting δ converge to 0, we deduce that

$$\{f(x)\}=\bigcap_{\varepsilon>0} \overline{\text{co}}\, f(((x+\varepsilon B)\cap E_n)\backslash N)\subset\phi(x). \qquad \square$$

2. Qualitative Properties of the Set of Trajectories of Convex-Valued Differential Inclusions

We shall study the properties of the map $\mathscr{T}_T(x^0)$ that associates to a given initial condition $(0, x^0)$ the set of solutions through it. As it happens in the case of an ordinary differential equation, solutions through different initial points or even different solutions through the same initial point may have unequal intervals of existence. To avoid this difficulty we shall limit ourselves to local properties and make the following assumption.

Boundedness Assumption

Let Ω be an open subset of $R\times R^n$ such that $(0, x_0)\in\Omega$ and F from Ω into the compact and convex subsets of R^n be an upper semicontinuous map. There exist $b>0$, $T>0$ and $M>0$ such that the set

(1) $$Q\doteq[0, T]\times(x_0+(b+MT)B)$$

is contained in Ω and F maps Q into the ball of radius M. ▲

The proofs of *Theorem* 1.3 show that for every initial state $x_1\in x_0+b\mathring{B}$, there exists a solution to the differential inclusion $x'(t)\in F(t, x(t))$, $x(0)=x_1$ on the interval $[0, T]$.

We shall supply the space $\mathscr{C}(0, T; R^n)$ with the topology of uniform convergence and the space $L^\infty(0, T; R^n)$ with the weak *-topology. We denote by

(2) $$\mathscr{A}(0, T; R^n)\doteq\{x\in\mathscr{C}(0, T; R^n)\,|\,x'\in L^\infty(0, T; R^n)\}$$

the space of continuous functions whose derivatives (in the sense of distributions) belong to $L^\infty(0, T; R^n)$.

It is supplied with the initial topology which makes the map $x(\cdot) \to (x(\cdot), x'(\cdot))$ continuous from $(0, T; R^n)$ to $\mathscr{C}(0, T; R^n) \times L^\infty(0, T; R^n)$, the first space with its strong topology, the second one with its weak *-topology.

We denote by $\mathscr{T}_T(\xi)$ the set of trajectories of the differential inclusion

$$x'(t) \in F(t, x(t)), \qquad x(0) = \xi. \tag{3}$$

on the interval $[0, T]$.

We denote by

(4) $$\mathscr{T}_T'(\xi) \doteq \{x' \in L^\infty(0, T; X) \mid x(\cdot) \in \mathscr{T}_T(\xi)\},$$
the set of the derivatives of the trajectories.

The *attainable set* $\mathscr{A}_T(\xi)$ is defined by

$$\mathscr{A}_T(\xi) \doteq \{x(T) \mid x(\cdot) \in \mathscr{T}_T(\xi)\}. \tag{5}$$

Theorem 1. *Let F be an upper semicontinuous map from Ω to the compact convex subsets of R^n.*

We posit the Boundedness Assumption. *The maps $\xi \to \mathscr{T}_T(\xi)$ and $\xi \to \mathscr{A}_T(\xi)$ are strict upper semicontinuous maps from $x_0 + b\mathring{B}$ into the compact subsets of $\mathscr{A}(0, T; R^n)$ and R^n.* ▲

Corollary 1. *Under the assumptions of Theorem 1, the map $\xi \to \mathscr{T}_T'(\xi)$ is upper semicontinuous from $x_0 + b\mathring{B}$ into $L^\infty(0, T; R^n)$ supplied with the weak *-topology.* ▲

Proof. Let $b_0 < b$ and $\xi_h \in x_0 + b_0 B$, $x_h(\cdot) \in \mathscr{T}_T(\xi_h)$. Since $x_h(t) \in x_0 + (b + tM)B$, and $\|x_k'(t)\| \leq M$, assumptions of the Compactness Theorem hold; this implies that a subsequence (again denoted) $x_k(\cdot)$ converges to a solution $x(\cdot)$ in $\mathscr{C}(0, T; X)$ and $x_k'(\cdot)$ converges weakly to $x'(\cdot)$ in $L^\infty(0, T; X)$. Since $((t, x_n(t)), x_k'(t))$ belongs to the graph of F for almost all $t \in [0, T]$, the Convergence Theorem implies that $x'(t)$ belongs to $F(t, x(t))$ for almost all t.

Hence the graph of the map $\mathscr{T}_T(\cdot)$ is *compact* in $(x_0 + b_0 B) \times \mathscr{A}(0, T; R^n)$. This implies in particular that it is upper semicontinuous with compact images. □

Under the assumptions of *Theorem* 1.4, we can take $T = \infty$.

The maps $\xi \to \mathscr{T}_T(\xi)$ and $\xi \to \mathscr{A}_T(\xi)$ are upper semicontinuous but they are not, in general, convex valued. Still we shall show that they enjoy the same properties as the convex valued maps described in Chapter 1, namely the Approximate Selection Theorem, the theorem on the existence of fixed points and, in a sense to be made precise, the Approximation Theorem. As it is the case for ordinary differential equations, the images $\mathscr{A}_T(\xi)$ are connected sets: we shall prove that they are continuous images of acyclic sets.

We shall consider maps F, ϕ on Q; the corresponding maps $\mathscr{T}_T^F$ and $\mathscr{T}_T^\phi$ should be thought as maps from x_0+bB into the subsets of $\mathscr{C}(0, T; R^n)$.

Proposition 1. *Let F be upper semicontinuous with compact convex values, satisfying the Boundedness Assumption. Then for every $\varepsilon>0$ there exists $\delta>0$ such that: for every map ϕ satisfying the Boundedness Assumption, and*

$$\text{graph}(\phi)\subset\text{graph}(F)+\delta B \tag{6}$$

it follows

$$\text{graph}(\mathscr{T}_T^\phi)\subset\text{graph}(\mathscr{T}_T^F)+\varepsilon B. \tag{7}$$ ▲

Proof. Suppose there exists a sequence of maps ϕ^n, $\delta(\text{graph}\,\phi^n, \text{graph}\,F)\to 0$, and solutions x_n to

$$x_n'\in\phi^n(t, x_n), \qquad x_n(0)=\xi_n\in x_0+bB$$

such that

$$d((\xi_n, x_n(\cdot)), \text{graph}\,\mathscr{T}_T^F)>\varepsilon. \tag{8}$$

Since each ϕ_n satisfies Boundedness Assumption, we can assume that for some subsequence $\xi_n\to\xi^*\in x_0+bB$, $x_n(\cdot)\to x^*(\cdot)$ in $\mathscr{C}(0, T; R^n)$ and $x_n'(\cdot)\to x^{*\prime}(\cdot)$ weakly* in $L^\infty(0, T)$. Since $d((t, x_n(t)),\ x_n'(t),\ \text{graph}\,F)\leq \delta(\text{graph}(\phi^n), \text{graph}(F))$, the Convergence Theorem implies that $x^*(\cdot)$ is a solution to (3), i.e. that $(\xi^*, x^*(\cdot))\in\text{graph}\,\mathscr{T}_T^F$, a contradiction to (8).

Corollary 2 (The Approximate Selection Theorem). *Let F be upper semicontinuous with closed convex values, satisfying the Boundedness Assumption. Then for every $\varepsilon>0$ there exists a single-valued continuous s_ε: $x_0+bB\to C(0, T)$ such that*

$$\text{graph}\,s_\varepsilon\subset\text{graph}\,\mathscr{T}_T^F+\varepsilon B.$$ ▲

Proof. It is enough to set, in Proposition 1, ϕ to be a single-valued locally Lipschitzean approximation to F, as provided by Theorem 1.12.1. □

Corollary 3 (The Fixed Point Property). *Let F be upper semicontinuous with closed convex values, satisfying the Boundedness Assumption. Assume that there exists a compact convex set $S\subset x_0+MTB$ such that ξ in S implies $\mathscr{A}_T(\xi)\subset S$. Then there exists ξ^* in S, $\xi^*\in\mathscr{A}_T(\xi^*)$.* ▲

Proof. Let $\varepsilon_n\to 0$. Let $s_n(\cdot)$ be as in Corollary 1. In particular $\xi\to s_n(\xi)(T)$ maps S into $S+\varepsilon_n B$. By Theorem 1.12.2 there exist ξ_n: $d(\xi_n, s_n(\xi_n)(T))\leq\varepsilon_n$. We can assume $\xi_n\to\xi^*$ (hence, that $s_n(\xi_n)(T)\to\xi^*$).

$$\begin{aligned} &d((\xi^*, \xi^*), \text{graph}(\mathscr{A}_T)) \\ &\quad\leq d((\xi^*, \xi^*), (\xi_n, s_n(\xi_n)(T))+\delta((\xi_n, s_n(\xi_n)(T), \text{graph}(\mathscr{A}_T)) \end{aligned}$$

and by Corollary 1, the right hand side converges to zero. Since graph$(\mathscr{A}_T)$ is closed, $\xi^* \in \mathscr{A}_T(\xi^*)$. $\square$

Corollary 4 (Dependence on Parameters). *Let the map $(t, x, \lambda) \to F(t, x, \lambda)$ be upper semicontinuous from $Q \times I$ into the compact convex subsets of R^n and satisfy the Boundedness assumption for every λ in I. Then for every $\varepsilon > 0$ there exists $\delta > 0$ such that $|\lambda - \lambda_0| < \delta$, λ and λ_0 in I, implies*

$$\operatorname{graph}(\mathscr{T}_T^{\lambda}) \subset \operatorname{graph}(\mathscr{T}_T^{\lambda_0}) + \varepsilon B. \qquad \blacktriangle$$

In the proofs of *Theorems* 2 and 3 below we make use of properties of solutions to differential equations of the form

$$x'(t) = f(t, x(t))$$

where f satisfies the Carathéodory Conditions i.e. f is measurable in t for each x fixed and is continuous (or Lipschitzean) in x for each t fixed and there exists a map g in $L^1_{\text{Loc}}(I)$ such that $\|f(t, x)\| \le g(t)$ for all (t, x) in Dom(f).

Theorem 2. *Under the same assumptions as in Theorem 1, the sets $\mathscr{A}_T(\xi)$ are connected.* $\blacktriangle$

Proof. It follows from *Theorem* 1.2.1 and *Theorem* 3 below.

However, we provide a self-contained proof.

It is enough to show that for any two points ξ^0, ξ^1 in $\mathscr{A}_T(\xi)$ and for any $\varepsilon > 0$, there is a curve p, a continuous image of the interval $[0, 1]$, such that $x^0 = p(0)$, $x^1 = p(1)$ and, for λ in $[0, 1]$, $p(\lambda) \in \mathscr{A}_T(\xi) + \varepsilon B$. Proposition 1 assures that there exists $\eta \doteq \eta(\varepsilon)$ such that for any map $f(t, x)$ satisfying graph$(f) \subset$ graph$(F) + \eta B$, a solution x_f to $x'_f = f(t, x_f)$, $x_f(0) = \xi$, is such that $x_f(T) \in \mathscr{A}_T(\xi) + \varepsilon B$.

Hence to prove the theorem it is enough to show that:

a) Given two solutions $x^j(\cdot)$, $j = 0, 1$, of (3) such that $x^j(T) = x^j$, there are maps $f_n^j(t, x)$, measurable in t and Lipschitzean in x, such that $\mathfrak{d}(\operatorname{graph}(f_n^i), \operatorname{graph}(F)) \to 0$ and $x^j(\cdot)$ is a solution to $x' = f_n^j(t, x(t))$, all $n, j = 0, 1$;

b) Whenever n is sufficiently large, any convex combination $g_{n,\lambda}(t, x) \doteq \lambda f_n^1(t, x) + (1 - \lambda) f_n^0(t, x)$ is in graph$(F) + \eta B$.

If this is the case, in fact, setting $p(\lambda) = x_{n,\lambda}(T)$, where $x_{n,\lambda}(\cdot)$ is the unique solution, continuously depending on the parameter λ, of

$$x'(t) = g_{n,\lambda}(t, x(t)), \qquad x(0) = \xi$$

we obtain the required curve.

Since b) follows from a simple contradiction argument, let us prove a). Let x^* be a solution on I and call $S \doteq \{(t, x^*(t)) \mid t \in I\}$. Fix $\sigma > 0$ arbitrarily. Let $\delta(t) > 0$ (we assume $\delta(t) < \sigma/3$) be such that $(\tau, \xi) \in (t, x^*(t)) + 2\delta(t)\mathring{B} \Rightarrow F(\tau, \xi) \subset F(t, x^*(t)) + \sigma/3\,\mathring{B}$. The family $\{(t, x^*(t)) + \delta(t)\mathring{B}\}$ covers the compact S: let $\{(t^i, x^*(t^i))+\delta^i \mathring{B}\}$ be a finite subcover. Set δ_m be such that $S+\delta_m B \subset \bigcup_{i=1}^{m} \{t^i, x^*(t^i))+\delta^i \mathring{B}\}$. Let $f(t,x)$ be a continuous map, locally Lipschitzean in x, approximating F in the sense that $\operatorname{graph}(f) \subset \operatorname{graph}(F) + \delta_m B$. The map f is uniformly continuous on Q, hence for some $\delta^*(0<\delta^*<\delta_m)$, $d(x,\xi)\le\delta^*$ implies $d(f(t,x), f(t,\xi))\le\sigma/3$. Define $\rho(t,x)$ to be a locally Lipschitzean map, such that $\rho=0$ on Comp $(S+\delta^* \mathring{B})$, $\rho=1$ on S and $0\le\rho\le 1$ on $S+\delta^* \mathring{B}$. Set

(9) $$g(t,x) \doteq \rho(t,x)\, x^{*\prime}(t) + (1-\rho(t,x))\, f(t, x^*(t)) + f(t,x) - f(t, x^*(t)).$$

Then g is locally Lipschitzean in x for t fixed, measurable in t for x fixed. Moreover $g(t, x^*(t)) = x^{*\prime}(t)$, hence x^* is the solution to $x' = g(t,x)$, $x(0)=\xi$. We wish to show that $\operatorname{graph}(g) \subset \operatorname{graph}(F) + \sigma B$.

On Comp$(S+\delta^* \mathring{B})$, $g=f$; on S, $g(t, x^*(t)) \in F(t, x^*(t))$. Hence we can consider only (t,x) in $S+\delta^* \mathring{B}$. By the properties of f, there is (τ,ξ) in $(t, x^*(t)) + \delta_m B$ such that $f(t, x^*(t)) \in F(\tau,\xi) + \delta_m B$; by the definition of δ_m, there is some i: $(\tau,\xi) \in (t^i, x^*(t^i)) + \delta^i B$. Hence (recalling that $\delta_m + \sigma/3 < 2/3\,\sigma$)

(10) $$f(t, x^*(t)) \in F(t^i, x^*(t^i)) + 2/3\,\sigma B.$$

For the same i,

$$d((t, x^*(t)), (t^i, x^*(t^i))) \le d((t, x^*(t)), (\tau,\xi)) + d((\tau,\xi), (t^i, x^*(t^i))$$
$$\le \delta_m + \delta^i < 2\delta^i = 2\delta(t^i),$$

hence

(11) $$x^{*\prime}(t) \in F(t, x^*(t)) \subset F(t^i, x^*(t^i)) + \sigma/3 B.$$

Inclusions (10) and (11) and the convexity of F imply

$$\rho x^*(t) + (1-\rho)\, f(t, x^*(t)) \in F(t^i, x^*(t^i)) + 2/3\,\sigma B.$$

Since $d(x, x^*(t)) \le \delta$, we have $\|f(t,x) - f(t, x^*(t))\| \le \sigma/3$, hence we have

$$g(t,x) \in F(t^i, x^*(t^i)) + \sigma B,$$

and

$$d((t,x), (t^i, x^*(t^i))) \le d((t,x), (t, x^*(t))) + d(t, x^*(t)), (t^i, x^*(t^i)))$$
$$\le \sigma/3 + 2/3\,\sigma = \sigma$$

as it was required. □

Theorem 3 (The Approximation Theorem). *We posit the assumption of Theorem 1. There exists a sequence of convex compact subsets U^m and of continuous maps s_n from $(x_0 + b\mathring{B}) \times U^m$ to $\mathscr{C}(0,T;R^n)$ satisfying, for all $\xi \in x_0 + b\mathring{B}$:*

(12) $$\begin{cases} \text{i)}\ \forall m \ge 0, & \mathscr{T}_T(\xi) \subset \ldots \subset s_{m+1}(\xi, U^{m+1}) \subset s_m(\xi, U^m) \subset \ldots \\ \text{ii)}\ \forall \varepsilon > 0, & \exists N_{\xi,\varepsilon} \mid \forall m \ge N_{\xi,\varepsilon}, \quad s_m(\xi, U^m) \subset \mathscr{T}_T(\xi) + \varepsilon B \end{cases}$$

where B denotes the unit ball of $\mathscr{C}(0,T;R^n)$. Furthermore,

(13) $$\forall x \in \mathscr{T}_T(\xi), \quad \{u \in U^m \mid s_m(\xi,u) = x(\cdot)\} \quad \text{is convex.}$$ ▲

Proof. By the Approximation Theorem 1.13.1, F can be approximated by a decreasing sequence of set-valued maps F_m defined on Q by

(14) $$F_m(t,x) \doteq \sum_{i\in I_m} \psi_i^m(t,x)\, C_i^m \subset MB$$

where I_m is a finite subset of indexes, $\psi_i^m(t,x)$ are non negative locally Lipschitzean functions defined on Q and summing up to 1 and where C_i^m are convex compact subsets of R^n contained in the ball of radius M.

Let us denote by

(15) $\mathscr{T}_T^m(\xi)$ the set of trajectories of the differential inclusion
$$x'(t)\in F_m(t,x(t)), \qquad x(0)=\xi.$$

Since $F(x)\subset\ldots\subset F_{m+1}(x)\subset F_m(x)\subset\ldots$, we deduce at once that

$$\mathscr{T}_T(\xi)\subset\ldots\subset\mathscr{T}_T^{m+1}(\xi)\subset\mathscr{T}_T^m(\xi)\subset\ldots.$$

We shall prove that for all $\varepsilon>0$, there exists $N_{\varepsilon,\xi}$ such that for all $m\ge N_{\varepsilon,\xi}$, we have $\mathscr{T}_T^m(\xi)\subset\mathscr{T}_T(\xi)+\varepsilon B$, B denoting the unit ball of $\mathscr{C}(0,T,R^n)$. Assume the contrary: there exist $\varepsilon>0$ and a sequence of solutions $x_m(\cdot)\in\mathscr{T}_T^m(\xi)$ satisfying $d(x_m(\cdot),\mathscr{T}_T(\xi))>\varepsilon$.

We know that $x_m'(t)\in F_m(t,x_m(t))\subset MB$ by (14). Therefore,

$$\forall t\in[0,T], \qquad \|x_m'(t)\|\le M \quad\text{and}\quad \|x_m(t)-x_0\|\le b+MT.$$

On the other hand, since for all $(t,x(t))$ in Q, there exists $N_{\varepsilon,t}$ such that $F_m(t,x(t))\subset F(t,x(t))+\varepsilon B$ when $m\ge N_{\varepsilon,t}$, we obtain:

(16) $$\forall m\ge N_{\varepsilon,t}, \qquad (t,x_m(t),x_m'(t))\in\operatorname{graph}F+\varepsilon B.$$

Hence the Compactness Theorem and the Convergence Theorem imply that a subsequence of $(x_m(\cdot))_m$ convergence uniformly to a solution $x(\cdot)\in\mathscr{T}_T(\xi)$. This is a contradiction.

We set
$$U^m \doteq \prod_{i\in I_m} L^\infty(0,T;\,C_i^m).$$

which is a convex compact subsets of $L^\infty(0,T;\,R^n)^{I_m}$ supplied with the weak *-topology.

Let $u^m(\cdot)\doteq(u_i^m(\cdot))_{i\in I_m}$ be a function of U^m. Since the map

$$(t,x)\to\sum_{i\in I_m}\psi_i^m(t,x)\,u_i^m(t)$$

is measurable with respect to t, locally Lipschitzean with respect to x and bounded by M on Q (independently of $u^m(\cdot)$), we know that there exists a *unique solution* on $[0,T]$ to the differential equation

(17) $$x'(t)=\sum_{i\in I_m}\psi_i^m(t,x(t))\,u_i^m(t), \qquad x(0)=\xi.$$

We shall denote by $s_m(\xi,u^m)$ this solution. Hence, we have defined a map s_m from $(x_0+b\mathring{B})\times U^m$ to $\mathscr{C}(0,T;\,R^n)$. Since the solution of (17) can be written

(18) $$x(t)=\xi+\sum_{i\in I_m}\int_0^t \psi_i^m(\tau, x(\tau))\, u_i^m(\tau)\, d\tau$$

we see at once that s_m is a continuous map. Then, for any $u^m \in U^m$, $s_m(\cdot, u^m)$ provides a continuous selection of $\mathscr{T}_T^m(\cdot)$. Actually, we have

(19) $$\mathscr{T}_T^m(\xi)=s_m(\xi, U^m).$$

Indeed, let $x(\cdot)\in\mathscr{T}_T^m(\xi)$ be a solution to the differential inclusion $x'(t)\in\sum_{i\in I_m}\psi_i^m(t, x(t))\, C_i^m$.

By *Corollary* 1 of the *measurable selection Theorem*, there exists a measurable function $u^m(\cdot)$ taking its values in $\prod_{i\in I_m} C_i^m$ such that

(20) $$x'(t)=\sum_{i\in I_m}\psi_i^m(t, x(t))\, u_i^m(t), \qquad \text{i.e. } x=x_m(\xi, u^m).$$

The subset of such controls u^m is obviously convex. The Theorem is now proved. □

Corollary 5. *Let F be an upper semicontinuous map from Ω to the compact convex subsets of R^n.*

We posit the Boundedness Assumption. The map $\xi\to\mathscr{T}_T(\xi)$ from $x_0+b\mathring{B}$ to $\mathscr{C}(0, T; R^n)$ is σ-selectionable, as well as $\xi\to\mathscr{A}_T(\xi)$.

Furthermore, the images $\mathscr{T}_T(\xi)$ are a cyclic. ▲

Proof. Since, by (12) i) and iii),

(21) $$\mathscr{T}_T(\xi)=\bigcap_{m\geq 0} s_m(\xi, U^m)$$

where the set-valued maps $s_m(\cdot, U^m)$ are obviously selectionable, $\mathscr{T}_T(\cdot)$ and consequently, $\mathscr{A}_T(\cdot)$ are σ-selectionable.

Theorem 1.2.1 and Properties (12) i) and iii) imply that the set-valued map $\mathscr{T}_T(\cdot)$ has connected values. □

Application of the Connectedness of $\mathscr{A}_T(\xi)$: the Hukuhara Theorem

In the theory of optimal control one encounters the following situation. A system evolves along the solutions to the problem

(22) $$x'(t)\in F(t, x(t)), \qquad x(0)=x_0.$$

A point x_1 is given as a target and one is interested both in the existence of a minimal time T such that $x_1\in\mathscr{A}_T(x_0)$ and in the properties of the solutions that lead to this point. Under very general conditions one has that when such a T exists, x_1 belongs to the boundary $\partial\mathscr{A}_T(x_0)$. We are interested in knowing whether there exists a solution $\bar{x}(\cdot)$ defined on $[0, T]$ such that $\bar{x}(0)=x_0$, $\bar{x}(T)=x_1$ and such that

$$\bar{x}(t)\in\partial\mathscr{A}_t(x_0), \qquad \forall t\in[0, T]. \tag{23}$$

In some cases, in particular when F is parameterized by a set of controls $\mathscr{U}$, i.e., when there exists a single valued $f(t, x, u)$ and a set $\mathscr{U}$ such that $F(t, x) = f(t, x, \mathscr{U})$, one obtains in this way necessary conditions for a solution to be optimal. It was Hukuhara who first proved the above property for the case of an ordinary differential equation.

Theorem 4. *We posit the Boundedness Assumption. Let $\mathscr{A}_t(x_0)$ be the attainable set at t of problem (22). Let $\xi_1\in\partial\mathscr{A}_T(x_0)$. Then there exists $\bar{x}(\cdot)$, a solution to (22), such that (23) holds and $\bar{x}(T)=\xi_1$.*

For its proof we need the following lemma. ▲

Lemma 1. *We posit the Boundedness Assumption.*

Let $\bar{x}$ be a solution of (22) on $[0, T]$. Let $\bar{t}\in[0, T]$ be such that $\bar{x}(\bar{t})\in\operatorname{Int}(\mathscr{A}_{\bar{t}}(x_0))$. Then there exists $\delta>0$ such that for $t\in[\bar{t}, \bar{t}+\delta]$, $\bar{x}(t) +\delta B\subset\mathscr{A}_t(x_0)$. ▲

Proof. Let $\eta>0$ be such that $\bar{x}(\bar{t})+\eta B\subset\mathscr{A}_{\bar{t}}(x_0)$. Let c be a bound for $\|F(t, x)\|$, $(t, x)\in Q$. If the claim is false there is a sequence $\{t_n\}$, $t_n\downarrow\bar{t}$ as well as a sequence ξ_n, $\xi_n\to\bar{x}(\bar{t})$ such that ξ_n does not belong to $\mathscr{A}_{t_n}(x_0)$. A solution $y_n(\cdot)$ through (t_n, ξ_n) cannot be such that $y_n(\bar{t})\in\mathscr{A}_{\bar{t}}(x_0)$, otherwise, by piecing together the solution of (22) that leads to $y_n(\bar{t})$ and then y_n, by definition $\xi_n \doteq y_n(t_n)$ would belong to $\mathscr{A}_{t_n}(x_0)$. Hence on the one hand

$$\begin{aligned}\|\bar{x}(\bar{t})-y_n(\bar{t})\| &\le \|\bar{x}(\bar{t})-y_n(t_n)\|+\|y_n(t_n)-y_n(\bar{t})\|\\ &\le \|\bar{x}(\bar{t})-\xi_n\|+c|t_n-\bar{t}|\end{aligned}$$

while on the other

$$\|\bar{x}(\bar{t})-y_n(\bar{t})\|>\eta,$$

a contradiction. □

Proof of Theorem 6. a) For any n, set $t_0^{(n)}=0$, $t_i^{(n)}=t_{i-1}^{(n)}+T/n$, $i=1, 2, \dots, n$. We claim that it is enough to show that (23) holds for any sufficiently large n and for $t\in\{t_i^{(n)}\}_{i=1}^n$. If this is the case, in fact, by letting n tend to infinity we obtain a sequence $\{x_n(\cdot)\}$ of solutions, each satisfying the previous requirement, that, by the compactness of the set of solutions issued from x_0, must contain a subsequence converging uniformly to a solution $x^*(\cdot)$. Assume that at some $\bar{t}\in[0, T]$, $x^*(\bar{t})\notin\partial\mathscr{A}_{\bar{t}}(x_0)$. By the preceding Lemma, there is a positive δ such that in the interval $[\bar{t}, \bar{t}+\delta]$, $x^*(t)+\delta B\subset\mathscr{A}_t(x_0)$. For every $n\ge T/\delta$, in this interval there falls at least one point $t_i^{(n)}$ and also, for all n sufficiently large, $\|x_n(t_i^{(n)})-x^*(t_i^{(n)})\|\le\delta/2$. Since $x_n(t_i^{(n)})\in\partial\mathscr{A}_{t_i^{(n)}}(x_0)$, in $x^*(t_i^{(n)})+\delta B$ there are points of $\operatorname{Comp}(\partial\mathscr{A}_{t_i^{(n)}}(x_0))$, a contradiction. This proves our first claim.

b) All we have to show then is that given any two values t_0, t_1 in $[0, T]$, $t_0 < t_1$, $(t_1 - t_0)$ sufficiently small, and any $\xi^1 \in \partial \mathscr{A}_{t_1}(x_0)$, there is a solution x^* such that $x(t_1) = \xi_1$ and $x^*(t_0) \in \partial \mathscr{A}_{t_0}(x_0)$. Whenever $|t_0 - t_1|$ is sufficiently small, any solution to

(24) $$x'(t) \in F(t, x(t)), \quad x(t_1) \in \mathscr{A}_{t_1}(x_0)$$

does not leave Q on $[t_0, t_1]$. Hence on this interval the set of solutions is well defined and, in view of the preceding paragraph, its section $\mathscr{A}$ at $t = t_0$ is a compact and connected set. We claim $\mathscr{A} \cap \partial \mathscr{A}_{t0}(x_0)$ is nonempty.

Consider $\mathrm{Comp}(\mathscr{A}_{t_0}(x_0))$ and $\mathrm{Int}(\mathscr{A}_{t_0}(x_0))$, two disjoint and open sets. Then either they don't cover $\mathscr{A}$, and we have done, or one of them has empty intersection with $\mathscr{A}$. Assume that the claim is false. Then, since $\mathscr{A} \cap \mathrm{Int}(\mathscr{A}_{t_0}(x_0))$ is non-empty, we must have $\mathscr{A} \cap \mathrm{Comp}(\mathscr{A}_{t_0}(x_0))$ empty, i.e. $\mathscr{A} \subset \mathrm{Int}(\mathscr{A}_{t_0}(x_0))$. Moreover, since $\mathscr{A}$ is compact, there must exist a positive ε such that $(\mathscr{A} + \varepsilon B) \cap \mathrm{Comp}(\mathscr{A}_{t_0}(x_0)) = \emptyset$. To reach a contradiction it is enough to show that there is a solution $\bar{x}$ to $x' \in F(t, x)$ such that $\bar{x}(t_0) \in \mathscr{A}_{t_0}$ while $\bar{x}(t_1) \in \mathrm{Comp}(\mathscr{A}_{t_1}(x_0))$. In fact by piecing together a solution $x(\cdot)$ of (22), such that $x(t_0) = \bar{x}(t_0)$, up to $t = t_0$ and then $\bar{x}(\cdot)$ on $[t_0, t_1]$, we would contradict the definition of $\mathscr{A}_{t_1}(x_0)$. Since the map from ξ to the set of solutions on $[t_0, t_1]$ of $x' \in F(t, x)$, $x(t_1) = \xi$, is upper semicontinuous, there is a positive η such that solutions $y(\cdot)$ on $[t_0, t_1]$ such that $y(t_1) \in \xi_1 + \eta B$ are within an ε neighborhood of the set of solutions through ξ_1. In particular the section of this set at t_0 is contained in $\mathscr{A} + \varepsilon B \subset \mathscr{A}_{t_0}(x_0)$.

In $\xi_1 + \eta B$ there are points of $\mathrm{Comp}(\mathscr{A}_{t_1}(x_0))$. This is the desired contradiction. □

3. Nonconvex-Valued Differential Inclusions

We shall prove the existence of solutions to differential inclusions

$$x'(t) \in F(t, x(t)), \quad x(0) = x_0$$

when the values of F are no longer assumed to be convex.

As we have remarked, in existence theorems for differential inclusions emphasis has to be placed on the convergence of the derivatives of the approximate solutions, and for this purpose compactness is the essential tool. Compactness is easier to prove for weakened topologies; however the topology where convergence takes place has to imply at least pointwise convergence for some subsequence, otherwise there is no guarantee that the limit of approximate solutions will be a solution. In the Convergence Theorem, convexity was used to pass from the convergence in a weakened topology, obtained by an easy compactness argument, to strong convergence for a sequence of convex combinations. When we have no convexity, the compactness argument has to be used directly in a strong topology.

We shall use the Compactness Theorem on the space of bounded functions to prove the existence of local solutions when F is continuous with (nonconvex) compact images. Later, by assuming that F is absolutely continuous, the Ascoli-Arzelà Theorem will be sufficient for proving the existence of continuously differentiable solutions.

Theorem 1. *Let $\Omega \subset R \times R^n$ be an open subset containing $(0, x_0)$ and let F be a continuous map from Ω to the nonempty compact subsets of R^n. Then there exists $T>0$ and an absolutely continuous function $x(\cdot)$ defined on $[0, T]$, a solution to the differential inclusion*

$$x'(t) \in F(t, x(t)), \quad x(0) = x_0. \tag{1}$$

Moreover, the derivative $x'(\cdot)$ is regulated. ▲

Proof. a) A rectangle $\mathscr{K}$ with sides 2a and 2b, centered at $(0, x^0)$ is contained in Ω and there exists some constant M such that $\|F\| \leq M$ on $\mathscr{K}$. Set $T \doteq \min[a, b/M]$. Then any function $x(\cdot)$ Lipschitzean with constant M such that $x(t_0) = x_0$, does not leave K on $[0, T]$. We shall define the solution on the interval $I = [0, T]$.

The idea of the proof is to define piecewise linear approximations such that the sequence of their derivatives is equioscillating, and thus, by the *compactness* Theorem 0.3.5, is precompact in the space $\mathscr{B}(0, T; R^n)$ of bounded functions. Therefore, we have to provide a sequence of finite partitions of I such that the oscillation of $x'_n(\cdot)$ on each element of the k-th partition is bounded by some number ε_k (independent of n) and then show that the sequence $\{\varepsilon_k\}$ converges to zero.

Let $\eta_k = 2^{-k}$. Since F is uniformly continuous on $\mathscr{K}$, there exists δ_k such that $|t-t'| \leq \delta_k$ and $|x-x'| \leq M\delta_k$ imply $\mathfrak{d}(F(t,x), F(t',x')) \leq \eta_k$. Choose a number l_1 smaller than δ_1 and such that T/l_1 is an integer and define the first partition of I by the intervals $[il_1, (i+1)l_1[$. In general, the m-th partition is obtained by subdividing each interval of the preceding partition into a finite number of right-open intervals each of equal length l_m, with $l_m < \delta_m$. The points il_m are the nodal points of the m-th partition.

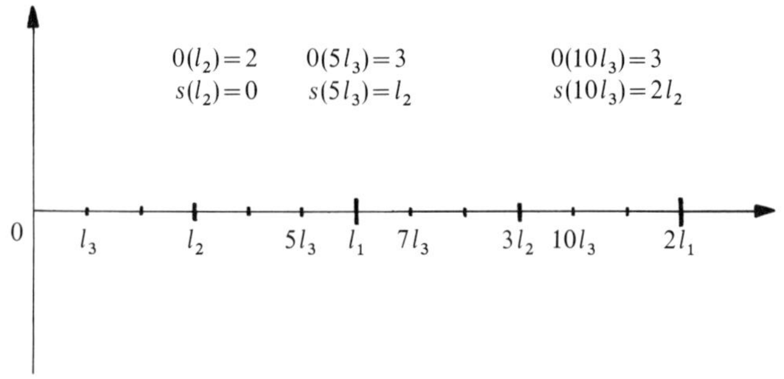

Fig. 10. Order and initial point associated to a nodal point

Since each partition is a refinement of the previous partitions, there are nodal points in the m-th partition that are also nodal points of preceding partitions. When τ is a nodal point, set $0(\tau)\in N$ to be the order of the partition where τ first appears as a nodal point. When $0(\tau)>1$, set $s(\tau)$ to be the initial point of the interval of the $(0(\tau)-1)$-th partition to which τ belongs.

b) The first polygonal line x_1 is a continuous function defined on I, linear on each interval of the first partition. On each $[il_1,(i+1)l_1]$ the derivative x_1' is constant, a point of $F(il_1, x_1(il_1))$. Then $x_1(\cdot)$ is an η_1 approximate solution, in the sense that $d(x_1'(t), F(t, x_1(t))\leq\eta_1$ (x_1' is only a right derivative at the points il_1): Let t belong to $[il_1,(i+1)l_1[$. Then, since $t-il_1<l_1$, hence, $d(x_1(il_1), x_1(t))\leq Ml_1$, we deduce from the uniform continuity of F that

$$d(x_1'(t), F(t, x_1(t))=d(x'(il_1), F(t, x_1(t))$$
$$\leq\mathfrak{d}(F(il_1, x_1(il_1)), F(t, x_1(t))\leq\eta_1.$$

Also, the oscillation of $x_1'(\cdot)$ is zero on the intervals of the first partition, and, a fortiori, on the intervals of all following partitions.

c) For each m, we whish to define a piecewise linear function $x_m(\cdot)$ with the following properties: for each nodal point t of the m-th partition,

$$(2)\quad\begin{cases}\text{i) } x_m(\cdot) \text{ is continuous on } [0,t],\ x_m(0)=x^0, \text{ and } x_m'(\cdot)\\ \quad\text{is constant on each interval of the } m\text{-th partitions of } [0,t],\\ \text{ii) At each nodal point } \tau\in[0,t[,\ x_m'(\tau)\in F(\tau, x_m(\tau)),\\ \text{iii) At each nodal point } \tau\in[0,t[,\ d(x_m'(\tau), x_m'(s(\tau)))\leq\eta_{0(\tau)-1}.\end{cases}$$

Each x_m would be a candidate for an approximate solution.

We prove the existence of x_m by induction on the nodal points of the m-th partition. For $t=0$, the constant x^0 is a continuous map defined on the closed interval $[0,0]$. There are no nodal points in $[0,0[$ and the right derivative is not defined. Let x_m be defined up to the nodal point t^*. Consider $0(t^*)$ and $s(t^*)$. Since

$$|t^*-s(t^*)|<l_{0(t^*)-1},\qquad d(x_m(t^*), x_m(s(t^*)))\leq Ml_{0(t^*)-1},$$

hence

$$\mathfrak{d}(F(t^*, x_m(t^*)), F(s(t^*), x_m(s(t^*))))\leq\eta_{0(t^*)-1}.$$

By assumption (2) ii), $x_m'(s(t^*))\in F(s(t^*), x_m(s(t^*)))$ and there exists $v\in F(t^*, x_m(t^*))$ such that $d(v, x_m'(s(t^*))\leq\eta_{0(t^*)-1}$. Set

$$(3)\qquad x_m(t)\doteq x_m(t^*)+(t-t^*)v \quad\text{on}\quad]t^*, t^*+l_m].$$

Then (2) i), ii) and iii) are satisfied up to $t+l_m$. By induction, the existence of x_m defined on $[0,T]$ has been proved.

d) We wish to compute the oscillation of x_m' on the interval $J\doteq[t_r, t_r+l_r[$ of the r-th partition. Let $t_*\in J$ be a nodal point. Then we must have $0(t_*)\geq 0(t_r)$, $0(t_*=0(t_r)$ if and only if $t_*=t_r$, and, when $d\doteq 0(t_*)-0(t_r)>0$, $s(t_*)\in J$. As a consequence, $t_*=s^0(t_*), s(t_*), \ldots, s^d(t_*)=t_r$ are in J.

To apply the above to computing the oscillation of x'_m on J, remark that on intervals of the m-th partition and, a fortiori, on intervals of subsequent partitions, x'_m is constant and its oscillation is zero. Hence we have to consider only $m>r$. Fix t in J and let $[t_*, t_*+l_m[$ be the interval of the m-th partition containing t. We have $x'_m(t)=x'_m(t_*)$ and

$$d(x'_m(t_*), x'_m(t_r)) \le d(x'_m(t_*), x'_m(s(t_*)) + \ldots + d(x'_m(s^{d-1}(t_*), x'_m(s^d(t_*))).$$

By construction,

$$d(x'_m(t_*), x'_m(s(t_*))) \le \eta_{0(t_*)-1},$$

$$d(x'_m(s(t_*)), x'_m(s(s(t_*)))) \le \eta_{0(t_*)-2}, \ldots,$$

$$d(x'_m(s^{d-1}(t_*)), x'_m(s(s^{d-1}(t_*)))) \le \eta_r,$$

hence

$$d(x'_m(t), x'_m(t_r)) \le \sum_{k=r}^{0(t)-1} \eta_k \le \sum_{k=r}^{m-1} \eta_k \le \sum_{k=r}^{\infty} \eta_k.$$

The oscillation of x_m on J, then, is at most $2\sum_{k=r}^{\infty} \eta_k = 1/2^{r-2}$. This estimate is independent of m.

e) Summarizing, we have proved that

(4) the sequence $\{x_m(\cdot)\}$ is equicontinuous

because

(5) the sequence $\{x'_m(\cdot)\}$ is bounded in $L^\infty(0, T; R^n)$

thanks to properties (2) i) and ii) of the functions $x_m(\cdot)$.

We also have proved that

(6) the sequence $\{x'_m(\cdot)\}$ is equioscillating.

Therefore, sequences $x_m(\cdot)$ and $x'_m(\cdot)$ converge uniformly to a continuous function $x_*(\cdot)$ and a regulated function $v_*(\cdot)$ related by the formula

$$x_*(t) = x^0 + \int_0^t v_*(\tau)\, d\tau. \tag{7}$$

Therefore $v_*(t)$ coincides almost everywhere with the derivative of $x_*(t)$.

The function $x(\cdot)$ satisfies $x(0)=x_0$ and is Lipschitzean of constant M, i.e. $(t, x(t)) \in \mathscr{K}$ for t in $[0, T]$.

Let t be a point such that $d(x'_m(t), v_*(t))$ converges to 0 and $x'(t)=v_*(t)$ (almost every t in I has the above properties); call t_m the initial point of the interval of the m-th partition to which t belongs. Then

$$\begin{aligned} d(v_*(t), F(t, x(t))) &\le d(v_*(t), x'_m(t)) + d(x'_m(t), x'_m(t_m)) + d(x'_m(t_m), F(t_m, x_m(t_m))) \\ &\quad + \mathfrak{d}(F(t_m, x_m(t_m)), F(t, x_m(t))) + \mathfrak{d}(F(t, x_m(t)), F(t, x(t))). \end{aligned}$$

Since: $x'_m(t)$ converges to $v_*(t)$, $x'_m(t)$ equals $x'_m(t_m)$, $x'_m(t_m)$ belongs to $F(t_m, x_m(t_m))$, $x_m(\cdot)$ is Lipschitzean, t_m converges to t, $x_m(\cdot)$ converges to $x(\cdot)$ and F is continuous, the right-hand side of the above inequality can be made as small as we please. Since F has closed values, we deduce that

$$x'(t) = v_*(t) \in F(t, x(t)). \qquad \square$$

In general more restrictive assumptions than continuity are needed to guarantee the existence of more regular solutions. The result that we wish to obtain and the tool we choose are strictly related. To have that the limit of the derivatives of the approximate solutions is continuous, the compactness theorem we have used in the preceding paragraph is not enough. We should need to construct a sequence of approximate solutions whose derivatives are equibounded and equicontinuous and exploit the Ascoli-Arzelà Theorem. The choice of the tool in a sense dictates the additional conditions to be imposed on F.

Definition 1. A map F from $K \subset R^n$ into the subsets of R^n is called "*absolutely continuous*" on K if for every $\varepsilon > 0$ there exists $\delta > 0$ such that

$$(8) \qquad \Sigma_i \mathfrak{d}(F(x_i), F(x_{i+1})) \leq \varepsilon$$

for every finite sequence of points $\{x_i\}$ with the property that

$$(9) \qquad \Sigma_i d(x_i, x_{i+1}) \leq \delta. \qquad \blacktriangle$$

An absolutely continuous map is continuous, actually uniformly continuous on K: the preceding definition applied to all the sequences consisting of the two points x_0, x yields that, given ε, there exists δ such that $d(x_0, x) \leq \delta$ implies $\mathfrak{d}(F(x_0), F(x)) \leq \varepsilon$. The classical theory of real variable provides examples of (single-valued) maps that are continuous but not absolutely continuous. Lipschitzean maps provide the easiest examples of absolutely continuous maps.

Theorem 2. *Let F, from an open subset $\Omega \subset R^n$ into the compact subsets of R^n, be absolutely continuous. Let x_0 be in Ω, and v_0 in $F(x_0)$, Then there exists an interval $I \doteq [-T, T]$ and a continuously differentiable functions $x(\cdot)$, defined on I, a solution to the problem*

$$(10) \qquad x'(t) \in F(x(t)), \quad x(0) = x_0 \text{ and } x'(0) = v_0. \qquad \blacktriangle$$

Proof. There exists some ball $B_r \doteq B(x_0, r)$ centered at x^0 with radius r, contained in Ω. Since F is continuous on B_r and has compact values, there exists M such that $\|F\| \leq M$ on B_r. Set $T \doteq r/M$.

The idea of the proof is to define approximate solutions essentially by polygonal lines: to have equicontinuity on the set of derivatives, care is taken to choose at each step a velocity nearest to the previous choice.

We associate with each integer n a subdivision of $[0, T]$ defined by the points $t_n^i \doteq iT/n$. Define $x_n(t_n^0) \doteq x_n^0$ and $v_n(t_0) \doteq v_0$. Assume that we have defined $x_n(t_n^i)$ and $v_n(t_n^i)$ up to $i = \nu - 1 < n$ and define $x_n(t_n^\nu)$ and $v_n(t_n^\nu)$ as follows.

We set

$$x_n(t_n^\nu) \doteq x_n(t_n^{\nu-1}) + \frac{T}{n} v_n(t_n^{\nu-1}) \tag{11}$$

and we choose for "velocity" $v_n(t_n^\nu)$ a projection of best approximation of $v_n(t_n^{\nu-1})$ by elements of $F(x_n(t_n^\nu))$: $v_n(t_n^\nu)$ belongs to $F(x_n(t_n^\nu))$ and

$$d(v_n(t_n^{\nu-1}), v_n(t_n^\nu)) = d(v_n(t_n^{\nu-1}), F(x_n(t_n^\nu))). \tag{12}$$

We then define the functions x_n and v_n on $[0, T]$ by interpolating linearly the sequences $x_n(t_n^i)$ and $v_n(t_n^i)$ at the nodal points t_n^i. We observe that $v_n(\cdot)$ is no longer the derivative of x_n, which is not continuous, being x_n' piecewise constant, but is a close by continuous function.

We wish to show that the sequence $\{v_n(\cdot)\}$ is equibounded and equicontinuous.

For any t and any n, $v_n(t)$ is a convex combination of vectors belonging to a sphere of radius M. Fix $\varepsilon > 0$. Let $\delta \doteq \delta(\varepsilon)$ be as provided by the assumption of absolute continuity.

Set $\delta_1 = \delta/M$ and consider those n larger than $3T/\delta_1$ (hence $T/n < \delta_1/3$). Let t and t' in I with $|t - t'| < \delta_1/3$ and let, for each n, $i \doteq i(n)$ and $j \doteq j(n)$ be such that

$$t_n^i \le t < t_n^{i+1} < \ldots < t' \le t_n^j.$$

In particular, $t_n^j - t_n^i \le \delta_1/3 + \delta_1/3 + \delta_1/3 = \delta_1$. We have that

$$d(v_n(t), v_n(t')) \le d(v_n(t), v_n(t_n^{i+1})) + \sum_{k=i+2}^{j-1} d(v_n(t_n^{k-1}), v_n(t_n^k)) + d(v_n(t_n^{j-1}), v_n(t')).$$

Since on $[t^i, t^{i+1}]$ and on $[t^{j-1}, t^j]$, v_n is linear

$$d(v_n(t), v_n(t_n^{i+1})) = \frac{t_n^{i+1} - t}{t_n^{i+1} - t_n^i} d(v_n(t_n^i), v_n(t_n^{i+1}) \le d(v_n(t_n^i), v_n(t_n^{i+1}))$$

and analogously

$$d(v_n(t_n^{j-1}), v_n(t')) \le d(v_n(t_n^{j-1}), v_n(t_n^j)).$$

It follows that

$$d(v_n(t), v_n(t')) \le \sum_{k=i+1}^{j} d(v_n(t_n^{k-1}), v_n(t_n^k)).$$

For each k, $d(v_n(t_n^{k-1}), v_n(t_n^k) = d(v_n(t_n^{k-1}), F(x_n(t_n^k))$, and since $v_n(t_n^{k-1}) \in F(x(t_n^{k-1}))$, we have

$$d(v_n(t_n^{k-1}), v_n(t_n^k)) \le \delta(F(x_n(t_n^{k-1}), F(x_n(t_n^k))).$$

Also
$$\sum_{h=i+1}^{j} d(x_n(t_n^{h-1}), x_n(t_n^h)) \leq M \sum_{h=i+1}^{j} |t_n^{h-1} - t_n^h| \leq M\,\delta_1 = \delta$$

and, by our choice of δ, this implies that

(13) $$d(v_n(t), v_n(t')) \leq \varepsilon.$$

This estimate, independent of n, does not in general hold for the finite number of n's that do not satisfy $n > 3T/\delta_1$. However each of these $v_n(\cdot)$ is uniformly continuous, hence for some δ_2, $|t-t'| \leq \delta_2$ implies $d(v_n(t), v_n(t')) \leq \varepsilon$, $n \leq 3T/\delta_1$. Hence by choosing $\min(\delta_1, \delta_2)$, (13) holds for any n, and we have proved the equicontinuity of the sequence $\{v_n(\cdot)\}$.

Some subsequence (we still call it $\{v_n\}$) converges uniformly to a continuous $v_*(\cdot)$. Define $y_*(\cdot)$ to be

(14) $$y_*(t) = x^0 + \int_0^t v_*(s)\,ds.$$

We claim that y_* is a solution to our problem, i.e. that

(15) $$v_*(t) \in F(y_*(t)).$$

We show first that the sequence $\{x_n(\cdot)\}$ uniformly converges to $y_*(\cdot)$:

$$y_*(t) - x_n(t) = \int_0^t (v_*(s) - x_n'(s))\,ds$$

and

$$\|y_x(t) - x_n(t)\| \leq \int_0^t \|v_*(s) - v_n(s)\|\,ds + \int_0^t \|v_n(s) - x_n'(s)\|\,ds;$$

since v_n and x_n' agree at the points t_n^i (where, actually x_n' is only a left derivative) and x_n' is constant on each subinterval while v_n is linear, the integral of their difference is at most $\frac{1}{2}(t_n^k - t_n^{k-1})^2 M$ on each $[t_n^{k-1}, t_n^k]$, i.e. at most $\frac{1}{2}MT/n$ on $[0, \delta]$. This proves the convergence of $\{x_n(\cdot)\}$ to $y_*(\cdot)$.

In the following final estimate we denote by (t_n^j, t_n^{j+1}), the interval to which t belongs. The index j depends actually on n.

(16) $$\begin{aligned} &d(v_*(t), F(y_*(t))) \\ &\leq d(v_*(t), v_*(t_n^j)) + d(v_*(t_n^j), v_n(t_n^j)) + d(v_n(t_n^j), F(x_n(t_n^j))) \\ &\quad + \mathfrak{d}(F(x_n(t_n^j)), F(y_*(t_n^i))) + \mathfrak{d}(F(y_*(t_n^i)), F(y_*(t))). \end{aligned}$$

We know that v_* is continuous at t, that v_n converges uniformly to v_*, that $v_n(t_n^i)$ belongs to $F(x_n(t_n^j))$, that x_n converges uniformly to the continuous function y_* and that F is uniformly continuous on B_r; hence the right-hand side of the inequality (16) can be made arbitrarily small.

Since F has closed values, $v_*(t) \in F(y_*(t))$, i.e. y_* is a continuously differentiable solution to the differential inclusion (10). □

Remark. The proof of the compactness of the set of derivatives of the approximate solutions was, in the preceding theorem, partially based on the knowledge that F was locally uniformly bounded, i.e. that the approximate solutions are Lipschitzean with constant M. This a priori estimate is no longer available when we drop the assumption of boundedness on F. In this case at each step of the construction we have to estimate at once the slope of the derivatives of the approximate solutions and their size, i.e. the slope of the approximate solutions themselves. A careful construction yields the following result.

Theorem 3. *Let $\Omega \subset R^n$ be open, F from Ω into the closed nonempty subsets of R^n be Lipschitzean on Ω, with constant K.*

Let $x^0 \in \Omega$ and $v^0 \in F(x^0)$ be given.

Let T be such that

$$x^0 + [\|v^0\| (e^{KT} - 1)/K] B \subset \Omega. \tag{17}$$

Then there exists a continuously differentiable function x from $]-T, +T[$ to Ω, a solution to the differential inclusion

$$x'(t) \in F(x(t)), \quad x(0) = x_0 \text{ and } x'(0) = v_0. \tag{10}$$ ▲

Remark. One word of warning about the dependence on time. At it is well known, results for the autonomous problem

$$x'(t) \in F(x(t)), \qquad x(0) = x_0 \tag{18}$$

can be used for the non autonomous case

$$x'(t) \in F(t, x(t)), \quad x(t_0) = x_0 \tag{19}$$

by introducing the vector (x, s) and solving the autonomous system

$$(x(t), s(t))' \in (F(s(t), x(t)), 1) \quad (x(0), s(0)) = (x_0, t_0). \tag{20}$$

This procedure demands the same conditions on the space and time variables. In ordinary differential equations conditions on the time variable can be considerably relaxed without changing the essential properties of the solutions: uniqueness is assured, for instance by assuming Lipschitzeanity with respect to x, while continuity or measurability in t is enough. The same is in general true for differential inclusions but there are some cases, in particular the theorems on the existence of continuously differentiable solutions, where clearly this cannot be done. Set-valued maps F exist, Lipschitzean in x but only continuous with respect to t, such that the problem

$$x'(t) \in F(t, x(t)), \quad x(t^0) = x^0 \tag{21}$$

admits no continuously differentiable solution. In fact consider the set-valued map ϕ provided by Example 1.6.1. Depending only on t, ϕ is certainly Lipschitzean in x. ϕ admits no continuous selection, hence no C^1-function $x(\cdot)$ can satisfy (21) with $F \doteq \emptyset$ as right-hand side.

4. Differential Inclusions with Lipschitzean Maps and the Relaxation Theorem

In the preceding section, an existence theorem for solutions having a regulated derivative was provided for inclusions with continuous right-hand side and the existence of continuously differentiable solutions was proved when F is absolutely continuous. Either result applies to the class of Lipschitzean maps. For this class of maps, however, we wish to have a qualitative theory, in particular to study the connection between the set of solutions of the two differential inclusions

$$x'(t) \in F(t, x(t)) \quad \text{and} \quad x'(t) \in \overline{\text{co}}\, F(t, x(t)).$$

To do so we must accept as solutions functions that are only absolutely continuous. It is only within this class that the following results, in particular the analog of Gronwall's inequality, do hold.

Proposition 1 Gronwall Inequality. *Let $I \doteq [a, b]$, $\alpha(\cdot)$ and $\phi(\cdot)$ be continuous and $k(\cdot)$ be a nonnegative integrable function, all defined on I. Assume that*

$$\phi(t) \leq \alpha(t) + \int_a^t k(s)\phi(s)\,ds, \qquad t \in I. \tag{1}$$

Then

$$\phi(t) \leq \alpha(t) + \int_a^t k(s)\,\alpha(s)\, e^{\int_s^t k(u)\,du}\, ds. \tag{2}$$

▲

For simplicity, we begin by examining its consequences in the case when we are given an ordinary differential equation $x' = f(t, x)$ with f continuous and satisfying a generalized Lipschitz condition with respect to x, i.e., such that there exists an integrable function $k(t)$ with the property that for (t, x) and (t, x') in the domain of f, $\|f(t, x) - f(t, x')\| \leq k(t)\,\|x - x'\|$. Let x_1 and x_2 be two solutions, defined on the same interval $[a, b]$. For the norm of their difference we have

$$\|x_1(t) - x_2(t)\| \leq \|x_1(a) - x_2(a)\| + \int_a^t k(s)\,\|x_1(s) - x_2(s)\|\,ds. \tag{3}$$

This is a special case of (1) with $\alpha(t) \doteq \|x_1(a) - x_2(a)\|$.

After an integration has been carried on the right-hand side of (2) we obtain

$$\|x_1(t) - x_2(t)\| \leq \|x_1(a) - x_2(a)\|\, e^{\int_a^t k(s)\,ds} \tag{4}$$

an estimate of the difference of the two solutions given the difference in the data.

The inequality, in its full power, takes into account the case when x_2, say, is not a solution, but differs from being a solution by some function, in the sense that $\|x_2'(t)-f(t,x_2(t))\| \le p(t)$, with $p(\cdot)$ a non-negative integrable function. The difference $\phi(t) \doteq \|x_1(t)-x_2(t)\|$ has to satisfy

$$\|x_1(t)-x_2(t)\| \le \|x_1(a)-x_2(a)\| + \int_a^t k(s)\,\|x_1(s)-x_2(s)\| + \int_a^t p(s)\,ds$$

and we can apply the inequality with $\alpha(t) \doteq \delta + \int_a^t p(s)\,ds$ and $\delta \doteq \|x_1(a) -x_2(a)\|$. Also in this case, though, it is worth working on the right-hand side to have it in a more suitable form.

Note that $k(s)\,e^{\int_s^t k(x)\,dx} = -\frac{d}{ds}\left(e^{\int_s^t k(x)\,dx}\right)$ so that

$$\phi(t) \le \delta + \int_a^t p(s)\,ds + \int_a^t \left(k(s)\left[\delta + \int_a^s p(u)\,du\right] e^{\int_s^t k(u)\,du}\right) ds$$

$$= \int_a^t p(s)\,ds + \delta e^{\int_a^t k(u)\,du} + \int_a^t k(s)\left(\int_a^s p(u)\,du\right) e^{\int_s^t k(v)\,dv}\,ds$$

and that, by interchanging the order of integration, the last integral becomes

$$\int_a^t \left\{\int_u^t \left(-\frac{d}{ds}\,e^{\int_s^t k(v)\,dv}\right) ds\right\} p(u)\,du = \int_a^t p(u)\left[e^{\int_u^t k(v)\,dv}\right] du - \int_a^t p(u)\,du$$

so that inequality (2) takes the form

$$\|x_1(t)-x_2(t)\| \le \delta e^{\int_a^t k(u)\,du} + \int_a^t p(s)\,e^{\int_s^t k(u)\,du}\,ds. \tag{5}$$

In this formulation we can explicitly see the contribution of the error in the initial condition, δ, and in not being a solution, p.

In what follows we shall prove a corresponding inequality for a Lipschitzean differential inclusion. Before proceeding let us remark explicitly that we shall deal with solutions that are not, in general, continuously differentiable functions and that the statement and the proof will be definitely longer than in the ordinary differential case. In fact, given an almost solution y, we shall have to prove the existence of at least one solution x satisfying the desired inequality (other solutions with the same initial data need not, obviously, satisfy any reasonable inequality). Hence the following is also an existence result.

Theorem 1. *Let there be given an interval $I \doteq [a,b]$, an absolutely continuous function $y\colon I \to R^n$, a positive constant β, and call Q the subset of $I \times R^n$ defined by $(t,x) \in Q$ if $t \in I$ and $\|x-y(t)\| \le \beta$. Assume that F, from Q into the nonempty and closed subsets of R^n, is continuous and satisfies*

$$\mathfrak{d}(F(t,x), F(t,y)) \le k(t)\,\|x-y\|. \tag{6}$$

Assume moreover that

(7) $$\|y(a)-x_0\|=\delta\leq\beta, \qquad d(y'(t), F(t, y(t)))\leq p(t) \quad a.e.$$

with $p\in L^1(I)$. *Set*

(8) $$\xi(t)\doteq\delta e^{\int_a^t k(s)\,ds}+\int_a^t e^{\int_s^t k(u)\,du}\,p(s)\,ds$$

and let $J\doteq[a,\omega]$ *be a nonempty interval such that* $t\in J$ *implies* $\xi(t)\leq\beta$. *Then there exists a solution* $x(\cdot)$ *on* J *to the problem*

(9) $$x'(t)\in F(t, x(t)), \qquad x(a)=x_0$$

such that

(10) $$|x(t)-y(t)|\leq\xi(t)$$

and

(11) $$|x'(t)-y'(t)|\leq k(t)\,\xi(t)+p(t) \quad a.e.$$ ▲

Remark. Let x_0, y_0 two initial points, $|x_0-y_0|=\delta\leq b$ and take $p=0$ in the preceding theorem. Then to any solution $y(\cdot)$ such that $y(0)=y_0$ we can associate a solution $x(\cdot)$ such that

$$x(0)=x_0 \quad\text{and}\quad |x(t)-y(t)|\leq|x_0-y_0|\,e^{\int_0^t k(s)\,ds}.$$

Hence we have the following

Corollary 1. *The map* $\mathcal{T}_I$ *from* R^n *to nonempty subsets of* $C(I, R^n)$ *that associates to an initial point the set of solutions on* I *issued from that point, is Lipschitzean with constant* $e^{\int_I k(s)\,ds}$. ▲

Proof. We shall construct a Cauchy sequence of successive approximations, x_n, such that their derivatives x'_n form also a Cauchy sequence on J.

Set for convenience $m(t)\doteq\int_a^t k(s)\,ds$. Consider $y(t)$. Since (in general) it is not a solution, $y'(t)\notin F(t, y(t))$. By the measurable selection Theorem there exists a measurable selection v_0 from $F(\cdot, y(\cdot))$ such that for almost all t,

$$\|v_0(t)-y'(t)\|=d(y'(t), F(t, y(t)))\leq p(t).$$

Let us call x_1 the absolutely continuous function given by $x_0+\int_a^t v_0(s)\,ds$. We note that $\|x_1(t)-y(t)\|\leq\delta+\int_a^t p(s)\,ds\leq\beta$.

We claim that we can define a sequence of absolutely continuous functions x_i with the following properties:

(12)
$$x_i(t)=x_0+\int_a^t v_{i-1}(s)ds, \qquad \text{with } v_{i-1}(s)\in F(s,x_{i-1}(s))$$
for almost all $t\in J$ and $(t,x_i(t))\in\Omega$ and for $i\geq 1$

and

(13)
$$\|x_i'(t)-x_{i-1}'(t)\|\leq k(t)\left\{\delta\frac{(m(t))^{i-2}}{(i-2)!}+\int_a^t\frac{(m(t)-m(s))^{i-2}}{(i-2)!}p(s)\,ds\right\}$$
for almost all t and for $i\geq 2$.

Remark that, as a consequence of (13), we have also

(14)
$$\begin{aligned}\|x_i(t)-x_{i-1}(t)\|&\leq\int_a^t\frac{\delta}{(i-2)!}k(s)\,(m(s))^{i-2}\,ds\\&\quad+\int_a^t\left(\int_a^s\frac{(m(s)-m(u))^{i-2}}{(i-2)!}k(s)\,p(u)\,du\right)ds\\&=\int_a^t\frac{\delta}{(i-1)!}\frac{d}{ds}(m(s))^{(i-1)}\,ds\\&\quad+\frac{1}{(i-1)!}\int_a^t p(u)\left(\int_u^t\frac{d}{ds}(m(s)-m(u))^{i-1}\,ds\right)du\\&=\frac{\delta}{(i-1)!}(m(t))^{i-1}+\frac{1}{(i-1)!}\int_a^t p(u)\,(m(t)-m(u))^{i-1}\,du.\end{aligned}$$

Assume we have defined on J our functions x_i up to $i=n$, satisfying (12) and (13), and with values such that $(t,x_n(t)\in Q$. Consider $x_n(\cdot)$. F is well defined along it and there exists a measurable selection v_n from $F(\cdot,x_n(\cdot))$ such that $\|v_n(t)-x_n'(t)\|=d(x_n',F(t,x_n(t)))$ a.e. in J. Define x_{n+1} to be $x_0+\int_a^t v_n(s)\,ds$. Then x_{n+1} is defined on J. Moreover

$$\begin{aligned}\|x_{n+1}'-x_n'\|=\|v_n-x_n'\|=\|v_n-v_{n-1}\|&\leq\delta((F(t,x_n(t)),F(t,x_{n-1}(t)))\\&\leq k(t)\,\|x_n(t)-x_{n-1}(t)\|.\end{aligned}$$

By (14) this last term is bounded by

$$k(t)\left\{\frac{1}{(n-1)!}(m(t))^{n-1}+\frac{1}{(n-1)!}\int_a^t p(u)\,(m(t)-m(u))^{n-1}\,du\right\}$$

and this is part (13) of the induction.

By adding the inequalities (14) we obtain that

$$\begin{aligned}\|x_{n+1}(t)-y(t)\|&\leq\|y(t)-x_1(t)\|+\ldots+\|x_n(t)-x_{n+1}(t)\|\\&\leq\sum_{k=0}^{n}\left(\delta\frac{(m(t))^k}{k!}+\int_a^t\frac{(m(t)-m(s))^k}{k!}p(s)\,ds\right)\end{aligned}$$

and this last term is bounded by $\xi(t)$. Hence x_{n+1} not only is defined on J, but it is such that $\|x_{n+1}(t)-y(t)\|\leq\xi(t)\leq\beta$ on it. This completes part (12) of the induction.

Inequalities (14) and (13) imply that $\{x_n(\cdot)\}$ is a Cauchy sequence of continuous functions, converging uniformly to a continuous function $x(\cdot)$, and that for almost all $t \in J$, $\{v_n(t)\}$ is a Cauchy sequence, hence that $v_n(\cdot)$ converges pointwise almost everywhere to a measurable function $v(\cdot)$. Moreover, from (12)

$$x_{i+1}(t) = x^0 + \int_a^t v_i(s)\, ds$$

so that

(15) $$x(t) = x^0 + \int_a^t v(s)\, ds.$$

Finally, since a.e. on J, $v_i(s) \in F(s, x_i(s))$ and F has a closed graph, $v(s) \in F(s, x(s))$, a.e. on J. This completes the proof. □

The Relaxation Theorem

The next result plays a fundamental role in the qualitative theory of differential equations, and concerns the relations between the set of solutions of the two problems $x' \in F(x)$ and $x' \in \overline{\text{co}}(F(x))$. Certainly solutions of the first are also solutions of the second: we wish to study, however, to what extent the operation of convexifying the right-hand side really introduces new solutions. In other words, under what conditions will the set of solutions to the differential inclusion

(16) $$x'(t) \in F(x(t))$$

be *dense* in the set of solutions of the "convexified" differential inclusion

(17) $$x'(t) \in \overline{\text{co}}(F(x(t)))?$$

This problem is particularly relevant in control theory; solutions to the convexified problem are often called *relaxed solutions*, and the problem we have mentioned, the *problem of relaxation*. We shall prove that the relaxation property holds when F is Lipschitzean with compact values, while it does not necessarily hold when F is only continuous.

In the theory of control, one encounters the following problem. It is given an affine differential equation,

$$x' = Ax + bu$$

a set of controls $\mathscr{U}$, a compact convex subset of R^n, an initial condition and a time interval $[0, T]$. One considers the attainable set $\mathscr{A}_T$ at time T, the images at T of all solutions issued at $t=0$ from ξ using all controls $u(\cdot)$, measurable on $[0, T]$, such that $u(t) \in \mathscr{U}$ a.e. on $[0, T]$.

The question can then be raised as to whether it would be possible to have the same attainable set economizing on the set of controls, hence to have the same results with controls that are much simpler to build. It is a famous result (the Bang-Bang principle) that one can actually reduce the set $\mathscr{U}$ to $\text{extr}(\mathscr{U})$, the set of its extremal points, even when this set is not closed.

For a differential inclusion one focuses on the set of solutions through an initial point. Since for a nonconvex right-hand side this set is, in general, not closed (even if its section at any given time might be), we should consider the problem of the possible equivalence of the closure of the set of solutions of a nonconvex problem with the set of solutions of the convexified one. In other words one can look at the set of solutions of a nonconvexified problem and ask for conditions to insure the equivalence of its closure with the set of solutions of the convexified. A second way of looking at the question, more related to the Bang-Bang principle, would be to begin with a differential inclusion $x' \in F(x)$ with compact convex values and to ask for a subset of $F(x)$ in order essentially to retain the solutions of the original problem. This second question is by far more difficult and so far has no complete answer.

Here is the Relaxation Theorem.

Theorem 2 (Filippov-Ważewski). *Let F, from $Q \doteq \{x \in R^n \mid \|x - \xi_0\| \le b\}$ into the compact subsets of R^n, be Lipschitzean. Set $I \doteq [-T, +T]$ and let $x: I \to Q$ be a solution to*

$$(17) \qquad x'(t) \in \overline{\text{co}}(F(x(t))), \quad x(0) = \xi_0$$

such that, for $t \in I$, $\|x(t) - \xi_0\| < b$. Then for every positive ε, there exists y: $I \to Q$, a solution to

$$(16) \qquad y'(t) \in F(y(t)), \quad y(0) = \xi_0$$

such that for $t \in I$, $\|y(t) - x(t)\| \le \varepsilon$. ▲

Proof. Since the mapping $t \to \|x(t) - \xi_0\|$ is continuous on the compact I, there exists a positive η such that on I, $\|x(t) - \xi_0\| + \eta \le b$. By taking a smaller ε if needed, we can assume as well $\varepsilon < \eta$. By *Theorem 1* it is enough to prove the existence of a quasi-solution $z(\cdot)$ such that, on I,

$$\|z(t) - x(t)\| \le \varepsilon/2, \quad z(0) = \xi_0$$

and

$$d(z'(t), F(z(t))) \le \varepsilon K [2(e^{KT} - 1)]^{-1} \quad \text{a.e.},$$

where K is the Lipschitz constant of F. Indeed, *Theorem 1* with $\delta \doteq 0$ and $p(t) \doteq \dfrac{\varepsilon K}{2(e^{KT} - 1)}$ implies the existence of a solution $y(\cdot)$ to $y' \in F(y)$ such that, on the intersection of its domain with I, $\|y(t) - z(t)\| \le \varepsilon/2$ and thus $\|y(t) - x(t)\| \le \varepsilon$. This means that z never reaches the boundary of Q at a time smaller (in absolute value) than T, i.e., that z is defined on I. Hence we proceed to the construction of such a function $z(\cdot)$.

It is convenient to partition the interval $I \doteq [-T, +T]$ into $2n$ subintervals $I_i \doteq [t_i, t_{i+1}]$ of length T/n.

Let M be the norm of the bounded subset $F(Q)$.

The mapping $x \to \overline{\text{co}}\, F(x)$ is Lipschitzean with the same constant K as F (see Proposition 1.1.6) so that both $t \to \overline{\text{co}}(F(x(t)))$ and $t \to F(x(t))$ are Lipschitzean with constant KM. Set $\phi_i \doteq \overline{\text{co}}\, F(x(t_i))$. Then, for $t \in I_i$, $\overline{\text{co}}\,(F(x(t)))$ is contained in an (KMT/n) neighborhood of ϕ_i. In particular, so is $x'(t)$. Fix any such i and partition the set $\{\overline{\text{co}}(F(x(t))) | t \in I_i\}$ into a finite number of (not necessarily open or closed, but borel) subsets S_j having diameter not larger than ζ. Set $E_j = (x')^{-1}(S_j)$, let $\chi_j(\cdot)$ be its characteristic function and ξ_j be some point in S_j. Consider the simple function:

(18) $$\xi \doteq \Sigma_j\, \xi_j\, \chi_j.$$

It has the property that a.e. on I_i, $\|\xi(t) - x'(t)\| \leq \zeta$ and that each of its values ξ_j differs from ϕ_i by at most KMT/n. By definition of a closed convex hull, for each j there exist finitely many points $z_{jk} \in F(x(t_i))$ and positive constants α_{jk}, such that

$$\Sigma_k\, \alpha_{jk} = 1 \quad \text{and} \quad \|\xi_j - \Sigma_k\, \alpha_{jk}\, z_{jk}\| \leq 2KM^2/n.$$

Now we wish to partition each measurable set E_j into measurable subsets E_{jk} such that $\text{meas}(E_{jk}) = \alpha_{jk}\, \text{meas}(E_j)$. This can be done as follows. Let $[\alpha, \omega]$ be an interval containing E_j and consider the function $\psi\colon [\alpha, \omega] \to R$ defined by $\psi(t) = \int_a^t \chi_j(s)\, ds$. It is a continuous and monotonic mapping with values that range from 0 to $\text{meas}(E_j)$. Set

$$\tau_k \doteq \sup\left\{ t \in [\alpha, \omega] \,\middle|\, \psi(t) \leq \text{meas}(E_j) \sum_{h=1}^{k} \alpha_{jh} \right\},$$

$J_1 \doteq [\alpha, \tau_1]$, $J_k \doteq]\tau_{k-1}, \tau_k]$ and define E_{jk} to be $E_j \cap J_k$. Then $\text{meas}(E_{jk}) = \psi(\tau_k) - \psi(\tau_{k-1}) = \text{meas}(E_j)\, \alpha_{jk}$ and the collection $\{E_{jk}\}$ is a partition of E_j. Set χ_{jk} to be the characteristic function of E_{jk} and define $\rho\colon I \to R^n$ to be the simple function whose restriction to E_j, ρ_j, is given by

(19) $$\rho_j(t) \doteq \Sigma_k\, z_{jk}\, \chi_{jk}(t).$$

Finally set

(20) $$z(t) = \xi_0 + \int_0^t \rho(s)\, ds.$$

We shall now impose conditions on n in order to insure that z is the quasi-solution sought which approximates x. First of all note that the derivative of z has values in the image through F of Q; hence z is Lipschitzean with the same constant M as x. Then by choosing the intervals sufficiently small, to approximate x uniformly over I by z, it is enough to approximate it at the nodal points t_i.

Set $\gamma \doteq \min\{1, [2K(e^{KT} - 1)^{-1}]\}$. Since any $t \in I$ belongs to some I_i and $\|z(t) - x(t)\| \leq \|z(t_i) - x(t_i)\| + \|z(t) - z(t_i)\| + \|x(t) - x(t_i)\|$ in order to have $\|z(t) - x(t)\| \leq \gamma\, \varepsilon/3$ it is enough to have both $\|z(t_i) - x(t_i)\| \leq \gamma\, \varepsilon/6$ and

$1/n \leq \gamma\,\varepsilon/6MT$, i.e., $n \geq 6MT/\gamma\,\varepsilon$. Moreover, on each interval I_i, we have, on one hand

$$\int_{I_i} \xi(s)\,ds = \Sigma_j\, \mu(E_j)\, \xi_j$$

while on the other hand

$$\int_{I_i} \rho(s)\,ds = \int_{I_i} [\Sigma_{j,k}\, z_{jh}\, \chi_{jh}(s)]\,ds$$
$$= \Sigma_{j,k} \int_{I_i} z_{jk}\, \chi_{jh}(s)\,ds = \Sigma_{j,h}\, \mathrm{meas}(E_j)\, \alpha_{jh}\, z_{jh}.$$

Hence

$$\|\int_{I_i} \rho(s)\,ds - \int_{I_i} \xi(s)\,ds\| \leq \Sigma_j\, \mathrm{meas}(E_j)\, [\Sigma_h\, \alpha_{jh}\, z_{jh} - \xi_j]$$
$$\leq \Sigma_j\, \mathrm{meas}(E_j)\, 2KMT/n \leq \mathrm{meas}(I_i)\, 2KMT/n.$$

For any point t_j of the subdivision then

$$\left\| z(t_j) - \int_0^{t_j} \xi(s)\,ds \right\| = \left\| \int_0^{t_j} (\rho(s) - \xi(s))\,ds \right\|$$
$$= \sum \|\int_{I_i} (\rho(s) - \xi(s))\,ds\| \leq 2KMT^2/n.$$

Since, on the other hand

$$\left\| x(t_j) - \int_0^{t_j} \xi(s)\,ds \right\| \leq \zeta\, T$$

in order to have $\|z(t_j) - x(t_j)\| \leq \gamma\,\varepsilon/6$ it is enough to have $2KMT^2/n \leq \gamma\,\varepsilon/12$ (a second condition on n) and to choose $\zeta \doteq \gamma\,\varepsilon/12T$.

Moreover $z'(t)$, whenever it exists, belongs to $F(x(t_i))$, hence $d(z'(t_i), F(x(t))) \leq KMT/n$ and $\mathfrak{d}(F(x(t)), F(z(t))) \leq KT\varepsilon/2$ that implies

$$d(z'(t), F(z(t))) \leq K\,\gamma\,\varepsilon/2 + KMT/n = \frac{\varepsilon}{4(e^{KT}-1)} + KMT/n.$$

If we choose n so large that, besides satisfying the two preceding conditions, it is also such that $KMT/n \leq \dfrac{\varepsilon}{4(e^{KT}-1)}$ we finally have that at the same time for $t \in I$,

$$\|z(t) - x(t)\| \leq \gamma\,\varepsilon/2 \leq \varepsilon/2$$

and for a.e. $t \in I$,

$$d(z'(t), F(z(t))) \leq \frac{\varepsilon}{2(e^{KT}-1)}$$

fulfilling the claim. □

Remark. What is essential in the preceding result is the Lipschitz condition on F since it is the estimate provided by the generalized Gronwall's inequality that, once an almost solution of the nonconvex problem has been built, guarantees the existence of a true solution nearby. This situation is far from being typical of the other cases where we can prove existence, like the absolute continuity or the continuity. The following well-known example is due to Pliss.

Example. Let F from R^2 into the compact subsets of R^2 be defined by

$$F(x_1, x_2) \doteq \{(v_1, v_2) | v_1 \in \{-1, +1\}, \ v_2 = \sqrt{|x_2|} + |x_1|\}. \tag{21}$$

Set $x \doteq (x_1, x_2)$ and consider the problem

$$x'(t) \in F(x(t)), \quad x(0) = 0. \tag{22}$$

Consider $x_1(\cdot)$. There is no interval on which it can be identically zero, since otherwise its derivative would be identically zero. Hence there is a sequence $\{\tau_n\}$ converging to zero and such that $|x_1(\tau_n)|$ is positive. Since x_1 is continuous, there is an interval $[\tau_n, \omega_n]$ and $\varepsilon_n > 0$ such that $|x_1(t)| \geq \varepsilon_n$ on it. On the interval $[0, \tau_n]$, x_2 is nonnegative while on $[\tau_n, \omega_n]$ it satisfies the inequality

$$x_2' \geq \sqrt{|x_2|} + \varepsilon_n, \quad x_2(\tau_n) > 0.$$

By a basic result on differential inequalities (see Ladas-Lakshmikantham) on this interval x_2 is larger than the maximal solution to

$$x_2' = \sqrt{|x_2|}, \quad x_2(\tau_n) = 0,$$

i.e., $x_2 \geq \frac{1}{4}(t - \tau_n)^2$.

On the right of ω_n, x_2 satisfies the inequalities

$$x_2' \geq \sqrt{|x_2|}, \quad x_2(\omega_n) \geq \tfrac{1}{4}(\omega_n - \tau_n)^2$$

hence it is not smaller than the minimal solution to the corresponding problem with equalities replacing the inequalities. This problem, however, has the unique solution $x = \frac{1}{4}(t - \tau_n)^2$, so that $x_2(t) \geq \frac{1}{4}(t - \tau_n)^2$ holds for all $t \geq \tau_n$. Since τ_n converges to 0, we conclude that $x_2(t) \geq \frac{1}{4} t^2$.

Consider, on the other hand, the convexified problem

$$x'(t) \in \overline{\text{co}}\, F(x(t)), \quad x(0) = 0. \tag{23}$$

To this problem, $x \equiv 0$ is a solution. The distance of this solution to any solution of (22) is at least $\frac{1}{4} t^2$; hence, on any given interval, it cannot be made arbitrarily small. □

5. The Fixed-Point Approach

For an ordinary differential equation a tool that has been widely used to prove existence of solutions has been the fixed point theorem of Schauder, stating that any continuous map from a compact convex subset of a Banach space into itself has a fixed point. In fact finding solutions to the Cauchy problem

$$x'(t)=f(t,x(t)), \quad x(t_0)=x_0 \tag{1}$$

is equivalent to finding fixed points of the map T: $C(I)\to C(I)$, the integral operator, that associates to each continuous function $x(\cdot)$, defined on I, the function Tx defined by

$$Tx(t) \doteq x_0+\int_{t_0}^{t} f(s,x(s))\,ds. \tag{2}$$

Knowing that $\|f\|$ is locally bounded by some constant M yields the information that it is enough to consider maps that are Lipschitzean with Lipschitz constant M on the interval I, chosen so that no Lipschitzean map with Lipschitz constant M issuing from x_0 at t_0 can leave the ball around x_0 where $\|f\|$ is bounded by M. The subset of $C(I)$ hence defined is compact and convex, and mapped into itself by T. Any fixed point is a solution to the differential equation (1).

It is our purpose to extend this approach to differential inclusions. In this section F is assumed to be defined on some open subset Ω, a subset of $R\times R^n$ and we consider a point (t_0,x_0) in Ω.

We posit the following assumption

$$\begin{cases} \text{There exist } I\doteq[t_0-T,t_0+T] \text{ and } M \text{ such that} \\ \text{i) } I\times\{x_0+TMB\} \text{ is contained in } \Omega, \\ \text{ii) } \|F(t,x)\|\le M \text{ on } I\times\{x_0+TMB\}. \end{cases} \tag{3}$$

Definition 1. Set $\mathscr{K}\doteq\{x\in C(I)\,|\,x$ is Lipschitzean with constant M and $x(t_0)=x_0\}$. For any x in $\mathscr{K}$ set the *integral operator* $\mathscr{I}$ to be

$$\mathscr{I}(x)=\{z\in\mathscr{K}\,|\,z'(t)\in F(t,x(t)) \text{ a.e. on } I\}. \tag{4}$$ ▲

Proposition 1. *Let F be upper semicontinuous from Ω into the nonempty compact subsets of R^n. Then:*

(5)
i) *There exist I and M such that* (3) *holds,*
ii) *For every $x\in\mathscr{K}$, $\mathscr{I}(x)$ is nonempty.* ▲

Proof. By *Proposition* 1.1.3, F is bounded by M on some neighborhood of (t_0,x_0) and by choosing T small, property (3) holds. The map $t\to F(t,x(t))$ is upper semicontinuous for every continuous $x(\cdot)$, hence it satisfies the assumptions of the Measurable Selection Theorem because of *Proposition* 1.1.4. For any measurable selection v, $t\to x_0+\int_{t_0}^{t} v(s)\,ds$ belongs to $\mathscr{I}(x)$. □

In order to obtain a fixed point of $\mathscr{I}$ we shall have either to exploit the convexity of F or, when F is non convex, to carefully take advantage of the properties of $\mathscr{I}$.

Theorem 1. *Assume that F is upper semicontinuous with compact convex values. Then:*

$$(6)\qquad \begin{cases} \text{i) for } x \text{ in } \mathcal{K},\ \mathcal{J}(x) \text{ is convex,} \\ \text{ii) the map } x \to \mathcal{J}(x) \text{ is upper semicontinuous,} \\ \text{iii) there exists a fixed point } x^* \in \mathcal{J}(x^*). \end{cases}$$

▲

Proof. a) Let z_1 and z_2 be in $\mathcal{J}(x)$. Then $z_1'(t)$ and $z_2'(t)$ belong to $F(t, x(t))$ for almost every t in I, and so does $\alpha z_1'(t) + (1-\alpha) z_2'(t)$ for α in $[0,1]$, i.e. $\alpha z_1 + (1-\alpha) z_2$ is in $\mathcal{J}(x)$.

b) Let us show that $\mathcal{J}$ has a closed graph. Let x_n be in $\mathcal{K}$, z_n in $\mathcal{J}(x_n)$ and assume $x_n \to \bar{x}$, $z_n \to \bar{z}$. Call y_n the maps $t \to z_n(t)$: they belong to the ball of radius M of $L^\infty(I)$. Hence, by Alaoglu's Theorem, a subsequence (kept with the same indices) converges to some $\bar{y}$ in $\sigma(L^\infty, L^1)$. In particular, for every map φ in $L^\infty(I)$,

$$\int_I \langle y_n, \varphi \rangle \to \int_I \langle \bar{y}, \varphi \rangle,$$

i.e. on one hand $\{y_n\}$ converges in $\sigma(L^1, L^\infty)$ and on the other

$$z_n(t) = \int_{t_0}^{t} y_n \to \int_{t_0}^{t} \bar{y},$$

i.e. $\bar{y} = \bar{z}'$. Applying the Convergence Theorem we infer that $\bar{z}'(t)$ belongs to $F(t, \bar{x}(t))$, hence the graph of $\mathcal{J}$ is closed. Since the values are contained in the compact $\mathcal{K}$, $\mathcal{J}$ is upper semicontinuous. □

Kakutani's Theorem (Corollary 1.12.1) yields the existence of a fixed point: this provides a fourth proof of Theorem 1.3, an analog to Peano's Theorem.

Corollary 1. *Under the assumptions of Theorem* 1, *the problem*

$$(7)\qquad x'(t) \in F(t, x(t)), \quad x(t_0) = x_0$$

admits at least one solution defined on I. ▲

Remark. In Theorem 1 convexity of $F(t, x)$ was essential both in proving that the images of $\mathcal{J}$ are convex and, through the Convergence Theorem, in proving that the map $x \to \mathcal{J}(x)$, from $C(I)$ into its subsets, is upper semicontinuous.

In the case of a map F having no convex values, hence, we face a double challenge. On one hand we have examples of maps that are continuous with compact but not convex images that fail to have a fixed point when mapping a compact and convex subset into its subsets (Example 1.6.2) on the other hand we cannot use the Convergence Theorem to prove the continuity of the map $x \to \mathcal{J}(x)$.

Still in this section we shall prove that the integral operator $\mathscr{J}$ does indeed have the fixed point property. This will be a consequence of a continuous selection argument that in its main lines follows the pattern of ideas of the selection theorem of Michael (Theorem 1.11.1) and of the measurable selection Theorem (1.14.1), i.e., that of defining a sequence of maps that are approximate selections (continuous or Borel-measurable) and such that the sequence is a Cauchy sequence, converging to the required selection. When continuity is required the real problem is how to connect continuously different values. Given vectors $y_1, \dots, y_n$ and continuous maps $\psi_1(\cdot), \dots, \psi_m(\cdot)$, a partition of unity, the map

$$u \to \Sigma\, \psi_i(t)\, y_i \tag{8}$$

is a continuous "interpolation" among the vectors y_i. However when these vectors are within a neighborhood of some set F, the vector $\Sigma\, \psi_i(t)\, y_i$ will be in the same neighborhood only when F is convex.

In some special cases the lack of convexity can be overcome by taking advantage of the structure of the space we are considering. Our vectors should not be viewed as points in some abstract space but, in a more concrete way, as maps from the interval I into R^n. From this point of view we can devise ways to continuously "interpolate" among points other than formula (8).

Definition 2. Let $\{\psi_i(\cdot)\}_1^m$ be a continuous partition of unity on some compact metric space $\mathscr{U}$ and let I be the interval $[t_0 - T, t_0 + T]$. Set $\tau_0 = t_0 - T$, $\tau_i(u) = \tau_{i-1}(u) + 2\psi_i(u)\, T$, $i = 1, \dots, m$ and call $J_i(u)$ the interval $[\tau_{i-1}(u), \tau_i(u)[$.

The collection $\{J_i(\cdot)\}$ is called a *continuous partition of* I, corresponding to $\{\psi_i(\cdot)\}$. ▲

Proposition 2. *Let $\{J_i(\cdot)\}_1^m$ be a continuous partition of I and call $\chi_i(u)\colon I \to R$ the characteristic function of the interval $J_i(u)$. Let v_i belong to the ball of radius M of $L^\infty(I)$. Define v by*

$$v(u)(t) \doteq \Sigma_i\, v_i(t)\, \chi_i(u)(t). \tag{9}$$

Then:

$$\begin{cases} \text{a) for every } \varepsilon > 0 \text{ there exists } \delta > 0 \text{ such that } d(u, w) \le \delta \\ \quad \text{implies } \operatorname{meas}\{t \in I \mid \|v(u)(t) - v(w)(t)\| > 0\} \le \varepsilon, \\ \text{b) the map } u \to v(u) \text{ is continuous from } \mathscr{U} \text{ into } L^1(I). \end{cases} \tag{10}$$

▲

Proof. Fix $\varepsilon > 0$. Let δ be so small that $d(u, w) < \delta$ implies $|\psi_i(u) - \psi_i(w)| \le \varepsilon/4m^2 T$, $i = 1, \dots, m$. In particular, $|\tau_i(u) - \tau_i(w)| \le |\tau_{i-1}(u) - \tau_{i-1}(w)| + \varepsilon/2m^2$; hence $\operatorname{meas}(J_i(u) \triangle J_i(w)) \le \varepsilon/m$. Since any t in I belongs to some $J_i(u)$, it either belongs to $J_i(u) \cap J_i(w)$ or to $J_i(u) \triangle J_i(w)$. i.e.

$$I \subset \left\{ \bigcup_{i=1}^m (J_i(u) \cap J_i(w)) \right\} \cup \left\{ \bigcup_{i=1}^m (J_i(u) \triangle J_i(w) \right\} \doteq E_1 \cup E_2. \tag{11}$$

On E_1, $v(u)(t)-v(w)(t)=0$, hence

$$\operatorname{meas}\{t\in I \mid \|v(u)(t)-v(w)(t)\|>0\}\leq \operatorname{meas} E_2\leq \varepsilon.$$

On E_2, $\|v(u)(t)-v(w)(t)\|$ is bounded by $2M$, hence $d(u,w)\leq\delta$ implies

$$\int_I \|v(u)(t)-v(w)(t)\|\,dt\leq \int_{E_2} \|v(u)(t)-v(w)(t)\|\,dt\leq 2\varepsilon M. \qquad \square$$

We intend to apply the above construction to the problem of defining a continuous selection for the integral operator $\mathscr{J}$ associated with a continuous map F having closed but not necessarily convex values. To do so we begin with the following Lemma.

Lemma 1. *Let F, from Ω into the nonempty compact subsets of R^n be continuous. Let I be as provided by Proposition 1. Then, for every $\varepsilon>0$, there exists a continuous map g^0: $\mathscr{K}\to L^1(I)$ such that, for every $u\in\mathscr{K}$, $d(g^0(u)(t), F(t,u(t)))\leq\varepsilon$ at almost every $t\in I$.* ▲

Proof. Fix ε. Since F is continuous, hence uniformly continuous on $I\times\{x_0+MTB\}$, there exists $\delta>0$ such that

(12) $$\mathfrak{d}(F(t,x),F(t,y))\leq\varepsilon$$

whenever $\|x-y\|\leq\delta$. Let $\{\mathscr{U}_i\}_1^m$ be a finite open covering of $\mathscr{K}$ such that $\operatorname{diam}(\mathscr{U}_i)\leq\delta$ for $i=1,\dots,m$, and let $\{\psi_i(\cdot)\}$ be a partition of unity subordinate to $\{\mathscr{U}_i\}$. For every i choose $u_i\in\mathscr{U}_i$ and set $v_i(\cdot)$ to be a measurable selection from the map $t\to F(t,u_i(t))$.

Consider the partition $\{J_i\}$ of I associated with $\{\psi_i\}$ and define

(13) $$g(u)(t)\doteq\Sigma_i\, v_i(t)\,\chi_i(u)(t).$$

Proposition 2 guarantees the continuity of g from $\mathscr{K}$ to $L^1(I)$.

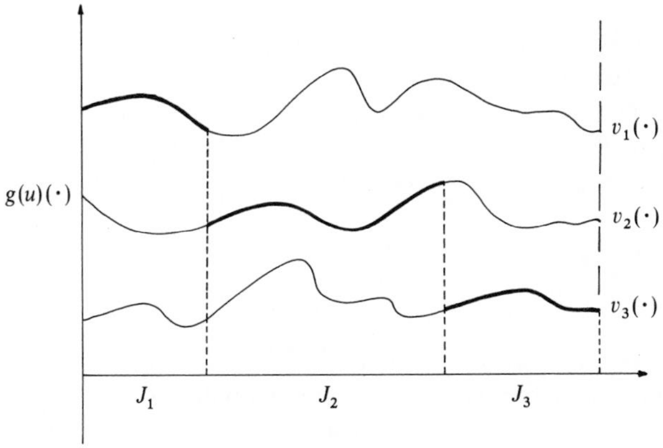

Fig. 11. Construction of the function $g(u)$

Fix $\bar{t} \in I$. There exists a unique $\bar{i}$ such that $\bar{t} \in J_{\bar{i}}$, and $g(u)(\bar{t}) = v_{\bar{i}}(\bar{t})$. Since $J_{\bar{i}}$ is nonempty, $\psi_{\bar{i}}(u) > 0$, i.e., $u \in \mathscr{U}_{\bar{i}}$. By construction, $\sup \|u(t) - u_{\bar{i}}(t)\| \leq \delta$, hence

$$\begin{aligned} d(g(u)(\bar{t}), F(\bar{t}, u(\bar{t}))) &= d(v_{\bar{i}}(\bar{t}), F(t, u(\bar{t}))) \\ &\leq d(v_{\bar{i}}(\bar{t}), F(\bar{t}, u_i(\bar{t}))) + \mathfrak{d}(F(\bar{t}, u_i(\bar{t})), F(\bar{t}, u(\bar{t}))) \leq \varepsilon. \end{aligned}$$

□

The following is the basic theorem on the existence of continuous selections.

Theorem 2. *Let F from Ω into the compact subsets of R^n be continuous, and let (t_0, x_0) be in Ω. Let I be as provided by Proposition 1. Then there exists a continuous map $\varphi: \mathscr{K} \to \mathscr{K}$, a selection from $\mathscr{J}$.* ▲

Proof. By the preceding Lemma there exists g^0, a continuous map from $\mathscr{K}$ into $L^1(I)$, and a constant δ^0 such that for every u in $\mathscr{K}$,

$$d(g^0(u)(t), F(t, u(t))) \leq 1/2 \tag{14}$$

at almost every t in I, and that

$$\operatorname{meas}\{t \in I \mid \|g^0(u)(t) - g^0(w)(t)\| > 0\} < 1/2 \tag{15}$$

whenever u and w are in $\mathscr{K}$ with $|u - w| \leq \delta^0$.

We will use induction to show that for every $n \geq 0$ there exists a continuous map $g^n: \mathscr{K} \to L^1(I)$ and a positive constant δ^n such that:

$$\begin{cases} \text{i) at almost every } t \in I,\ d(g^n(u)(t), F(t, u(t))) \leq 1/2^{n+1}, \\ \text{ii) } \|u - w\| \leq \delta^n \text{ implies} \\ \qquad \operatorname{meas}\{t \in I \,|\, |g^n(u)(t) - g^n(w)(t)| > 0\} < 1/2^{n+1}, \\ \text{iii) } \operatorname{meas}\{t \in I \,|\, |g^n(u)(t) - g^{n-1}(u)(t)| > 1/2^n\} < 1/2^n \\ \qquad \text{for } n \geq 1. \end{cases} \tag{16}$$

Assume that g^n and δ^n have been defined up to $n = \nu - 1$ and let us define g^ν and δ^ν.

By the uniform continuity of F on $I \times \{x_0 + MTB\}$, there exists a positive $\eta \leq \delta^{\nu-1}$ such that x and y in $I \times \{x_0 + MTB\}$ with $\|x - y\| \leq \eta$ imply $\mathfrak{d}(F(t, x), F(t, y)) \leq 1/2^{\nu-1}$. Let $\{\mathscr{U}_i^\nu\}_{i=1}^m$ be a finite open convering of the compact $\mathscr{K}$ with $\operatorname{diam}(\mathscr{U}_i^\nu)$ smaller than η and let u_i^ν belong to $\mathscr{U}_i^\nu$. For $i = 1, \ldots, m$ choose v_i, a measurable function from I into R^n, such that at almost every t

$$\begin{cases} \text{i) } v_i^\nu(t) \in F(t, u_i^\nu(t)), \\ \text{ii) } \|g^{\nu-1}(u_i^\nu)(t) - v_i^\nu(t)\| = d(g^{\nu-1}(u_i^\nu)(t), F(t, u_i^\nu(t))). \end{cases} \tag{17}$$

This can be done in view of Theorem 1.12.2.

Let $\{\psi_i^\nu\}$ be a partition of unity subordinate to $\{\mathscr{U}_i^\nu\}$: each ψ_i^ν is continuous, hence uniformly continuous, on the compact $\mathscr{K}$. Define δ^ν, $0 < \delta^\nu \leq \eta$, to be such that u and w in $\mathscr{K}$, $\|u - w\| \leq \delta^\nu$, imply

(18) $$|\psi_i^\nu(u)-\psi_i^\nu(w)|\leq 1/2^{\nu+3}m^2T), \quad \text{for all } i.$$

Consider $\{J_i^\nu\}$, the partition of I corresponding to $\{\psi_i^\nu\}$, and χ_i^ν, the characteristic function of J_i^ν. Define g^ν as

(19) $$g^\nu(u)(t) \doteq \Sigma_i\, v_i^\nu(t)\,\chi_i^\nu(t).$$

To check (16) i) fix any t in I. There exists a unique i such that t belongs to J_i^ν: hence $\psi_i^\nu(u)>0$ and $u\in\mathscr{U}_i$, i.e. $g^\nu(u)(t)=v_i^\nu(t)$. Then

$$d(g^\nu(u)(t), F(t,u(t)))\leq d(v_i^\nu(t), F(t,u_i^\nu(t)))+\mathfrak{d}(F(t,u_i^\nu(t)), F(t,u(t)))$$
$$\leq 1/2^{\nu+1},$$

since $|u_i^\nu(t)-u(t)|\leq\eta$.

To prove (16) ii) remark that, as in the proof of Proposition 2, the choice of δ^ν implies that the measure of each $J_i^\nu(u)\triangle J_i^\nu(w)$ is at most $1/(2^{\nu+1}m)$, i.e.

$$\operatorname{meas}\{t\in I\,|\,|g^\nu(u)-g^\nu(w)|>0\}<\operatorname{meas}\cup(J_i^\nu(u)\triangle J_i^\nu(w)$$
$$<1/2^{\nu+1}.$$

Finally

$$\|g^\nu(u)(t)-g^{\nu-1}(u)(t)\|$$
$$\leq\|v_i^\nu(t)-g^{\nu-1}(u_i^\nu)(t)\|+\|g^{\nu-1}(u_i^\nu)(t)-g^{\nu-1}(u)(t)\|$$
$$=d(g^{\nu-1}(u_i^\nu)(t), F(t,u_i^\nu(t)))+\|g^{\nu-1}(u_i^\nu)(t)-g^{\nu-1}(u)(t)\|.$$

By (16) i) of the induction assumptions, the first term is bounded by $1/2^\nu$ almost everywhere while, by (16) ii), the measure of those t such that the second term is positive is bounded by $1/2^\nu$. Hence

$$\operatorname{meas}\{t\in I\,|\,|g^\nu(u)(t)-g^{\nu-1}(u)(t)|>1/2^\nu\}<1/2^\nu$$

and this is (16) iii).

Fix u in $\mathscr{K}$. By (16) iii), the sequence of measurable maps $\{g^n(u)(\cdot)\}$ is a Cauchy sequence in measure, converging in measure to some measurable function that we denote by $g(u)(\cdot)$. Moreover a subsequence converges to $g(u)(\cdot)$ pointwise almost everywhere. By (16) i) and the compactness of $F(t,u(t))$, we infer that, for almost every t, $g(u)(t)\in F(t,u(t))$.

We wish to show that the map $\varphi\colon\mathscr{K}\to\mathscr{K}$ defined by

(20) $$\varphi(u)(t)\doteq x_0+\int_{t_0}^{t} g(u)(s)\,ds$$

is continuous, i.e. that it is the required continuous selection from $\mathscr{J}$. Let us show directly that φ is uniformly continuous on $\mathscr{K}$, i.e. that, given $\varepsilon>0$, there exists a positive δ such that $\|u-w\|\leq\delta$ implies $\|\varphi(u)(t)-\varphi(w)(t)\|\leq\varepsilon$ for every t in I. It turns out to be convenient to choose n such that $(5M+T)/2^n<\varepsilon$ and to choose $\delta\doteq\delta^n$.

In fact on one hand we have, by (16) iii), that for every $u\in\mathscr{K}$ and every ν,

$$\int_I\|g^{\nu+1}(u)(s)-g^\nu(u)(s)\|\,ds\leq 2T/2^{\nu+1}+2M/2^{\nu+1}$$
$$=(T+M)/2^\nu$$

so that

$$\int_I \|g^{n+1}(u)(s)-g^n(u)(s)\|\,ds+\int_I \|g^{n+2}(u)(s)-g^{n+1}(u)(s)\|\,ds+\dots$$
$$\leq(1/2^n)(T+M)(1+\tfrac{1}{2}+\tfrac{1}{4}+\dots)=(T+M)/2^{n-1}$$

and, since $\int_I \|g^\nu(u)(s)-g(u)(s)\|\,ds$ converges to 0,

$$\|\varphi(u)(t)-\varphi(w)(t)\|\leq\int_{t_0}^{t}\|g(u)(s)-g(w)(s)\|\,ds$$
$$\leq\int_I \|g^n(u)(s)-g^n(w)(s)\|\,ds+\int_I \|g^n(u)(s)-g(u)(s)\|\,ds$$
$$+\int_I \|g^n(w)(s)-g(w)(s)\|\,ds$$
$$\leq\int_I \|g^n(u)(s)-g^n(w)(s)\|\,ds+4(T+M)/2^n.$$

On the other hand by (16) ii)

$$\int_I \|g^n(u)(s)-g^n(w)(s)\|\,ds\leq 2M/2^{n+1}=M/2^n.$$

Summarizing, $\|u-w\|\leq\delta$ implies, for every t in I,

$$\|\varphi(u)(t)-\varphi(w)(t)\|\leq(T+5M)/2^n\leq\varepsilon,$$

proving the continuity of φ. □

Corollary 2. *Let F, from Ω into the compact subsets of R^n, be continuous, let (t_0, x_0) be in Ω and let the interval I be defined as in Proposition 1. Then the problem*

(21) $$x'(t)\in F(t,x(t)),\quad x(t_0)=x_0$$

admits at least one solution, defined on I. ▲

Proof. The continuous φ maps the compact convex set $\mathscr{K}$ into itself, hence, by Schauder's theorem, φ has a fixed point $\bar{u}$ in $\mathscr{K}$: $\bar{u}$ is a Lipschitzean function on I, such that $u(t_0)=x_0$ and, for almost every t in I,

(22) $$\bar{u}'(t)=g(\bar{u})(t)\in F(t,\bar{u}(t)).$$ □

6. The Lower Semicontinuous Case

By the same technique as in Section 5, we shall prove the existence of solutions when the map $(t,x)\to F(t,x)$ is lower semicontinuous on some open $\Omega\subset R\times R^n$.

We posit the following assumption.

There exist $I \doteq [t_0 - T, t_0 + T]$ and M such that

$$(1) \qquad \begin{cases} \text{i)} \ I \times \{x_0 + TMB\} \text{ is contained in } \Omega, \\ \text{ii)} \ \|F(t,x)\| \le M \text{ on } I \times \{x_0 + TMB\}. \end{cases}$$

We define the compact $\mathscr{K} \subset C(I)$ and the integral operator $\mathscr{I}$ as in Definition 5.1. We shall show the existence of solutions to

$$x' \in F(t,x) \qquad x(t_0) = x_0$$

by the same selection technique as in the previous section. We begin by the following technical lemma.

Lemma 1. *Let H, from a compact subset K of $I \times R^n$ into the subsets of R^n be lower semicontinuous, and let $\varepsilon > 0$. Set*

$$(2) \qquad \eta^\varepsilon(t,x) \doteq \sup\{\eta \mid \bigcap_{(\tau,\xi) \in (t,x) + \eta B} H(\tau,\xi) + \varepsilon \mathring{B} \neq \emptyset\}.$$

Then: a) for some $\eta^\varepsilon > 0$ we have $\eta^\varepsilon(t,x) \ge \eta^\varepsilon$, all (t,x); b) for every continuous u, $(t,u(t))$ in K, there exists a measurable $v: t \to v(t)$, such that: $\|(t,x) - (t,u(t))\| < \eta^\varepsilon$ implies

$$d(v(t), H(t,x)) \le \varepsilon. \qquad \blacktriangle$$

Proof. a) The definition of lower semicontinuity implies that the set inside brackets in (2) is non-empty, so that $\eta^\varepsilon(t,x)$ is positive. We claim that it is a continuous function. Fix $\sigma > 0$ arbitrarily and remark that whenever $d((\tau_1,\xi_1),(\tau_2,\xi_2)) < \sigma/3$,

$$B_1 \doteq B[(\tau_1,\xi_1), \eta^\varepsilon(\tau_2,\xi_2) - 2\sigma/3] \subset B[(\tau_2,\xi_2), \eta^\varepsilon(\tau_2,\xi_2) - \sigma/3] \doteq B_2,$$

i.e.

$$\bigcap_{(\tau,\xi) \in B_2} H(\tau,\xi) + \varepsilon \mathring{B} \neq \emptyset \Rightarrow \bigcap_{(\tau,\xi) \in B_1} H(\tau,\xi) + \varepsilon \mathring{B} \neq \emptyset.$$

Whenever $d((t,x),(t^*,x^*)) < \sigma/3$, setting $(t,x) \doteq (\tau_1,\xi_1)$, $(t^*,x^*) \doteq (\tau_2,\xi_2)$ we obtain

$$\eta^\varepsilon(t,x) \ge \eta^\varepsilon(t^*,x^*) - 2\sigma/3$$

while interchanging (t,x) and (t^*,x^*) we have

$$\eta^\varepsilon(t^*,x^*) \ge \eta^\varepsilon(t,x) - 2\sigma/3.$$

Hence $(t,x) \to \eta^\varepsilon(t,x)$ is a continuous and positive map defined on a compact set.

b) Call Φ the map

$$(3) \qquad (t,x) \to \Phi(t,x) \doteq \bigcap_{(\tau,\xi) \in (t,x) + \eta^\varepsilon B} H(\tau,\xi) + \varepsilon \mathring{B}.$$

Then Φ is lower semicontinuous. In fact let y^* be in $\Phi(t^*,x^*)$, so that for every (τ,ξ) in $(t^*,x^*) + \eta^\varepsilon B$, $d(y^*, H(\tau,\xi)) = \varepsilon - \eta^\varepsilon(t,\xi)$, $\eta^\varepsilon(\tau,\xi) > 0$, or equivalently, there exists $y(\tau,\xi)$ in $H(\tau,\xi)$ so that $d(y^*, y(\tau,\xi)) \le \varepsilon - \sigma/2$.

By the lower semicontinuity of H, there exists $\delta \doteq \delta(\tau, \xi)$ so that (τ', ξ') in $\mathring{B}[(\tau, \xi), \delta]$ implies $d(y(\tau, \xi), H(\tau', \xi')) < \sigma/2$, hence in particular $d(y^*, H(\tau', \xi')) < \varepsilon$. The open set

$$\mathscr{U} \doteq \bigcup_{(\tau, \xi) \in (t^*, x^*) + \eta^\varepsilon B} \mathring{B}[(\tau, \xi), \delta(\tau, \xi)]$$

contains the compact $(t^*, x^*) + \eta^\varepsilon B$, hence whenever $d((t, x), (t^*, x^*)) < \rho$ sufficiently small, $B[(t, x), \eta^\varepsilon] \subset \mathscr{U}$, and thus,

$$d(y^*, H(\tau, \xi)) < \varepsilon \quad \text{or} \quad y^* \in \Phi(t, x).$$

Finally, the map $t \to \bar{\Phi}(t, u(t))$ is lower semicontinuous and has closed values, hence the *measurable selection* Theorem 1.14.1 applies to yield the required $v(t)$. □

We can prove the following Theorem, the analogue to Theorem 5.2.

Theorem 1. *Let F from $\Omega \subset R \times R^n$ into the compact non empty subsets of R^n be lower semicontinuous and let (t_0, x_0) be in Ω. We posit Assumption* (1). *Then there exists a continuous map φ: $\mathscr{K} \to L^1(I)$, a selection from $\mathscr{J}$.* ▲

Proof. We shall first show the existence of a finite number $m(0)$ of measurable maps v_i from I into MB; of a continuous partition of I into $\mathscr{I}_i^0 \doteq [\tau_{i-1}^0, \tau_i^0[$, with characteristic functions χ_i^0 such that setting

$$g^0(u)(t) \doteq \sum \chi_i^0(t)\, v_i(t)$$

we have:

$$\text{(4)} \qquad \text{for every } t, \quad d(g^0(u)(t), F(t, u(t))) < 1.$$

In fact, set in Lemma 1, H to be F, ε to be 1 and let η_0 be the constant provided by a). Let $\mathscr{U}^i = u^i + \eta_0 \mathring{B}$ be a finite open covering of the compact $\mathscr{K}$, v_i the corresponding measurable functions as provided by b). Fix u and t: whenever $\chi_i^0(u)(t) > 0$, u is in $u^i + \eta_0 \mathring{B}$, and $d(v_i(t), F(t, u(t))) < 1$ and (4) holds.

We claim that for $n = 0, 1, \ldots$, we can define: $m(n)$ measurable functions $v_i^{(n)}$ from I into MB; a continuous partition of I, $\mathscr{I}_i^{(n)} \doteq [\tau_{i-1}^{(n)}(u), \tau_i^{(n)}(u)[$ having characteristic functions $\chi_i^{(n)}$, such that setting

$$g^n(u)(t) \doteq \sum \chi_i^{(n)}(t)\, v_i^{(n)}(t)$$

we have

$$\begin{cases} \text{i) for every } t, \quad d(g^n(u)(t), F(t, u(t))) < 1/2^n, \\ \text{ii) except on a finite number of intervals, having total length } 1/2^n, \\ \quad d(g^n(u)(t), g^{n-1}(u)(t)) < 1/2^{n+1}, \qquad n \geq 1. \end{cases}$$

Assume the above to hold up to $n = \nu - 1$ and let us prove it to hold for $n = \nu$.

There exists an open S^ν such that all the maps $t \to v_i^{(\nu-1)}(t)$ are continuous on $I \backslash S^\nu$, and the measure of S^ν is smaller than $1/2^{\nu+1}$. Let $\delta > 0$ be such that $\|w-u\|<\delta$ implies that for each i, $|\tau_i^{(v-1)}(u) - \tau_i^{(v-1)}(w)| < (2^{-v}(4m(v-1)))^{-1}$. A finite number of balls $\hat{u}_j+\delta\mathring{B}$ covers $\mathscr{K}$. For each j call E_j the finite union of open intervals $|t-\tau_i(\hat{u}_j)| < (2^{-v}(4(v-1)))^{-1}, i=1, \ldots, m(v-1)$. Then whenever u is in $\hat{u}_j+\delta\mathring{B}$, when t is in any of the closed intervals whose union is $I\backslash E_j$, $g^{v-1}(u)(t)=g^{v-1}(\hat{u}_j)(t) = v_i^{v-1}(t)$ for some i. Hence when t belongs to the closed $(I\backslash E_j)\backslash S^v$, the map $t \to g^{v-1}(u)(t)$ is continuous.

Set $\rho_j^v(t)$ to be $2M$ on the open $(E_j \cup S^v)$ and to be $1/2^{v-1}$ on the closed $I\backslash(E_j \cup S^v)$. The map $(t,x) \to F_j(t,x)$ defined by

$$F_j(t,x) \doteq (g^{v-1}(\hat{u}_j)(t) + \rho_j^v(t)\,\mathring{B}) \cap F(t,x)$$

is strict for (t,x) in $(t,\hat{u}_j(t))+\delta B$. In fact, when t is in $(E_j \cup S^v)$ it is enough to remark that both g^{v-1} and F take values in MB. Let (t,x): t in $I\backslash(E_j \cup S^n)$, $d(x,\hat{u}_j(t))<\delta$. Then a translate $u(\cdot)$ of $\hat{u}_j(\cdot)$ is in $\hat{u}_j(\cdot)+\delta B$ and is such that $u(t)=x$. For this u, $g^{v-1}(u)(t) = g^{v-1}(\hat{u}_j)(t)$ and, by point i) of the induction,

$$d(g^{v-1}(u)(t), F(t,x)) < 1/2^{v-1}.$$

Since $t \to \rho^v(t)$ is lower semicontinuous and F_j is strict, Proposition 1.1.5 implies that $(t,x) \to F_j(t,x)$ is lower semicontinuous.

Set in Lemma 1, H to be F_j, ε to be $1/2^{v+1}$ and call η^j be the constant provided by point a). A finite number of balls $u_j^i + \eta_j\mathring{B}$ covers the compact $\mathscr{K} \cap (\hat{u}_j+\delta B)$. By Lemma 1, b), there exists for each i a measurable v_j^i such that $d((t,x),(t,u_j^i(t))) < \eta_i$ implies

$$d(v_j^i, F_j(t,x)) \leq 1/2^{v+1} < 1/2^v. \tag{5}$$

The collection of open sets $\mathscr{U}_j^i \doteq (\hat{u}_j+\delta\mathring{B}) \cap (u_j^i+\eta_j\mathring{B})$ covers $\mathscr{K}$. Let χ_j^i be the characteristic functions of the corresponding continuous partition $\{\mathscr{I}_j^i\}$ of I. Set

$$g^v(u)(t) \doteq \sum_{i,j} \chi_j^i(t)\, v_j^i(t).$$

We claim that the functions v_j^i and the map g^v satisfy our induction assumptions.

Fix u and t. Whenever t belongs to $\mathscr{I}_j^i(u)$, $g^v(u)(t) = v_j^i(t)$ and u belongs to $u_j^i + \eta_j\mathring{B}$, and by (5)

$$d(v_j^i(t), F_j(t,u(t))) < 1/2^v. \tag{6}$$

Since $F_j(t,u(t)) \subset F(t,u(t))$, (6) checks point i).

To check ii) assume t in $I\backslash(E_j \cup S^v)$. Then $\rho^v(t) = 1/2^{v-1}$, $F_j(t,x) \subset g^{v-1}(\hat{u}_j)(t) + (1/2^{v-1})\,\mathring{B} = g^{v-1}(u)(t) + (1/2^{v-1})\,\mathring{B}$, hence

$$\begin{aligned} &d(v_j^i(t), g^{v-1}(u)(t) + (1/2^{v-1})\,\mathring{B}) < 1/2^v \quad \text{or:} \\ &d(g^v(u)(t), g^{v-1}(u)(t)) < 1/2^{v+1} \end{aligned} \tag{7}$$

except on an open set $(E_j \cup S^v)$ with measure at most $1/2^v$.

The remainder of the proof follows exactly as the proof of Theorem 5.2.

□

Corollary 1. *Let F, from Ω into the compact subsets of R^n be lower semicontinuous. Let (t^0, x^0) be in Ω and let M and I be defined as in Hypothesis (H). Then the problem*

$$x'(t) \in F(t, x(t)), \qquad x(t_0) = x_0$$

has at least one solution defined on I. ▲

Chapter 3. Differential Inclusions with Maximal Monotone Maps

Introduction

We devote this chapter to a very important class of differential inclusions

$$(1) \qquad x'(t) \in -A(x(t))$$

where $A(x) \doteq -F(x)$ is a so-called "maximal monotone" set-valued map.

They include the case where $F(x) = -\partial V(x)$ is the subdifferential of a convex lower semicontinuous function; the end of this chapter is devoted to a comprehensive study of this special case and to some applications.

Before giving a precise definition of this class of maps, we mention their main properties: For any initial state $x_0 \in \operatorname{Dom} A$, there exists a unique solution $x(\cdot)$ to (1) issued from x_0. The map $x_0 \to x(\cdot)$ is non-expansive (Lipschitzean with Lipschitz constant one). The velocity $x'(t) \in -A(x(t))$ is the velocity of minimal norm (we say that $x(\cdot)$ is a *slow solution*).

Uniqueness and non-expansivity of the map $x_0 \to x(\cdot)$ depend upon the *monotonicity of A*, defined by

$$(2) \qquad \forall x, y \in D(A),\ \forall p \in A(x),\ \forall q \in A(y), \quad \langle p-q, x-y \rangle \geq 0.$$

Among the monotone maps, we single out the ones that are maximal; by Minty's theorem, these are the monotone maps such that $1+A$ is surjective.

The values $A(x)$ of such maps are closed and convex, but A is not necessarily upper semicontinuous (the graph of A is closed in $X \times X$ when one of the factors of this product is supplied with the weak topology). Therefore, we cannot use the above theorems dealing with upper semicontinuous set-valued maps. Nevertheless, we overcome this lack of continuity thanks to the special structure of maximal monotone maps. They enjoy an approximation property by Lipschitzean maps analogous to the existence of approximate selections of upper semicontinuous maps: we prove that the *Yosida approximation* $A_\lambda \doteq \frac{1}{\lambda}(1-(1+\lambda A)^{-1})$ is a Lipschitzean map from X to X such that $(x, A_\lambda(x)) \in \operatorname{Graph}(A) + \lambda \| m(A(x)) \| B$. Therefore the approach for solving the differential inclusion is clear: we solve the ordinary differential equation

$$(3) \qquad x'_\lambda(t) = -A_\lambda(x_\lambda(t))$$

we check that the sequence of solutions $x_\lambda(\cdot)$ is a Cauchy sequence and that its limit $x(\cdot)$ is actually a solution to the differential inclusion.

We proceed our investigations with the asymptotic behavior of the trajectories of the differential inclusion $x'(t) \in -A(x(t))$ when $t \to \infty$. We wish that they converge weakly to an *equilibrium* $\bar{x}$, *solution to the inclusion* $0 \in A(\bar{x})$.

This is the case when there exists at least an equilibrium and when all the cluster points of the trajectory are equilibria.

This latter condition is not always satisfied by the trajectory $x(\cdot)$ itself. It happens that it is always satisfied by its Cesaro means

(4) $$\sigma(T) \doteq \frac{1}{T} \int_0^T x(\tau)\, d\tau .$$

We shall prove the *Ergodic Theorem*, stating that the function $\sigma(T)$ converges weakly to an equilibrium when $T \to \infty$.

We also prove that the sequence of elements

(5) $$x_n = (1+A)^{-n} x$$

converges to an equilibrium when $n \to \infty$.

All these convergence results assume the existence of at least an equilibrium. For proving them, we use the properties of asymptotic centers of bounded sequences.

Since the subdifferential $\partial V(x)$ of a proper lower semicontinuous convex function defines a maximal monotone map $x \to \partial V(x)$, the gradient inclusion

(6) $$x'(t) \in -\partial V(x), \quad x(0) = x_0$$

has a unique solution $x(t)$ such that

(7) $$t \to V(x(t)) \quad \text{is convex and nonincreasing.}$$

Furthermore, if V achieves its minimum at some point, the trajectory $x(t)$ as $t \to \infty$, converges weakly to a point achieving the minimum of V.

1. Maximal Monotone Maps

In this chapter, X always denotes a *Hilbert space* (identified with its dual X^*).

Definition 1. A set-valued map A from X into X is called *monotone* if and only if

(1) $$\forall x_1, x_2 \in \mathrm{Dom}(A),\ \forall v_i \in A(x_i), \quad i = 1, 2,\ \langle v_1 - v_2, x_1 - x_2 \rangle \geq 0.$$

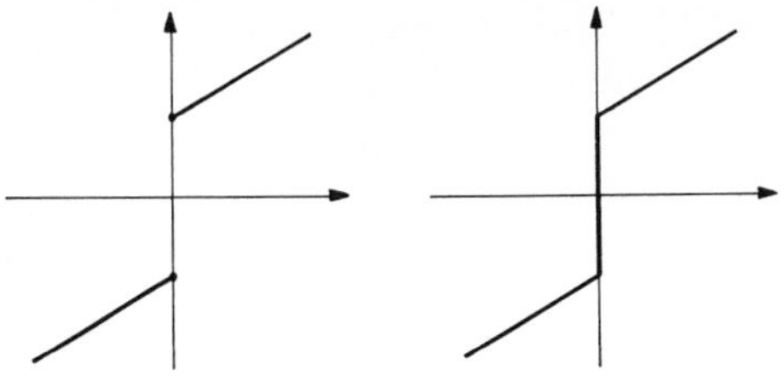

a b

Fig. 12. *a* Graph of a monotone map; *b* maximal monotone map

A *monotone* set-valued map is *maximal* if there is no other monotone set-valued map $\tilde{A}$ whose graph contains strictly the graph of A. ▲

We begin by pointing out the following remarks:

A set-valued map is monotone (resp. maximal monotone) if and only if its inverse A^{-1} is monotone (resp. maximal monotone).

By Zorn's lemma, the graph of any monotone set-valued map is contained in the graph of a maximal monotone set-valued map, because the union of an increasing family of graphs of monotone set-valued maps is obviously the graph of a monotone set-valued map.

Actually, we will use the following formulation of the definition of a maximal monotone set-valued map.

Proposition 1. *A set-valued map A is maximal monotone if and only if the following statements* a) *and* b) *are equivalent:*

$$(2) \qquad \begin{cases} \text{a) } \textit{For every } (y,v)\in\text{Graph}(A),\ \langle u-v, x-y\rangle \geq 0 \\ \text{b) } u\in A(x) \end{cases}$$ ▲

This provides a useful and manageable way for recognizing that u belongs to $A(x)$.

Proposition 2. *Let A be maximal monotone. Then:*

a) *Its images are closed and convex.*

b) *Its graph is strongly-weakly closed in the sense that if x_n converges to x and if $u_n\in A(x_n)$ converges weakly to u, then $u\in A(x)$.* ▲

Proof. a) By the above proposition, $A(x)$ is the intersection of the closed half spaces $\{u\in X \mid \langle u-v, x-y\rangle \geq 0\}$ when (y,v) ranges over the graph of A. Hence $A(x)$ is closed and convex.

b) Let x_n converge to x and $u_n\in A(x_n)$ converge weakly to u. Let us choose (y,v) in the graph of A. Inequalities

$$\langle u_n - v, x_n - y\rangle \geq 0$$

imply, by passing to the limit, inequalities

$$\langle u-v, x-y\rangle \geq 0.$$

Hence $u\in A(x)$ by Proposition 1. □

Proposition 3. *Any monotone single valued map A from the whole space X to X, which is continuous from X supplied with the strong topology to X supplied with the weak topology, is maximal monotone.* ▲

Proof. Let $x \in X$ and $u \in X$ such that

$$(3) \qquad \langle u - A(y), x - y \rangle \geq 0 \quad \text{for all } y \in X.$$

To show that A is maximal monotone, we have to check that $u = A(x)$ [by Proposition 1]. For that purpose, we take $y \doteq x - \lambda(z - x)$ where $\lambda \in]0, 1[$ and $z \in X$. Inequality (3) becomes

$$\langle u - A(x - \lambda(z - x)), z - x \rangle \geq 0 \quad \text{for all } z \in X.$$

By letting λ converge to 0 and by the continuity of A, we deduce that $\langle u - A(x), z - x \rangle \geq 0$ for all $z \in X$, i.e., $u = A(x)$.

Remark. An alternative proof.

Assume A is not maximal. There exists $\tilde{A}$, a proper monotone extension of A, i.e. for some x there exists $u \neq A(x)$, $u \in \tilde{A}(x)$. Consider $u - A(x)$ and for λ positive, call y_λ the vector $y_\lambda = x + \lambda(u - A(x))$. The map A is defined (by assumption) on y_λ and, by the monotonicity of $\tilde{A}$, we must have:

$$\langle y_\lambda - u, A(y_\lambda) - u \rangle \geq 0, \quad \text{i.e. } \langle u - A(x), A(y_\lambda) - u \rangle \geq 0.$$

By the continuity of A, passing to the limit for $\lambda \to 0$, it follows

$$\langle u - A(x), A(x) - u \rangle \geq 0,$$

a contradiction. □

We shall now characterize those monotone maps which are maximal.

Theorem 1 (Minty). *Let A be a monotone set-valued map from X to X. It is maximal if and only if $1 + A$ is surjective.* ▲

Proof. a) Assume that A is maximal.

Let $\hat{y} \in X$; we have to prove that there exists x such that $\hat{y} \in x + A(x)$. It is sufficient to choose $\hat{y} = 0$, since this amounts to replacing A by $x \to -\hat{y} + A(x)$, which is also maximal monotone. By Proposition 1 we must prove that there exists $\bar{x}$ satisfying

$$(4) \qquad \forall y \in D(A), \ \forall v \in A(y), \quad \langle -\bar{x} - v, \bar{x} - y \rangle \geq 0.$$

For convenience, set $\phi(x; (y, v)) \doteq \langle x + v, x - y \rangle$, or equivalently $\phi(x; (y, v)) \doteq \|x\|^2 + \langle x, v - y \rangle - \langle v, y \rangle$. We have to prove that there exists $\bar{x}$ such that

$$(5) \qquad \forall (y, v) \in \operatorname{graph}(A), \quad \phi(\bar{x}; (y, v)) \leq 0.$$

Fix (y_0, v_0) to be any point in the graph of A. If a solution to (5) exists, it certainly belongs to

$$L \doteq \{x \in \mathrm{Dom}(A) \mid \phi(x;(y_0,v_0)) \leq 0\}.$$

The mapping $x \to \phi(x;(y_0,v_0))$ is quadratic, hence the set $\{x \mid \phi(x;(y_0,v_0)) \leq 0\}$ is convex, closed and bounded, hence weakly compact. Its intersection L with $\mathrm{Dom}(A)$ need not be so, but so is $\overline{\mathrm{co}}(L)$. Also, the maps $x \to \phi(x;(y,v))$ are convex and (strongly) continuous. Therefore their lower sections are convex and closed, hence weakly closed. It follows then that they are weakly lower semicontinuous.

We use Theorem 3.10.1 of Aubin [1979] a), Chap. 3, § 10, p. 119): Set S^n to be the simplex of R^n and $\mathscr{S}$ to be the family of all finite subsets $K \doteq \{(y_1,v_1), \ldots, (y_n,v_n)\}$ of graph (A); there exists $\bar{x} \in \overline{\mathrm{co}}(L)$ such that

$$\sup_{(y,v)\in \mathrm{graph}(A)} \phi(\bar{x};(y,v))$$

$$(6) \qquad \begin{aligned} &\leq \sup_{K\in\mathscr{S}} \inf_{x\in\mathrm{co}(L)} \max_{i=1,\ldots,n} \phi(x;(y_i,v_i)) \\ &\leq \sup_{K\in\mathscr{S}} \inf_{x\in\mathrm{co}(y_1,\ldots,y_n)} \max_{i=1,\ldots,n} \phi(x;(y_i,v_i) \\ &\leq \sup_{K\in\mathscr{S}} \inf_{x\in\mathrm{co}(y_1,\ldots,y_n)} \sup_{\mu\in S^n} \sum_{j=1}^{n} \mu_j \phi(x;(y_j,v_j) \\ &< \sup_{K\in\mathscr{S}} \inf_{\lambda\in S^n} \sup_{\mu\in S^n} \phi_K(\lambda,\mu) \end{aligned}$$

where we set

$$(7) \qquad \phi_K(\lambda,\mu) \doteq \sum_{j=1}^{n} \mu_j \phi(\beta(\lambda);(y_j,v_j)), \qquad \beta(\lambda) \doteq \sum_{i=1}^{n} \lambda^i y_i.$$

This function is continuous with respect to λ and

$$\begin{aligned} \phi_K(\mu,\mu) &= \sum_{i,k=1}^{n} \mu^i \mu^k \langle \beta(\xi)+v_i, y_k - y_i\rangle \\ &= \sum_{i,k=1}^{n} \mu^i \mu^k \langle \beta(\mu), y_k - y_i\rangle + \sum_{i,k=1}^{n} \mu^i \mu^k \langle v_i, y_k - y_i\rangle. \end{aligned}$$

The first term is zero for reasons of symmetry, while the second can be written

$$\begin{aligned} &\tfrac{1}{2} \sum_{i,k=1}^{n} \mu^i \mu^k \langle v_i, y_k - y_i\rangle + \tfrac{1}{2} \sum_{j,l=1}^{n} \mu^j \mu^l \langle -v_j, y_j - y_l\rangle \\ &= \tfrac{1}{2} \sum_{i,k=1}^{n} \mu^i \mu^k \langle v_i - v_k, y_k - y_i\rangle \leq 0 \end{aligned}$$

since A is monotone. Hence assumptions of Ky Fan's inequality are satisfied (see Chapter 5, Section 2). Therefore, for all K,

$$\inf_{\lambda\in S^n} \sup_{\mu\in S^n} \phi_K(\lambda,\mu) \leq 0$$

and thus, by (6), inequality (5) is satisfied for $\bar{x}$. We have shown that $\bar{x}$ is a solution to $0 \in \bar{x} + A(\bar{x})$.

b) Assume that A is monotone and that $1 + A$ is surjective.

Let $x \in X$ and $u \in X$ satisfy

$$(8) \qquad \forall y \in \operatorname{Dom} A,\ \forall v \in A(y), \quad \langle u-v, x-y\rangle \geq 0.$$

By Proposition 1, we have to prove that $u \in A(x)$. Since $1+A$ is surjective, we can choose in (8) y to be a solution y_0 of the set-valued equation $u+x \in y_0 + A(y_0)$. Let $v_0 \in A(y_0)$ such that $u+x=y_0+v_0$. Then $\|x-y_0\|^2=\langle x-y_0, x-y_0\rangle = -\langle u-v_0, x-y_0\rangle \leq 0$. Hence $x=y_0$ and thus, $u=v_0 \in A(y_0)=A(x)$. □

Yosida Approximation

We show that a maximal monotone A can be approximated in some sense by single valued Lipschitzean maps A_λ from X to X that are maximal monotone. These maps, called Yosida approximations, play an important role.

Theorem 2. *Let A be maximal monotone. Then, for all $\lambda > 0$,*

$$(9) \quad J_\lambda \doteq (1+\lambda A)^{-1} \quad \text{is a non-expansive single valued map from } X \text{ to } X$$

and the map $A_\lambda \doteq \frac{1}{\lambda}(1-J_\lambda)$ *satisfies*

$$(10) \quad \begin{array}{l} \text{i) } \forall x \in X,\ A_\lambda(x) \in A(J_\lambda x), \\ \text{ii) } A_\lambda \text{ is Lipschitzean with constant } \frac{1}{\lambda} \text{ and maximal monotone.} \end{array}$$

We also have

$$(11) \qquad \forall x \in \operatorname{Dom}(A), \quad \|A_\lambda(x) - m(A(x))\|^2 \leq \|m(A(x))\|^2 - \|A_\lambda(x)\|^2$$

and, for all $x \in \operatorname{Dom}(A)$

$$(12) \qquad \begin{cases} \text{i) } J_\lambda x \text{ converges to } x, \\ \text{ii) } A_\lambda x \text{ converges to } m(A(x)), \end{cases}$$

where $m(A(x)) \doteq \pi_{A(x)}(0)$ is the element of $A(x)$ of minimal norm. ▲

Definition 2. The maps J_λ and A_λ are called "resolvent" of A and the "Yosida approximation" of A respectively. ▲

Proof. a) Let x_i $(i=1,2)$ be solutions to the set-valued equation

$$(13) \qquad y_i \in x_i + \lambda A(x_i) \quad (i=1,2)$$

(which exist since $1+\lambda A$ is surjective by Theorem 1). So $y_i = x_i + \lambda v_i$ where $v_i \in A(x_i)$. We obtain $\|y_1-y_2\|^2 = \|x_1-x_2+\lambda(v_1-v_2)\|^2 = \|x_1-x_2\|^2 + \lambda^2\|v_1-v_2\|^2 + 2\lambda\langle v_1-v_2, x_1-x_2\rangle \geq \|x_1-x_2\|^2 + \lambda^2\|v_1-v_2\|^2$. Hence

$$(14) \qquad \begin{cases} \text{i) } \|x_1-x_2\| \leq \|y_1-y_2\|, \\ \text{ii) } \|v_1-v_2\| \leq \frac{1}{\lambda}\|y_1-y_2\|. \end{cases}$$

By taking $y_1=y_2$, (14) i) proves the uniqueness of the solution. We note that

(15) $$x_i = J_\lambda y_i \quad \text{and} \quad v_i = A_\lambda(y_i).$$

Hence inequalities (14) prove that J_λ and A_λ are Lipschitzean with constants 1 and $1/\lambda$ respectively.

b) By the very definitions of J_λ and A_λ, we have

(16) $$A_\lambda(y) = \frac{1}{\lambda}(y - J_\lambda(y)) \in A(J_\lambda y) \quad \text{for all } y \in X.$$

Therefore, since $y_i = J_\lambda(y_i) + \lambda A_\lambda(y_i)$, we obtain

$$\begin{aligned}\langle A_\lambda(y_1) - A_\lambda(y_2), y_1 - y_2\rangle &= \langle A_\lambda(y_1) - A_\lambda(y_2), J_\lambda(y_1) - J_\lambda(y_2)\rangle \\ &\quad + \lambda \|A_\lambda(y_1) - A_\lambda(y_2)\|^2 \geq \lambda \|A_\lambda(y_1) - A_\lambda(y_2)\|^2 \geq 0.\end{aligned}$$

Hence A_λ is monotone (and, by Proposition 3, maximal monotone).

c) Let $x \in D(A)$. We compute

$$\begin{aligned}\|A_\lambda(x) - m(A(x))\|^2 &= \|A_\lambda(x)\|^2 + \|m(A(x))\|^2 - 2\langle A_\lambda(x), m(A(x))\rangle \\ &= -\|A_\lambda(x)\|^2 + \|m(A(x))\|^2 - 2\langle A_\lambda(x), m(A(x)) - A_\lambda(x)\rangle.\end{aligned}$$

But since A is monotone, $m(A(x)) \in A(x)$ and $A_\lambda(x) \in A(J_\lambda x)$, we obtain

$$\langle A_\lambda(x), m(A(x)) - A_\lambda(x)\rangle = \frac{1}{\lambda}\langle x - J_\lambda(x), m(A(x)) - A_\lambda(x)\rangle \geq 0.$$

Therefore, we have proved inequality

(17) $$\|A_\lambda(x) - m(A(x))\|^2 \leq \|m(A(x))\|^2 - \|A_\lambda(x)\|^2.$$

d) Then, when $x \in D(A)$, we have

$$\|x - J_\lambda(x)\| = \lambda \|A_\lambda(x)\| \leq \lambda \|m(A(x))\|.$$

Therefore, $J_\lambda(x)$ converges to x when λ converges to 0.

e) Since $J_\lambda(x) = x - \lambda A_\lambda(x)$ and since $A_\lambda(x) \in A(J_\lambda x)$, we see that $y = A_\lambda(x)$ is a solution to the inclusion $y \in A(x - \lambda y)$. Conversely, any such solution y is $A_\lambda(x)$: indeed, set $z \doteq x - \lambda y$; this equation becomes $x \in z + \lambda A(z)$. Hence $z = J_\lambda(x)$ and $y = \frac{1}{\lambda}(x - J_\lambda(x)) = A_\lambda(x)$.

f) This remark implies that

(18) $$A_{\mu+\lambda}(x) = (A_\mu)_\lambda(x).$$

Indeed, $y = A_{\mu+\lambda}(x)$ is a solution to the equation $y \in A(x - \lambda y - \mu y)$; then $y \in A_\mu(x - \lambda y)$. By applying again the above remark to the Yosida approximation A_μ, which is maximal monotone, we deduce that $y = (A_\mu)_\lambda(x)$.

g) Now, we use inequality (17) where A is replaced by A_μ. Since $m(A_\mu(x)) = A_\mu(x)$, we obtain

$$\|A_{\lambda+\mu}(x) - A_\mu(x)\|^2 \leq \|A_\mu(x)\|^2 - \|A_{\lambda+\mu}(x)\|^2.$$

Then the sequence $\|A_\mu(x)\|^2$ is an increasing sequence of real numbers bounded above by $\|m(A(x))\|^2$. Hence it converges to some real number α.

This implies that

$$\lim_{\lambda,\mu\to 0} \|A_{\lambda+\mu}(x)-A_\mu(x)\|^2 \le \alpha-\alpha=0.$$

Consequently, $A_\lambda(x)$ is a Cauchy sequence that converges to some element v in X.

Since $A_\lambda(x)$ belongs to $A(J_\lambda(x))$ and since the graph of A is closed, we deduce that $v\in A(x)$. Also,

$$\|v\| = \lim_{\lambda\to 0} \|A_\lambda(x)\| \le \|m(A(x))\|.$$

Since $A(x)$ is closed and convex, the projection of 0 to $A(x)$ is unique and, consequently, $v=m(A(x))$. Therefore $A_\lambda(x)$ converges to $m(A(x))$ for all $x\in\mathrm{Dom}(A)$.

Remark. The Yosida approximation A_λ of A is an approximate selection in the following sense:

(19) $$\forall x\in\mathrm{Dom}(A),\quad (x, A_\lambda x)\in\mathrm{Graph}\, A+\lambda\,\|m(A(x))\|\, B.$$

(Compare with Theorem 1.12.1.) Indeed, $(x, A_\lambda x)=(J_\lambda x, A_\lambda x)+(x-J_\lambda x, 0)$ $\in\mathrm{Graph}\, A+(x-J_\lambda x, 0)$. Also, $\|x-J_\lambda x\|\le\lambda\|m(A(x))\|$.

Remark. We also deduce from the above theorem the following "tangential" property of $m(A)$ with respect to $\mathrm{Dom}(A)$

(20) $$\forall x\in\mathrm{Dom}(A),\quad \lim_{\lambda\to 0+}\frac{d_{\mathrm{Dom}(A)}(x-\lambda m(A(x)))}{\lambda}=0$$

Indeed,

$$d_{\mathrm{Dom}(A)}(x-\lambda m(A(x)))\le\|x-J_\lambda(x)-\lambda m(A(x))\|=\lambda\,\|A_\lambda(x)-m(A(x))\|.$$ □

Proposition 4. *Let A be a monotone (resp. maximal monotone) set-valued map from X to X. Let $\mathscr{A}$ be the set-valued map from $D(\mathscr{A})\subset L^2(0,T;X)$ to $L^2(0,T;X)$ defined by $(\mathscr{A}\,x)(t)\doteq A(x(t))$ a.e.*

Then $\mathscr{A}$ is a monotone (resp. maximal monotone) set-valued map. ▲

Proof. It is clear that $\mathscr{A}$ is monotone: if $p_i(\cdot)\in\mathscr{A}\,x_i(\cdot)$ for $i=1,2$, then

$$\int_0^T \langle p_1(t)-p_2(t), x_1(t)-x_2(t)\rangle\, dt\ge 0.$$

If A is maximal monotone, then $J_1=(1+A)^{-1}$ is Lipschitzean from X to X. Hence, if $y(\cdot)$ is given in $L^2(0,T;X)$, the function $x(\cdot)$ defined by

$$x(t)=J_1\, y(t)\quad \text{a.e.}$$

is obviously measurable. Also

$$\|x(t)-J_1(0)\|\le\|y(t)\|$$

and thus, $x(\cdot)$ belongs to $L^2(0,T;X)$ and is obviously a solution of $y(\cdot)\in(1+\mathscr{A})(x(\cdot))$. □

2. Existence and Uniqueness of Solutions to Differential Inclusions with Maximal Monotone Maps

We devote this section to the main theorem of this chapter.

Theorem 1. *Let A be a maximal monotone set-valued map from X to X and $\text{Dom}(A)$ be its domain. Consider the problem*

(1) $$x'(t) \in -A(x(t)), \quad x(0) = x_0 \in D(A).$$

Then there exists a unique solution $x(\cdot)$ defined on $[0, \infty[$, which is the slow solution:

(2) $$\text{for almost all } t, \quad x'(t) = m(A(x(t)))$$

Moreover

i) *$t \to \|x'(t)\|$ is nonincreasing.*

ii) *Let $x(\cdot)$ and $y(\cdot)$ be the solutions issued from x_0 and y_0.*

(3) *Then for $t \geq 0$,* $\|x(t) - y(t)\| \leq \|x_0 - y_0\|$.

iii) $\forall t \geq 0$, $x'(t) = \lim_{h \to 0+} \dfrac{x(t+h) - x(t)}{h}$ *and $x'(\cdot)$ is continuous from the right.* ▲

Proof. The proof is based on defining approximate solutions as solutions to ordinary differential equations when A is replaced by the Yosida approximation of A. To show the convergence of a sequence of approximate solutions we cannot use an extension to the infinite dimensional space X of the Ascoli Arzelà theorem. In fact, although the estimate (3) i) applied to the approximate solutions shows that they are an equicontinuous family of functions from an interval to X with the usual (strong) topology, the images of these approximate solutions are contained in a set that is only weakly compact. We shall instead show that they form a Cauchy sequence. Properties (3) i) and ii) have been pointed out explicitly in the statement of the theorem for their importance in this respect. Property (3) i) that would give a bound on $\|A(x(t))\|$, if we knew we had a solution $x(\cdot)$, yields the same bound on the set of approximate solutions. Property (3) ii), which shows that the solution is unique, also shows that the distance among two approximate solutions is small, i.e., that they are a Cauchy sequence.

We shall divide the proof into the following steps:

Step a: We prove (3) ii) (i.e., that the map $x_0 \to x(\cdot)$ is a contraction) and derive uniqueness.

Step b: We prove that $t \to \|x'(t)\|$ is nonincreasing.

Step c: We consider the solutions $x_\lambda(\cdot)$ to

(4) $$x'_\lambda(t) = -A_\lambda x_\lambda(t); \quad x_\lambda(0) = x_0$$

where A_λ is the Yosida approximation of A and prove that $x_\lambda(\cdot)$ is a Cauchy sequence that converges to some $x(\cdot)$ in $\mathscr{C}(0, \infty; X)$.

Step d: We prove that $x(\cdot)$ is a solution to the differential inclusion (1).

Step e: We show that $t \to \|m(A(x(t)))\|$ is nonincreasing and that $x(t)$ belongs to $\text{Dom}(A)$ for all $t \geq 0$.

Step f: We demonstrate that $t \to m(A(x(t)))$ is continuous from the right.

Step g: We conclude by establishing that $x'(t) = -m(A(x(t)))$ (i.e., that $x(\cdot)$ is the slow solution) and that $x'(\cdot) = \frac{d}{dt} x(t)$ is the derivative from the right.

a) Assume that $x(\cdot)$ and $y(\cdot)$ are solutions of the initial value problem

(5) $$\begin{aligned} &\text{i)} \ x'(t) \in -A\, x(t); \quad x(0) = x_0 \\ &\text{ii)} \ y'(t) \in -A\, y(t); \quad y(0) = y_0 \end{aligned}$$

Therefore, since A is monotone,

$$\frac{d}{dt} \frac{1}{2} \|x(t) - y(t)\|^2 = \langle x'(t) - y'(t), x(t) - y(t) \leq 0$$

and, by integrating from 0 to t, we deduce that

(6) $$\sup_{t \geq 0} \|x(t) - y(t)\| \leq \|x_0 - y_0\|.$$

In particular, when $x_0 = y_0$, this implies that the solution of (1), if any, is unique.

b) Consider now $y(t) = x(t+h)$ where $h > 0$, and where $x(\cdot)$ is a solution of (5) i). It satisfies

$$y'(t) \in -A\, y(t), \quad y(0) = x(h).$$

Therefore, (6) implies that

$$\|x(t+h) - x(t)\| \leq \|x(h) - x(0)\|.$$

By dividing by $h > 0$ and letting h tend to 0, we deduce that

(7) $$\|x'(t)\| \leq \|x'(0)\| \quad \text{for all } t \geq 0.$$

c) We consider the approximate solutions $x_\lambda(\cdot)$ as solutions to the differential equations

(4) $$x'_\lambda(t) = -A_\lambda x(t); \quad x_\lambda(0) = x_0.$$

Since A_λ is a Lipschitzean map from X into X (see Theorem 1.2) it has a unique continuously differentiable solution $x_\lambda(\cdot)$ defined on $[0, \infty[$. We shall prove that $x_\lambda(\cdot)$ is a Cauchy sequence in $\mathscr{C}(0, \infty; X)$ and check that $x_\lambda(\cdot)$ converge to the solution to the differential inclusion (1).

Since A_λ is also monotone, we deduce from (7) (with $x(\cdot)$ replaced by $x_\lambda(\cdot)$) that

$$\|A_\lambda x_\lambda(t)\| = \|x'_\lambda(t)\| \le \|x'_\lambda(0)\| = \|A_\lambda x_0\| \le \|m(A(x_0))\|. \tag{8}$$

(We use Theorem 1.2.)

We compute $\frac{1}{2}\|x_\lambda(t)-x_\mu(t)\|^2 \doteq a(t)$

$$\begin{aligned}a(t) = \tfrac{1}{2}\|x_\lambda(t)-x_\mu(t)\|^2 &= \int_0^t \frac{d}{dt}\|x_\lambda(\tau)-x_\mu(\tau)\|^2\,d\tau\\ &= -\int_0^t \langle A_\lambda x_\lambda(\tau)-A_\mu x_\mu(\tau), x_\lambda(\tau)-x_\mu(\tau)\rangle\,d\tau.\end{aligned}$$

We use now the relation $1-J_\lambda = \lambda A_\lambda$, the fact that $A_\lambda x \in AJ_\lambda(x)$ and the monotonicity of A:

$$\begin{aligned}a(t) = &-\int_0^t \langle A_\lambda x_\lambda(\tau)-A_\mu x_\mu(\tau), \lambda A_\lambda x_\lambda(\tau)-\lambda A_\mu x_\mu(\tau)\rangle\,d\tau\\ &-\int_0^t \langle A_\lambda x_\lambda(\tau)-A_\mu x_\mu(\tau), J_\lambda x_\lambda(\tau)-J_\mu x_\mu(\tau)\rangle\,d\tau\\ \le &-\int_0^t \langle A_\lambda x_\lambda(\tau)-A_\mu x_\mu(\tau), \lambda A_\lambda x_\lambda(\tau)-\mu A_\mu x_\mu(\tau)\rangle\,d\tau \qquad (9)\\ = &\int_0^t \lambda\langle A_\lambda x_\lambda(\tau), A_\mu x_\mu(\tau)\rangle\,d\tau + \int_0^t \mu\langle A_\mu x_\mu(\tau), A_\lambda x_\lambda(\tau)\rangle\,d\tau\\ &-\int_0^t (\lambda\|A_\lambda x_\lambda(\tau)\|^2 + \mu\|A_\mu x_\mu(\tau)\|^2)\,d\tau.\end{aligned}$$

We note that

$$\begin{aligned}\lambda\langle A_\lambda x_\lambda(\tau), A_\mu x_\mu(\tau)\rangle &\le \lambda\|A_\lambda x_\lambda(\tau)\|\,\|A_\mu x_\mu(\tau)\|\\ &\le \lambda\|A_\lambda x_\lambda(\tau)\|^2 + \frac{\lambda}{4}\|A_\mu x_\mu(\tau)\|^2\end{aligned}$$

and, in the same way, that

$$\mu\langle A_\mu x_\mu(\tau), A_\lambda x_\lambda(\tau)\rangle \le \mu\|A_\mu x_\mu(\tau)\|^2 + \frac{\mu}{4}\|A_\lambda x_\lambda(\tau)\|^2.$$

Therefore, using this and (9), we obtain

$$\begin{aligned}a(t) &\le \frac{1}{4}\int_0^t (\lambda\|A_\mu x_\mu(\tau)\|^2 + \mu\|A_\lambda x_\lambda(\tau)\|^2)\,d\tau\\ &\le t(\lambda+\mu)/4 \quad \|m(A(x_0))\|^2.\end{aligned}$$

Hence $x_\lambda(\cdot)$ is a Cauchy sequence in $\mathscr{C}(0,\infty;X)$ and thus, converges to some continuous function $x(\cdot)$ uniformly over compact intervals.

The inequality

$$\|x_\lambda(t)-J_\lambda x_\lambda(t)\| = \lambda\|A_\lambda x_\lambda(t)\| \le \lambda\|m(A(x_0))\| \tag{10}$$

yields that $J_\lambda x_\lambda(\cdot)$ converges to $x(\cdot)$ uniformly on compact intervals.

d) We note that $x'_\lambda(\cdot)$ remains in a bounded set of $L^\infty(0,\infty;X)$ and thus, that a subsequence $x'_{\lambda_n}(\cdot)$ converges weakly to some function, which is equal almost everywhere to $x'(\cdot)$. So, for any $T>0$, $x'_{\lambda_n}(\cdot)$ converges weakly to $x'(\cdot)$ in $L^2(0,T;X)$ and $x_{\lambda_n}(\cdot)$ converges strongly to $x(\cdot)$ in $L^2(0,T;X)$. Proposition 1.4 implies that the map $x(\cdot)\to(\mathscr{A}x)(\cdot)=A(x(\cdot))$ is maximal monotone in $L^2(0,T;X)$; we deduce that $x'(\cdot)\in -A(x(\cdot))$ in $L^2(0,T;X)$ by Proposition 1.2. Hence $x'(t)$ belongs to $-Ax(t)$ almost everywhere.

e) Let us prove that $t\to\|m(A(x(t)))\|$ is non increasing.

Since $\|A_\lambda x_\lambda(t)\|\le\|m(A(x_0))\|$, there exists a subsequence $A_{\lambda_n}x_{\lambda_n}(t)$ that converges weakly in X to some $v(t)$. But $A_{\lambda_n}x_{\lambda_n}(t)\in A(J_{\lambda_n}x_{\lambda_n}(t))$; therefore, we deduce from Proposition 1.2 that $v(t)\in A(x(t))$ (This implies in particular that $x(t)\in\operatorname{Dom}(A)$ for all $t\ge 0$).

Let $t>t_0$. Since

$$\|x'_\lambda(t)\|=\|A_\lambda x_\lambda(t)\|\le\|m(A(x(t_0)))\|$$

we deduce that

$$\|v(t)\|\le\liminf_{\lambda\to 0}\|x'_\lambda(t)\|\le\|m(A(x(t_0)))\| \tag{11}$$

and that the solutions $x_\lambda(\cdot)$ are uniformly Lipschitzean. Hence $x(\cdot)$ is also Lipschitzean. Moreover

$$\|m(A(x(t)))\|\le\|v(t)\|\le\|m(A(x(t_0)))\|. \tag{12}$$

Then $t\to\|m(A(x(t)))\|$ is nonincreasing.

f) Let us check that $t\to m(A(x(t)))$ is continuous from the right. Let $t_n>t_0$ converge to t_0; then $x(t_n)$ converges to $x(t_0)$ and, since

$$\|m(A(x(t_n)))\|\le\|v(t_n)\|\le\|m(Ax(t_0))\|,$$

by (12) we deduce that some subsequence (again denoted) $m(A(x(t_n)))$ converges weakly to some y in X.

Proposition 1.2 implies that $y\in A(x(t_0))$. But

$$\|y\|\le\liminf_{t_n\to\infty}\|m(A(x(t_n)))\|\le\|m(A(x(t_0)))\|.$$

Hence $y=m(A(x(t_0)))$ and $m(A(x(t_0)))$ is the weak limit of the sequence $m(A(x(t_n)))$. Since

$$\|m(A(x(t_0)))\|=\lim_{n\to\infty}\|m(A(x(t_n)))\|,$$

we deduce that $m(A(x(t_n)))$ converges strongly to $m(A(x(t_0)))$ when $t_n\to t_{0+}$. Hence $t\to m(A(x(t)))$ is continuous from the right.

g) Let N be the subset of $[0,\infty[$ where either $x(\cdot)$ is not differentiable or $x'(t)\notin -A(x(t))$. Let $t_0\notin N$.

Inequality

$$\|x(t_0+h)-x(t_0)\|\le\int_{t_0}^{t_0+h}\|x'(\tau)\|\,d\tau\le\int_{t_0}^{t_0+h}\|m(A(x(t_0)))\|\,d\tau$$
$$=h\,\|m(A(x(t_0)))\|\quad\text{(By (9))}$$

implies that

$$\|x'(t_0)\| = \lim_{h\to 0+}\left\|\frac{x(t_0+h)-x(t_0)}{h}\right\| \le \|m(A(x(t_0)))\|.$$

Since $x'(t_0)\in -A(x(t_0))$, we deduce that $x'(t_0) = -m(A(x(t_0)))$. Integrating from t_0 to t_0+h, we deduce that

$$\frac{x(t_0+h)-x(t_0)}{h} = -\frac{1}{h}\int_{t_0}^{t_0+h} m(A(x(\tau)))\,d\tau.$$

Since $m(A(x(\cdot)))$ is continuous from the right, we deduce that $\frac{d}{dt}x(t_0) = -m(A(x(t_0)))$. □

3. Asymptotic Behavior of Trajectories and the Ergodic Theorem

It is quite convenient to begin by introducing the concept of semigroup associated to differential inclusions with maximal monotone maps.

Semi-Groups of Nonexpansive Maps

Let A be a maximal monotone set-valued map from X to X.

We set

(1) $$\forall x\in \mathrm{Dom}(A), \quad S(t)x \doteq x(t)$$

where $x(\cdot)$ is the unique solution to the differential inclusion

(2) $$x'(t)\in -A(x(t)), \quad x(0)=x.$$

So, the maps $S(t)$ satisfy, by Theorem 2.1, the following properties

(3) $$\begin{aligned}&\text{i)}\ S(0)=I,\ \forall t,s\ge 0,\ S(t+s)=S(t)S(s),\\ &\text{ii)}\ \forall t\ge 0,\ \forall x,y\in D(A),\ \|S(t)x-S(t)y\|\le\|x-y\|.\end{aligned}$$

$S(t)$ is the "transition map", which carries the initial value x to $x(t) = S(t)x$, the value of the corresponding solution at time $t>0$. We observe that $S(t)$ can be extended to a non-expansive map $\bar{S}(t)$ from the closure $\overline{\mathrm{Dom}A}$ of $\mathrm{Dom}A$ to $\overline{\mathrm{Dom}A}$, since non-expansive maps are uniformly continuous. We shall identify $\bar{S}(t)$ with $S(t)$ and, from now on, assume that $\{S(t)\}_{t\ge 0}$ is a *semi-group of non-expansive maps* $S(t)$ from the closed convex subset $\overline{\mathrm{Dom}(A)}$ to itself.

A point x_* is a *stationary* point of a semi-group of maps $S(t)$ if and only if:

(4) $$\forall t\ge 0, \quad S(t)x_* = x_*.$$

It is clear that $x_* \in \operatorname{Dom} A$ is an equilibrium of A if and only if it is a stationary point of the semi-group $\{S(t)\}_{t>0}$.

We shall now devote some attention to the behavior of $S(t)x$ when $t \to \infty$ as well as to the behavior of "averages" $\sigma(T)x$ of $S(t)x$. For instance, the usual "time averages" are the Cesaro means

$$\sigma(T)x = \frac{1}{T}\int_0^T S(t)x\,dt. \tag{5}$$

More generally, we introduce the following definition.

Definition 1. We denote by D either $\mathbb{N}$ or R_+ and by K a closed convex subset of X. A family of maps $\{S(t)\}_{t\in D}$ from K to K is called a *semi-group of non-expansive maps* if

$$\begin{cases} \text{i)}\ S(0)=I,\ \forall t,s\in D,\ S(t+s)=S(t)S(s), \\ \text{ii)}\ \forall t\in D,\ \forall x,y\in K,\ \|S(t)x-S(t)y\| \le \|x-y\|. \end{cases} \tag{6}$$ ▲

If x_* is a stationary point of the semi-group, the function $t \to \|S(t)x - x_*\|$ is non-increasing: Indeed, when $t \ge s$, we have

$$\|S(t)x - x_*\| = \|S(t-s)S(s)x - S(t-s)x_*\| \le \|S(s)x - x_*\|. \tag{7}$$

Therefore, if there exists at least one stationary point x_*, the function $t \to \|S(t)x - x_*\|^2$ has a limit as $t \to \infty$: Hence

$$t \to S(t)x \text{ is bounded and thus it has an asymptotic center } x_\infty. \tag{8}$$

(See Definition 0.4.1).

Proposition 1. *If all the weak cluster point of $S(t)x$ are stationary points, then $S(t)x$ converges weakly to its asymptotic center as $t \to \infty$.* ▲

Proof. Let N be the set $\{y \in X$ such that $\lim_{t\to\infty} \|S(t)x - y\|$ exists$\}$ and F denote the set of stationary points x_* of the semi-group; property (7) implies that

$$F \subset N. \tag{9}$$

Therefore, Proposition 1 follows from Proposition 0.4.1. □

The following statement provides a sufficient condition for cluster points to be stationary points.

Proposition 2. *Let S be a non-expansive map from a closed convex subset K to itself. Assume that a sequence of elements $x_T \in K$ satisfies the condition*

$$\lim_{T\to\infty} \|x_T - Sx_T\| = 0. \tag{10}$$

Then any weak cluster point of this sequence is a fixed point of S. ▲

Proof. Let x_* be the weak limit of a generalized subsequence $\{x_{T'}\}_{T'}$ of the sequence $\{x_T\}_T$ and let us set $x_\lambda \doteq (1-\lambda)x_* + \lambda S(x_*)$, where $\lambda \in]0,1]$. Since S is non-expansive, we deduce that

$$\begin{aligned}&\langle x_\lambda - S x_\lambda - (x_{T'} - S x_{T'}), x_\lambda - x_{T'}\rangle \\ &\quad \geq \|x_\lambda - x_{T'}\|^2 - \|S x_\lambda - S x_{T'}\| \, \|x_\lambda - x_{T'}\| \geq 0.\end{aligned}$$

By letting $x_{T'}$ converge to x_*, we deduce from this inequality that

$$\langle x_\lambda - S x_\lambda, x_\lambda - x_* \rangle \geq 0$$

for $x_{T'} - S x_{T'}$ converge *strongly* to 0 by assumption (10). This latter inequality can be written, after division by $\lambda > 0$,

$$\langle (1-\lambda)x_* + \lambda S(x_*) - S[(1-\lambda)x_* + \lambda S(x_*)], S(x_*) - x_* \rangle \geq 0.$$

By letting λ converge to 0, we deduce that $\|S(x_*) - x_*\|^2 \leq 0$, i.e., that $x_* = S(x_*)$. □

Using this proposition, we need mechanisms yielding sequences of elements x_T that satisfy the condition

$$\text{(11)} \qquad \forall t \geq 0, \quad \lim_{T\to\infty} \|x_T - S(t)x_T\| = 0$$

for proving the weak convergence of x_T to some stationary point.

We turn to the Ergodic Theorem for $S(t)x$, that is the weak convergence of averages of $S(t)x$. The averages usually chosen are the Cesaro means

$$\text{(12)} \qquad \frac{1}{T}\sum_{t=0}^{T-1} S(t)x \quad \text{or} \quad \frac{1}{T}\int_0^T S(t)x\,dt$$

according to whether $D = \mathbb{N}$ or $D = R_+$. If we set

$$Q(T,t) \doteq \begin{cases} \dfrac{1}{T} & \text{when } 0 \leq t \leq T \\ 0 & \text{when } t > T \end{cases}$$

the Cesaro means can be written $\int_D Q(T,t)S(t)x\,d\mu(t)$ when μ is either the measure defined by $\int_D \phi\,d\mu = \sum_{t=1}^{\infty} \phi_t$ when $D = \mathbb{N}$ or the Lebesgue measure when $D = R_+$.

More generally, we consider a function $Q\colon D \times D \to R_+$ satisfying the following properties

$$\text{(13)} \qquad \begin{aligned}&\text{i) } \forall T \in D, \; s \to Q(T,s) \text{ is } \mu\text{-measurable,} \\ &\text{ii) } \forall T \in D, \; \int_D Q(T,s)\,d\mu(s) = 1.\end{aligned}$$

Assuming that $t \to S(t)x$ is strongly μ-measurable, we define the Q-average $\sigma_Q(T)x$ by

(14) $$\sigma_Q(T)x = \int_D Q(T,t)\,S(t)\,x\,d\mu(t).$$

Propositions 1 and 2 imply the following consequence.

Lemma 1. *Let us assume that there exists at least one stationary point of the semi-group and that*

(15) $$\forall t \geq 0, \quad \lim_{T\to\infty} \|\sigma_Q(T)x - S(t)\,\sigma_Q(T)x\| = 0.$$

When $T\to\infty$, $\sigma_Q(T)x$ converges weakly to the asymptotic center of $t\to S(t)x$, which is a stationary point of the semi-group. ▲

Proof. For simplicity, we set $y_T = \sigma_Q(T)x$ and $x_T = S(T)x$.

Proposition 2 implies that all the weak cluster points of the sequence $\{y_T\}$ are stationary points of the semi-group S; By (7), these weak cluster points belong to the subset

$$N \doteq \{y\in X \mid \lim_{t\to\infty} \|S(t)x - y\| \text{ exists}\}.$$

Property (13) imply that these weak cluster points belong to the closed convex hull C of the weak cluster points of $\{x_T\}$. So, the weak cluster points of $\{y_T\}_T$ are contained in $N\cap C$ and our lemma ensues, thanks to Proposition 0.4.4. □

To prove the Ergodic Theorem it is left to show that condition (15) is satisfied. For that purpose, we need the following result.

Lemma 2. *Set $y_T \doteq \sigma_Q(T)x$. Then*

(16) $$\begin{aligned}\|S(t)\,y_T - y_T\|^2 \leq & \int_{D\cap[0,t]} Q(T,s)\,\|S(s)x - S(t)\,y_T\|^2\,d\mu(s)\\ & + \int_D |Q(T,s+t) - Q(T,s)|\,\|S(s)x - y_T\|^2\,d\mu(s).\end{aligned}$$ ▲

Proof. Since $\|S(t+s)x - S(t)y\| \leq \|S(s)x - y\|$, we deduce that

$$\begin{aligned}0 &\leq \|S(s)x - S(t)y + S(t)y - y\|^2 - \|S(t+s)x - S(t)y\|^2\\ &\leq \|S(s)x - S(t)y\|^2 + \|S(t)y - y\|^2 - \|S(t+s)x - S(t)y\|^2\\ &\quad + 2\langle S(s)x - S(t)y, S(t)y - y\rangle.\end{aligned}$$

We multiply by $Q(T,s)\geq 0$ and we integrate on D with respect to s; we obtain by using property (13),

$$\begin{aligned}0 \leq & \int_D Q(T,s)(\|S(s)x - S(t)y\|^2 - \|S(t+s)x - S(t)y\|^2)\,d\mu(s)\\ & + \|S(t)y - y\|^2 + 2\langle y_T - S(t)y, S(t)y - y\rangle.\end{aligned}$$

We now take $y = y_T$. The above inequality becomes

$$\|S(t)\,y_T - y_T\|^2 \leq \int_D Q(T,s)(\|S(s)x - S(t)\,y_T\|^2 - \|S(t+s)x - S(t)\,y_T\|^2)\,d\mu(s).$$

Since

$$\int_D Q(T,s)\,\|S(s)x-S(t)y_T\|^2\,d\mu(s) = \int_{D\cap[0,t]} Q(T,s)\,\|S(s)x-S(t)y_T\|^2\,d\mu(s) + \int_D Q(T,s+t)\,\|S(t+s)x-S(t)y_T\|^2\,d\mu(s),$$

the above inequality can be written

$$\|S(t)y_T-y_T\|^2 \le \int_{D\cap[0,t]} Q(T,s)\,\|S(s)x-S(t)y_T\|^2\,d\mu(s) + \int_D |Q(T,s)-Q(T,s+t)|\,\|S(t+s)x-S(t)y_T\|^2\,d\mu(s).$$

We estimate the last integral by using

$$\|S(t+s)x-S(t)y_T\|^2 \le \|S(s)x-y_T\|^2$$

Hence inequality (16) ensues. □

Theorem 1. *Let us consider a μ-measurable semi-group of non-expansive maps $S(t)$: $K\to K$ which has at least one stationary point. Let Q be any function from $D\times D$ to R_+, measurable with respect to the second variable, of bounded variation, satisfying*

$$\text{(17)}\quad \begin{aligned} &\text{i) } \forall T>0,\ \int_D Q(T,s)\,d\mu(s)=1,\\ &\text{ii) } \forall t\in D,\ \lim_{T\to\infty} \int_{D\cap[0,t]} Q(T,s)\,d\mu(s)=0,\\ &\text{iii) } \forall t\in D,\ \lim_{T\to\infty} \int_D |Q(T,s+t)-Q(T,s)|\,d\mu(s)=0. \end{aligned}$$

Then the average $\sigma_Q(T)x$ converge weakly as $T\to\infty$ to the asymptotic center of $t\to S(t)x$ which is a stationary point of the semi-group. ▲

Proof. By Proposition 2, we have to check that

$$\text{(18)}\qquad \forall t\ge 0,\quad \lim_{T\to\infty} \|S(t)y_T-y_T\|^2=0.$$

For that purpose, we use Lemma 2. We observe that if $x_*\in K$ is a stationary point of the semi-group, we can estimate $\|S(s)y-z\|=\|S(s)y-S(s)x_*+x_*-z\|$ by $\|S(s)y-S(s)x_*\|+\|x_*-z\|\le\|x_*-y\|+\|x_*-z\|$. We observe also that $\|x_*-y_T\|=\|\int_D Q(T,s)(S(s)x_*-S(s)x)\,d\mu(s))\|\le\|x_*-x\|$. By taking $y=x$ and $z=S(t)y_T$, we deduce that $\|S(s)x-S(t)y_T\|\le\|x_*-x\|+\|x_*-S(t)y_T\|\le\|x_*-x\|+\|x_*-y_T\|\le 2\|x_*-x\|$ and by taking $y=x$ and $z=y_T$, we deduce that $\|S(s)x-y_T\|\le 2\|x_*-x\|$. Hence Lemma 2 implies that

$$\|S(t)y_T-y_T\|^2 \le 4\|x_*-x\|^2 \Big(\int_{D\cap[0,t]} Q(T,s)\,d\mu(s) + \int_D |Q(T,t+s)-Q(T,s)|\,d\mu(s)\Big).$$

The right-hand side of this inequality converges to 0 as $T \to \infty$ by assumption (17) i) and ii).

Corollary 1. *Let us consider a semi-group of non-expansive maps $S(t)\colon K \to K$ which has at least one stationary point. Then*

(19) $$\forall x \in K, \quad \frac{1}{T} \sum_{t=0}^{T-1} S(t)\,x$$

converges weakly to a stationary point.

If the semi-group $S(\cdot)$ is measurable, then

(20) $$\forall x \in K, \quad \frac{1}{T} \int_0^T S(\tau)\,x\,d\tau$$

converges weakly to a stationary point. ▲

An Ergodic Theorem for Products of Resolvents

Let A be a maximal monotone set-valued map. For $\lambda > 0$, we consider its resolvent $J_\lambda \doteq (1 + \lambda A)^{-1}$, which is a nonexpansive map from X to X.

Given $x_0 \in X$ and a sequence of positive numbers λ_n, we define the sequence of elements $x_n \in X$ iteratively by

(21) $$x_n \doteq J_{\lambda_{n-1}} x_{n-1} \qquad (n = 1, 2, \ldots)$$

and we associate with this sequence the sequence of averages $y_n \in X$ defined by

(22) $$y_n \doteq \left(\sum_{k=0}^{n} \lambda_k \right)^{-1} \sum_{k=0}^{n} \lambda_k x_k \qquad (n = 1, 2, \ldots).$$

We shall prove that this sequence of averages of products of resolvents converges weakly to a stationary point of A.

Theorem 2. *Let A be a maximal monotone set-valued map that has at least one equilibrium. Assume that $\sum_{k=0}^{\infty} \lambda_k = \infty$. Then the sequence of averages y_n defined by (22) converges weakly to the asymptotic center of the sequence of elements x_n defined by (21), which is an equilibrium of A.* ▲

Proof. Since there exists an equilibrium x_* of A, the sequence of elements x_n is bounded because

(23) $$\|x_n - x_*\| = \|J_{\lambda_{n-1}} x_{n-1} - J_{\lambda_{n-1}} x_*\| \leq \|x_{n-1} - x_*\| \leq \ldots \leq \|x_0 - x_*\|.$$

We denote by C the closed convex hull of the weak cluster points of $\{x_n\}_n$ and by N the subset of elements y such that $\lim_{n \to \infty} \|x_n - y\|^2$ exists. Note that the subset $A^{-1}(0)$ of equilibria is contained in N:

$$A^{-1}(0) \subset N.$$

Indeed, inequality (23) shows that the sequence of non-negative numbers $\|x_n - x_*\|^2$ is non-increasing, and thus that it converges.

We shall apply Proposition 0.4.4 to prove the theorem, i.e., show that the weak cluster points y_* of the (bounded) sequence $\{y_n\}_n$ are contained in $C \cap N$. (In this case, we know that y_n converges weakly to the asymptotic center of $\{x_n\}_n$.) First, it is clear that $y_* \in C$, since y_n is a convex combination of elements x_k, $k \leq n$. Since $A^{-1}(0) \subset N$, it suffices to show that $y_* \in A^{-1}(0)$. By doing so, we prove also that $x_\infty = y_*$ belongs to $A^{-1}(0)$. Let $(y, v) \in \operatorname{graph}(A)$. From the definition (21) of x_n, we have for all n,

$$\langle x_{n-1} - x_n, x_n - y \rangle \geq \lambda_n \langle v, x_n - y \rangle \tag{24}$$

and, thus,

$$\|x_n - y\|^2 + 2\lambda_n \langle v, x_n - y \rangle + \|x_n - x_{n-1}\|^2 \leq \|x_{n-1} - y\|^2. \tag{25}$$

Consequently,

$$2\lambda_n \langle v, x_n - y \rangle \leq \|x_{n-1} - y\|^2 - \|x_n - y\|^2$$

which implies

$$2\langle v, y_n - y \rangle \leq \left(\sum_{k=0}^{n} \lambda_k \right)^{-1} \|x_0 - y\|^2. \tag{26}$$

Let y_* be a weak cluster point, the weak limit of a subsequence $y_{n'}$. Passing to the limit through the subsequence $y_{n'}$, inequality (26) yields, recalling that $\sum_{k=0}^{\infty} \lambda_k = \infty$,

$$\langle v, y - y_* \rangle \geq 0 \quad \text{for all } (y, v) \in \operatorname{graph}(A). \tag{27}$$

Since A is maximal monotone, this proves that $(y_*, 0)$ belongs to graph (A), i.e., that $y_* \in A^{-1}(0)$. □

If we assume that $\sum_{k=0}^{\infty} \lambda_k^2 = +\infty$, we can prove that the sequence $\{x_n\}_n$ itself converges weakly.

Theorem 3. *Let A be a maximal monotone set-valued map that has at least one equilibrium. Assume that $\sum_{k=0}^{\infty} \lambda_k^2 = \infty$. Then the sequence of elements x_n defined by (21) converges weakly to an equilibrium of A as $n \to \infty$.* ▲

The particular case where $\lambda_n = \lambda > 0$ is worth mentioning.

Corollary 2. *Let A be a maximal monotone set-valued map that has at least one equilibrium and $\lambda > 0$. Given $x_0 \in X$, the sequence of elements $J_\lambda^n x_0$ converges weakly to an equilibrium of A as $n \to \infty$.* ▲

Proof of Theorem 3. We deduce it from Proposition 0.4.3. So, we have to prove that the weak cluster points x_* of the sequence x_n belong to the set N defined in the proof of Theorem 2. Since $A^{-1}(0) \subset N$, we shall prove that $x_* \in A^{-1}(0)$. This implies also that the weak limit of $\{x_n\}_n$ is an equilibrium of A.

From the definition of x_n it follows that $z_n \doteq \lambda_n^{-1}(x_{n-1} - x_n)$ belongs to $A(x_n)$. Since A is monotone, we deduce that

$$0 \leq \langle z_{n+1} - z_n, x_{n+1} - x_n \rangle = -\lambda_{n+1}^{-1} \langle z_{n+1} - z_n, z_{n+1} \rangle.$$

Therefore, the sequence of nonnegative numbers $\|z_n\|^2$ is non-increasing. Let $\bar{x} \in A^{-1}(0)$ be an equilibrium. Taking $y \doteq \bar{x}$ and $v \doteq 0$ in inequality (25) yields

(28) $$\lambda_n^2 \|z_n\|^2 \leq \|x_{n-1} - \bar{x}\|^2 - \|x_n - \bar{x}\|^2.$$

Therefore, by summing over n, we obtain

(29) $$\|z_n\|^2 \sum_{k=0}^{n} \lambda_k^2 \leq \sum_{k=0}^{n} \lambda_k^2 \|z_k\|^2 \leq \|x_0 - \bar{x}\|^2.$$

Since $\sum_{k=0}^{\infty} \lambda_k^2 = \infty$, we deduce that $\|z_n\| \to 0$ as $n \to \infty$. Let x_* be a weak cluster point of $\{x_n\}_n$, the weak limit of a subsequence $x_{n'}$. From the monotonicity of A, we have

(30) $$\forall (y, v) \in \operatorname{graph}(A), \quad \langle v - z_{n'}, y - x_{n'} \rangle \geq 0$$

and, passing to the limit $n' \to \infty$, we deduce that

(31) $$\forall (y, v) \in \operatorname{graph}(A), \quad \langle v, y - x_* \rangle \geq 0$$

which, by the maximality of A, implies that $x_* \in A^{-1}(0)$. □

4. Gradient Inclusions

Equations of the form

(1) $$x'(t) = -\nabla V(x(t)), \quad x(0) = x_0,$$

sometimes called "gradient equations", have been extensively studied for their importance in mechanics and stability theory. It is classically assumed that $V: \Omega \to R$ is a differentiable, and often convex, function defined on some open subset Ω of X. In this section we extend these results to the case where $V: X \to R \cup \{+\infty\}$ is only a lower semicontinuous convex function whose domain $\operatorname{Dom} V \doteq \{x \in X \mid V(x) < +\infty\}$ is non-empty, by replacing the above gradient equation by a "gradient inclusion"

(2) $$x'(t) \in -\partial V(x(t)), \quad x(0) = x_0.$$

Indeed, we shall prove that the set-valued map $x \to \partial V(x)$ is maximal monotone. So, existence and uniqueness of a solution to a gradient inclusion follows from Theorem 2.1. Solutions to gradient inclusions enjoy further properties.

a) They verify equality $\frac{d}{dt} V(x(t)) + \|x'(t)\|^2 = 0$.

b) If V achieves its minimum at some point, then $x(t)$ converges to a point that minimizes V when $t \to \infty$.

c) The solution to a gradient inclusion minimizes a functional on a certain space (variational principle).

Proposition 1. *Let V be a proper lower semicontinuous convex function from a Hilbert space X to $R \cup \{+\infty\}$. Then the set-valued map $x \to \partial V(x)$ is maximal monotone.* ▲

Proof. Since V is convex, the map $\partial V(\cdot)$ is monotone because if $p \in \partial V(x)$ and $q \in \partial V(y)$, inequalities $V(x) - V(y) \le \langle p, x-y \rangle$ and $V(y) - V(x) \le \langle q, y-x \rangle$ imply that $\langle p-q, x-y \rangle \ge 0$.

Theorem 0.7.3 implies that $1 + \partial V(\cdot)$ is surjective. Hence Theorem 1.1 states that $\partial V(\cdot)$ is maximal monotone. □

Hence we obtain the following corollary.

Theorem 1. *Let V be a proper lower semicontinuous convex function from a Hilbert space X to $R \cup \{+\infty\}$. For any $x_0 \in \mathrm{Dom}(\partial V)$, there exists a unique solution to the differential inclusion*

$$x'(t) \in -\partial V(x(t)), \quad x(0) = x_0. \tag{2}$$

Furthermore, $t \to V(x(t))$ is a convex non-increasing function that satisfies

$$\frac{d}{dt} V(x(t)) + \|x'(t)\|^2 = 0. \tag{3}$$

▲

Proof. Since $x'(t) \in -\partial V(x(t))$, we deduce that, for $h > 0$

$$V(x(t)) - V(x(t+h)) \le -\langle x'(t), x(t) - x(t+h) \rangle$$

and

$$V(x(t+h)) - V(x(t)) \le -\langle x'(t+h), x(t+h) - x(t) \rangle.$$

We recall that $t \to x'(t) = m(\partial V(x(t))$ is continuous from the right. (See Theorem 2.1.) By dividing by $h > 0$ the two sides of the second inequality, we obtain equality (3).

Then $\frac{d}{dt} V(x(t)) = -\|x'(t)\|^2$ is non-increasing by Theorem 2.1. Hence $t \to V(x(t))$ is convex and non-increasing. □

If V achieves its minimum, the trajectory $x(t)$ converges to such a minimizer when t goes to ∞.

Theorem 2. *Let V be a proper lower semicontinuous convex function from a Hilbert space X to $R \cup \{+\infty\}$. Assume that V achieves its minimum at some point. Then, for all x_0 in $\mathrm{Dom}(\partial V)$, the trajectory of the gradient inclusion* (2) *converges weakly to a point which minimizes V when $t \to \infty$.* ▲

Proof. We apply Proposition 3.1 to the semi-group of non-expansive maps defined by $S(t)x_0 \doteq x(t)$, where $x(t)$ is the solution to the gradient inclusion (2). We assumed that there exists a point achieving the minimum of V, which is a stationary point of the semi-group. We also know that $\sup_{t \geq 0} \|x'(t)\|$ is finite. It remains to check that all the weak cluster points x_* of $x(t)$ when $t \to \infty$ achieve the minimum of V.

Integrating inequality (3), we deduce that

$$V(x(t)) - V(x(s)) + \int_s^t \|x'(\tau)\|^2 \, d\tau \leq 0 \tag{4}$$

and thus, that $\lim_{t,s \to \infty} \int_s^t \|x'(\tau)\|^2 \, d\tau \leq \lim_{t,s \to \infty} (V(x(s)) - V(x(t)) = 0$. Hence the Cauchy criterion implies that

$$\int_0^\infty \|x'(\tau)\|^2 \, d\tau = \lim_{t \to \infty} \int_0^t \|x'(\tau)\|^2 \, d\tau < +\infty. \tag{5}$$

Let us denote by A_ε the subset of those $t > 0$ such that $\|x'(t)\| < \varepsilon$. Then $\mathrm{meas}\,(A_\varepsilon) = \infty$.

Otherwise, the measure of A_ε would be finite and the measure of its complement $\complement A_\varepsilon$ would be infinite. Thus,

$$\int_0^\infty \|x'(t)\|^2 \, dt \geq \varepsilon^2 \, \mathrm{meas}(\complement A_\varepsilon) = \infty \tag{6}$$

which is a contradiction of the property (5). Hence, for any $t \in A_\varepsilon$, we deduce from the inclusion $-x'(t) \in \partial V(x(t))$ that, for all $y \in X$,

$$\begin{aligned} \inf_{t>0} V(x(t)) &\leq V(x(t)) \leq V(y) + \langle -x'(t), x(t) - y \rangle \\ &\leq V(y) + \|x'(t)\| \, \|x(t) - y\| \leq V(y) + \varepsilon M \\ &\text{where } M \doteq \sup_{t>0} \|x(t) - y\|. \end{aligned}$$

Therefore, by letting $\varepsilon \to 0$, we obtain $\inf_{t \geq 0} V(x(t)) = \inf_{y \in X} V(y)$.

Therefore, because $t \to V(x(t))$ is non-increasing, any cluster point x_* satisfies

$$V(x_*) \leq \lim_{t_n \to \infty} V(x(t_n)) = \inf_{t \geq 0} V(x(t)) = \inf_{y \in X} V(y). \tag{7}$$

Proposition 3.1 states that $x(t)$ converges weakly to its asymptotic center, which thus, achieves the minimum of V. □

Variational Principle

Let $T<+\infty$ and $\mathscr{K}$ be the subset of functions defined by

(8) $$\mathscr{K} \doteq \{y \in L^2(0,T;X) \text{ such that } y' \in L^2(0,T;X) \text{ and } y(0)=x_0\}$$

We introduce the functional ϕ defined on $\mathscr{K}$ by

(9) $$\phi(y)=\int_0^T [V(y(t))+V^*(-y'(t))]\,dt+1/2\,\|y(T)\|^2-1/2\,\|x_0\|^2$$

where V^* is the conjugate function defined in Section 0.7 (Definition 0.7.1).

The functional ϕ *is nonnegative on* $\mathscr{K}$: indeed, since $\frac{d}{dt}\|y(t)\|^2 = 2\langle y'(t), y(t)\rangle$,

(10) $$\phi(y) \geq \int_0^T \langle -y(t), y'(t)\rangle\,dt+1/2\,\|y(T)\|^2-1/2\,\|x_0\|^2=0.$$

We shall prove that the solutions to the gradient inclusion (2) achieve the minimum of ϕ on $\mathscr{K}$.

Theorem 3. *Let $V: X \to R \cup \{+\infty\}$ be a lower semicontinuous convex function with non-empty domain. A function $x \in \mathscr{K}$ is a solution to the gradient inclusion $x'(t) \in \partial V(x(t))$ if and only if*

(11) $$\phi(x)=0\,(=\min_{y \in \mathscr{K}} \phi(y)).$$ ▲

Proof. a) If $x \in \mathscr{K}$ is a solution, then inequality (10) becomes an equality, since $V(x(t))+V^*(-x'(t))=\langle -x'(t), x(t)\rangle$ for almost all t.

b) *Conversely*, if $\phi(x)=0$, we obtain

$$\phi(x)=\int_0^T [V(x(t))+V^*(-x'(t))-\langle -x'(t), x(t)\rangle]\,dt=0.$$

Since $V(x(t))+V^*(-x'(t))-\langle -x'(t), x(t)\rangle \geq 0$ a.e., we deduce that for almost all t,

$$V(x(t))+V^*(-x'(t))=\langle -x'(t), x(t)\rangle,$$

i.e., that $x'(t) \in \partial V(x(t))$. □

Yosida Approximation of the Subdifferential

We associate with a proper lower semicontinuous convex function V from X to $R \cup \{+\infty\}$ the function V_λ defined by

$$\forall \lambda > 0, \quad V_\lambda(x) \doteq \inf_{y \in X} \left(V(y) + \frac{1}{2\lambda} \|y - x\|^2 \right) \tag{12}$$

Theorem 4. *The functions V_λ are differentiable convex functions from X to R approximating V in the sense that $\forall x \in \operatorname{Dom} V$, $V(x) = \lim_{\lambda \to 0} V_\lambda(x)$.*

The gradient $\nabla V_\lambda(x)$ is the Yosida approximation of $\partial V(x)$; the resolvent $J_\lambda(x) = (1 + \lambda \partial V)^{-1}(x)$ is the unique solution of the minimization problem (12) defining $V_\lambda(x)$. ▲

Proof. The functions V_λ are obviously convex. In Theorem 0.6.1, we have proved that the solution x_λ to the inclusion $x \in x_\lambda + \lambda \partial V(x_\lambda)$, is the unique minimum of problem (12). Then $x_\lambda \doteq J_\lambda x$ and $\frac{1}{\lambda}(1 - J_\lambda)(x) = \frac{1}{\lambda}(x - x_\lambda)$ defines the Yosida approximation of $\partial V(\cdot)$.

We prove first that

$$\begin{aligned}
V_\lambda(x) - V_\lambda(y) &= V(x_\lambda) - V(y_\lambda) + \frac{1}{2\lambda} \|x_\lambda - x\|^2 - \frac{1}{2\lambda} \|y_\lambda - y\|^2 \\
&\leq \left\langle \frac{1}{\lambda}(x - x_\lambda), x_\lambda - y_\lambda \right\rangle + \frac{1}{2\lambda} \|x_\lambda - x\|^2 - \frac{1}{2\lambda} \|y_\lambda - y\|^2 \\
&\leq \left\langle \frac{1}{\lambda}(x - x_\lambda), x - y \right\rangle + \frac{1}{\lambda} \langle (x - x_\lambda), x_\lambda - x - (y_\lambda - y) \rangle \\
&\quad + \frac{1}{2\lambda} \|x_\lambda - x\|^2 - \frac{1}{2\lambda} \|y_\lambda - y\|^2 \leq \left\langle \frac{1}{\lambda}(x - x_\lambda), x - y \right\rangle \\
&\quad - \frac{1}{2\lambda} \|x_\lambda - x\|^2 - \frac{1}{2\lambda} \|y_\lambda - y\|^2 + \frac{1}{\lambda} \|x_\lambda - x\| \, \|y_\lambda - y\| \\
&\leq \left\langle \frac{1}{\lambda}(x - x_\lambda), x - y \right\rangle.
\end{aligned}$$

Hence $\frac{1}{\lambda}(x - x_\lambda) \in \partial V_\lambda(x)$.

On the other hand, since $\frac{1}{\lambda}(y - y_\lambda) \in \partial V_\lambda(y)$, we obtain

$$\begin{aligned}
V_\lambda(x) - V_\lambda(y) &\geq \left\langle \frac{1}{\lambda}(y - y_\lambda), x - y \right\rangle \\
&\geq \left\langle \frac{1}{\lambda}(x - x_\lambda), x - y \right\rangle - \left\langle \frac{1}{\lambda}(x - x_\lambda) - \frac{1}{\lambda}(y - x_\lambda), x - y \right\rangle \\
&\geq \left\langle \frac{1}{\lambda}(x - x_\lambda), x - y \right\rangle - \left\| \frac{1}{\lambda}(x - x_\lambda) - \frac{1}{\lambda}(y - y_\lambda) \right\| \|x - y\| \\
&\geq \left\langle \frac{1}{\lambda}(x - x_\lambda), x - y \right\rangle - \frac{1}{\lambda} \|x - y\|^2
\end{aligned}$$

$\left(\text{because } x \to \frac{1}{\lambda}(x - x_\lambda)\right.$ is the Yosida approximation, which is Lipschitzean with constant $\left.\frac{1}{\lambda}\right)$. So we have proved that

$$-\frac{1}{\lambda}\|x-y\| \leq \frac{V_\lambda(x) - V_\lambda(y) - \left\langle \frac{1}{\lambda}(x - x_\lambda), x - y \right\rangle}{\|x-y\|} \leq 0.$$

This proves that $\frac{1}{\lambda}(x - x_\lambda) = \nabla V_\lambda(x)$, i.e., that $\nabla V_\lambda(\cdot)$ is the Yosida approximation of $\partial V(\cdot)$. □

5. Application: Gradient Methods for Constrained Minimization Problems

Let X and Y be Hilbert spaces, $U: X \to R \cup \{+\infty\}$ and $V: Y \to R \cup \{+\infty\}$ be proper lower semicontinuous convex functions and $A \in \mathscr{L}(X, Y)$ be a continuous linear operator from X to Y. We consider the minimization problem v:

Find $\bar{x} \in X$ such that

(1) $$v \doteq \inf_{x \in X} (U(x) + V(Ax)) = U(\bar{x}) + V(A\bar{x}).$$

An important example is provided by the case when $V \doteq \psi_M$ is the indicator of a non-empty closed convex subset M of Y: in this case, the minimization problem becomes

(2) $$w \doteq \inf_{Ax \in M} U(x) = U(\bar{x}) \quad \text{where } A\bar{x} \in M.$$

First we have to provide a sufficient condition for a solution to the problem v to exist:

Proposition 1. *Assume that*

(3) $$0 \in \operatorname{Int}(A^* \operatorname{Dom} V^* + \operatorname{Dom} U^*).$$

Then there exists a solution $\bar{x} \in X$ to the minimization problem v. ▲

Proof. It is sufficient to prove that the level subsets $F_\alpha \doteq \{x \in X \mid U(x) + V(Ax) \leq \alpha\}$ are weakly bounded. Being also closed and convex, they are therefore weakly compact, so that the intersection $\bigcap_{\alpha > v} F_\alpha$ is non-empty. This intersection is the set of solutions to the minimization problem v.

Let p belong to X. We have to prove that $\sup_{x \in F_\alpha} \langle p, x \rangle$ is finite. By assumption (3), $A^* \operatorname{Dom} V^* + \operatorname{Dom} U^*$ contains a ball of radius η. Hence there exist $q \in \operatorname{Dom} V^*$ and $r \in \operatorname{Dom} U^*$ such that $\frac{\eta p}{\|p\|} = A^* q + r$. Therefore, by Fenchel's inequality, we obtain for all $x \in F_\alpha$:

$$\frac{\eta}{\|p\|}\langle p, x\rangle = \langle q, Ax\rangle + \langle r, x\rangle \leq V^*(q) + V(Ax) - U^*(r) + U(x)$$
$$\leq \alpha + U^*(r) + V^*(q) < +\infty. \qquad \square$$

Then, Theorem 4.2 states that the trajectories of the gradient inclusion

(4) $$x'(t) \in -\partial(U + V \circ A)(x(t)), \quad x(0) = x_0$$

converge weakly to a solution to the minimization problem v. Theorem 0.7.4 tells us that the condition

(5) $$0 \in \operatorname{Int}(A \operatorname{Dom} U - \operatorname{Dom} V)$$

implies that $\partial(U + V \circ A) = \partial U + A^* \partial V \circ A$.

Theorem 1. *Let us assume that*

(6) $$\begin{cases} \text{i) } 0 \in \operatorname{Int}(A \operatorname{Dom} U - \operatorname{Dom} V), \\ \text{ii) } 0 \in \operatorname{Int}(A^* \operatorname{Dom} V^* + \operatorname{Dom} U^*). \end{cases}$$

Then, for all $x_0 \in \operatorname{Dom}(\partial V) \cap A^{-1} \operatorname{Dom}(\partial V)$, there exists a unique solution to the differential inclusion

(7) $$x'(t) \in -\partial U(x(t)) - A^* \partial V(Ax(t)), \quad x(0) = x_0$$

that converges to a solution $\bar{x}$ to the minimization problem v, which is a solution to the inclusion

(8) $$0 \in \partial U(\bar{x}) + A^* \partial V(A\bar{x}).$$

Actually, $x(\cdot)$ is a solution to the differential inclusion

(9) $$x'(t) = m(-\partial U(x(t)) - A^* \partial V(Ax(t))), \quad x(0) = x_0. \qquad \blacktriangle$$

When $A \in \mathscr{L}(X, Y)$ is surjective, its transpose A^* is injective with a closed range and thus, has an orthogonal left inverse $A^{*-} \doteq (AA^*)^{-1} A$. (See Aubin, [1979] a), chapter 4, Proposition 4, 5.3).

When G is a subset of Y^*, we denote by π_G the projector of best approximation onto G when Y^* is supplied with the initial norm $\|p\|_{Y^*} = \|A^* p\|_{X^*}$.

Then Proposition 0.6.5 implies that

(10) $$m(-\partial U(x) - A^* \partial V(Ax)) = m((1 - A^* \pi_{\partial V(Ax)} A^{*-})(-\partial U(x))).$$

In particular, when U is Gâteaux-differentiable on $A^{-1} \operatorname{Dom} V$, we have

(11) $$m(-\nabla U(x) - A^* \partial V(Ax)) = -\nabla U(x) - A^* \pi_{\partial V(Ax)}(-A^{*-} \nabla U(x)).$$

Remark. When $V \doteq \psi_{\{y\}}$ is the indicator of a point $\{y\}$, then $\partial \psi_{\{y\}}(y) = Y$ and formula (11) becomes

$$m(-\nabla U(x)-A^*\,\partial V(Ax))=-\nabla U(x)+A^*A^{*-}\,\nabla U(x)=-\pi_{\operatorname{Ker} A}\nabla U(x). \tag{12}$$

because $1-A^*A^{*-}$ is the orthogonal projector onto $(\operatorname{Im} A^*)^\perp=\operatorname{Ker} A$.

When $V\doteq\psi_M$ is the indicator of a closed convex subset M, then $\partial\psi_M(y)=N_M(y)$ is the normal cone to M at y and formula (11) becomes

$$\begin{aligned} m(-\nabla U(x)-A^*N_M(Ax)) &= -\nabla U(x)-A^*\pi_{N_M(Ax)}A^{*-}(-\nabla U(x)) \\ &= -\pi_{\operatorname{Ker} A}\nabla U(x)+A^*\pi_{T_M(Ax)}(-A^{*-}\,\nabla U(x)), \end{aligned} \tag{13}$$

because $1-\pi_{N_M(Ax)}$ is the projector onto the tangent cone $T_M(Ax)$.

Example. As a particular case, we consider proper lower semicontinuous convex functions V and U_i $(i=1,\dots,n)$ from a Hilbert space Y to $R\cup\{+\infty\}$ and the minimization problem

$$v\doteq\inf\left(\sum_{i=1}^n U_i(x_i)+V\left(\sum_{i=1}^n x_i\right)\right). \tag{14}$$

Corollary 1. *Assume that*

$$\begin{cases} \text{i) } 0\in\operatorname{Int}\left(\sum_{i=1}^n \operatorname{Dom} U_i-\operatorname{Dom} V\right), \\ \text{ii) } \exists\,\eta>0 \text{ such that } \forall\,p_i\in X^*,\ \exists\,p\in\operatorname{Dom} V^* \\ \quad \text{and } q_i\in\operatorname{Dom} U_i^* \text{ such that } \eta\,p_i/\|p_i\|=p-q_i. \end{cases} \tag{15}$$

For all $(x_{10},\dots,x_{n0})\in\prod_{i=1}^n\operatorname{Dom}(\partial U_i)$ *such that* $\sum_{i=1}^n x_{i_0}\in\operatorname{Dom} V$, *there exists a unique trajectory of the system of differential inclusions:*

$$\begin{cases} \text{i) } x_i'(t)\in-\partial U_i(x_i(t))-p(t),\ (i=1,\dots,n), \\ \text{ii) } x_i(0)=x_{i_0}, \\ \text{iii) } p(t)\in\partial V\left(\sum_{i=1}^n x_i(t)\right), \end{cases} \tag{16}$$

that converges weakly when $t\to\infty$ *to a solution* $(\bar{x}_1,\dots,\bar{x}_n)$ *to the minimization problem* v, *which is a solution to the system of inclusions*

$$\begin{cases} \text{i) } 0\in\partial U_i(\bar{x}_i)+\bar{p},\ (i=1,\dots,n), \\ \text{ii) } \bar{p}\in\partial V\left(\sum_{i=1}^n \bar{x}_i\right). \end{cases} \tag{17}$$

▲

Proof. We apply the above Theorem with

$$X\doteq Y^n,\quad Ax\doteq\sum_{i=1}^n x_i\quad\text{and}\quad U(x)\doteq\sum_{i=1}^n U_i(x_i).$$ □

We know that the system of differential inclusions (16) is actually a system of differential equations. For simplicity, let us assume that the functions U_i are Gâteaux-differentiable. Then the system (16) can be written

$$(18)\qquad \begin{cases} x_i'(t) = -\nabla U_i(x_i(t)) + P(x_1(t), \ldots, x_n(t))) \\ x_i(0) = x_{i_0} \end{cases}$$

where

$$(19)\qquad P(x_1, \ldots, x_n) \doteq -\pi_{\partial V\left(\sum_{i=1}^{n} x_i\right)}\left(-\frac{1}{n}\sum_{j=1}^{n} \nabla U_j(x_j)\right).$$

Indeed, we apply formula (11) and we observe that

$$A^* \pi_{\partial V(Ax)}(-A^{*-} \nabla U(x)) = P(x_1, \ldots, x_n).$$

A particular instance is provided by the case when $V = \psi_{\{y\}}$ is the indicator of a point. Then the minimization problem v becomes

$$(20)\qquad v \doteq \inf_{\sum_{i=1}^{n} x_i = y} \sum_{i=1}^{n} U_i(x_i)$$

and the map (19) is defined by

$$(21)\qquad P(x_1, \ldots, x_n) \doteq -\frac{1}{n}\sum_{j=1}^{n} \nabla U_j(x_j)$$

because $\partial\psi_{\{y\}}(y) = Y^*$. □

When $V \doteq \psi_M$ is the indicator of a closed convex subset M, then $\partial\psi_M(y) = N_M(y)$ is the normal cone to M at y. The differential equations (16) can be written

$$(22)\qquad \begin{aligned} x_i'(t) &= -\nabla U_i(x_i) - \pi_{N_M\left(\sum_{j=1}^{n} x_j\right)}\left(-\frac{1}{n}\sum_{j=1}^{n} U_j(x_j)\right) \\ &= \frac{1}{n}\sum_{j=1}^{n} \nabla U_j(x_j) - \nabla U_i(x_i) + \pi_{T_M\left(\sum_{j=1}^{n} x_j\right)}\left(-\frac{1}{n}\sum_{j=1}^{n} \nabla U_j(x_j)\right). \end{aligned}$$ □

The Dual Gradient Method

We saw that conditions

$$(6)\qquad \begin{cases} \text{i)}\ 0 \in \operatorname{Int}(A \operatorname{Dom} U - \operatorname{Dom} V) \\ \text{ii)}\ 0 \in \operatorname{Int}(A^* \operatorname{Dom} V^* + \operatorname{Dom} U^*) \end{cases}$$

imply the existence of a solution $\bar{x}$ to the minimization problem $v \doteq \inf_{x \in X}(U(x) + V(Ax))$, which is a solution to the inclusion

$$(23)\qquad 0 \in \partial U(\bar{x}) + A^* \partial V(A\bar{x}).$$

Hence there exists $\bar{p} \in \partial V(A\bar{x})$ such that $-A^* \bar{p} \in \partial U(\bar{x})$. This is equivalent to

$$(24)\qquad A\bar{x} \in \partial V^*(\bar{p}) \quad \text{and} \quad \bar{x} \in \partial U^*(-A\bar{p})$$

or, to

$$0 \in \partial V^*(\bar{p}) - A\,\partial U^*(-A\bar{p}). \tag{25}$$

Since $0 \in \operatorname{Int}(A^* \operatorname{Dom} V^* + \operatorname{Dom} U^*)$, Theorem 0.7.4 states that this inclusion is equivalent to

$$0 \in \partial(V^* + U^* \circ (-A^*))(\bar{p}) \tag{26}$$

i.e., that

$$\bar{p} \in Y^* \quad \text{minimizes } p \to V^*(p) + U^*(-A^* p) \quad \text{on } Y^*. \tag{27}$$

This is called the *dual problem* of the minimization problem v, denoted by

$$v_* \doteq \inf_{p \in Y^*} (V^*(p) + U^*(-A^* p)). \tag{28}$$

So, if we solve the dual problem and find a solution $\bar{p}$ to the inclusion (25), then any element $\bar{x}$ of $\partial U^*(-A^*\bar{p}) \cap A^{-1}\partial V^*(\bar{p})$ is a solution to the minimization problem v. Therefore, it may be advantageous to use the gradient method for the dual problem.

Theorem 2. *We posit the same assumptions as in* Theorem 1, *namely*

$$\text{(6)} \quad \begin{aligned} &\text{i) } 0 \in \operatorname{Int}(A \operatorname{Dom} U - \operatorname{Dom} V) \\ &\text{ii) } 0 \in \operatorname{Int}(A^* \operatorname{Dom} V^* + \operatorname{Dom} U^*). \end{aligned}$$

Then, for all $p_0 \in \operatorname{Dom} \partial V^* \cap (A^*)^{-1}(-\operatorname{Dom} U)$, *there exists a unique trajectory* $p(t)$ *of the differential inclusion*

$$p'(t) \in -\partial V^*(p(t)) + A\,\partial U^*(-A^* p(t))), \qquad p(0) = p_0 \tag{29}$$

that converges weakly when $t \to \infty$ *to a solution* $\bar{p}$ *to the dual minimization problem* (28). ▲

We know that the solutions to the differential inclusion (29) are actually the solutions to the differential equation

$$p'(t) \in m(A\,\partial U^*(-A^* p(t)) - \partial V^*(p(t))), \qquad p(0) = p_0. \tag{30}$$

They can be written

$$p'(t) = m((1 - \pi_{\partial V^*(p(t))})\,A\,\partial U^*(-A^* p(t))), \qquad p(0) = p_0. \tag{31}$$

Example (Continuation). We continue the study of our example. The dual problem of the problem

$$v \doteq \inf\left(\sum_{i=1}^{n} U_i(x_i) + V\left(\sum_{i=1}^{n} x_i\right)\right) \tag{14}$$

is defined by

$$v_* = \inf_{p \in Y^*}\left(V^*(p) + \sum_{i=1}^{n} U_i^*(-p)\right) \tag{32}$$

whose solutions are the solutions to the inclusion

$$0 \in \partial V^*(\bar{p}) - \sum_{i=1}^{n} \partial U_i^*(-\bar{p}). \tag{33}$$

When $\bar{p}$ is a solution to the dual problem, the solutions $\bar{x} = (\bar{x}_1, \ldots, \bar{x}_n)$ to

$$\bar{x}_i \in \partial U_i^*(-\bar{p})(i=1, \ldots, n) \quad \text{and} \quad \sum_{i=1}^{n} \bar{x}_i \in \partial V^*(\bar{p}) \tag{34}$$

are solutions to the minimization problem (14).

In particular, when U_i^* are Gâteaux-differentiable, solutions to the minimization problem (14) are obtained through formula

$$\bar{x}_i = \nabla U_i^*(-\bar{p}). \tag{35}$$

The dual gradient method can be written

$$p'(t) \in -\partial V^*(p(t)) + \sum_{i=1}^{n} \partial U_i^*(-p(t)), \qquad p(0) = p_0. \tag{36}$$

When the functions are Gâteaux-differentiable, we can write these equations in the form

$$\begin{aligned} &\text{i) } p'(t) \in -\partial V^*(p(t)) + \sum_{i=1}^{n} x_i(t), \qquad p(0) = p_0, \\ &\text{ii) } x_i(t) = \nabla U_i^*(-p(t)). \end{aligned} \tag{37}$$

Remark. When $V \doteq \psi_M$ is the indicator of a closed convex subset M, differential inclusion (37) can be written

$$\begin{aligned} &\text{i) } p'(t) \in \sum_{i=1}^{n} x_i(t) - M, \qquad p(0) = p_0 \\ &\text{ii) } \left\langle p(t), \sum_{i=1}^{n} x_i(t) - p'(t) \right\rangle = \sigma_M(p(t)) \\ &\text{iii) } x_i(t) = \nabla U_i^*(-p(t)) \end{aligned} \tag{38}$$

because $\partial V^*(p) = \{y \in M \mid \langle p, y \rangle = \sigma_M(p)\}$. □

Application. Convergence to Pareto Minima

We consider n *proper lower semicontinuous convex* functions U_i from X to $R \cup \{+\infty\}$, which are bounded below. Since we shall minimize them, we can assume without loss of generality that

$$\forall i = 1, \ldots, n, \inf_{x \in X} U_i(x) = 0. \tag{39}$$

We define *Pareto minima* on K as elements $\bar{x} \in K$ such that it is impossible to find elements $y \in K$ satisfying $U_i(y) < U_i(\bar{x})$ for all $i = 1, \ldots, n$.

It is known that when the functions U_i are convex, as we assumed they are, $\bar{x} \in K$ is a Pareto minimum if and only if there exists

$$\bar{\lambda} \in S^n = \left\{ \lambda \in R^n_+ \;\middle|\; \sum_{i=1}^n \lambda_i = 1 \right\}$$

satisfying

$$\sum_{i=1}^n \bar{\lambda}_i U_i(\bar{x}) = \min_{x \in K} \sum_{i=1}^n \bar{\lambda}_i U_i(x). \tag{40}$$

We also achieve Pareto minima by the best compromise method (see Aubin [1979]a), Chapter 3, Proposition 8.2, p. 49).

We associate with any $\lambda \in \mathring{R}^n_+$ the function:

$$U_\lambda(x) \doteq \max_{i=1,\dots,n} \frac{U_i(x)}{\lambda_i}. \tag{41}$$

Lemma 1. *If $\bar{x} \in K$ minimizes U_λ over K, then $\bar{x}$ is Pareto minimum. Conversely, if $\bar{x} \in K$ is a Pareto minimum satisfying $U_i(\bar{x}) > 0$ for all i, there exists $\bar{\lambda} \in \mathring{R}^n_+$ such that $\bar{x}$ minimizes $U_{\bar{\lambda}}(x)$ over K.* ▲

Proof. Indeed, if $\bar{x} \in K$ achieves the minimum of U_λ over K and is not a Pareto minimum, there would exist $y \in K$ such that $U_i(y) < U_i(\bar{x})$ for all $i = 1, \dots, n$.

Then, we deduce that $U_\lambda(y)$ would be strictly inferior to $U_\lambda(\bar{x})$, which is also impossible.

Conversely, let $\bar{x} \in K$ be a Pareto minimum. We take $\bar{\lambda}_i = U_i(\bar{x})$ for all $i \in 1, \dots, n$. Then $\bar{x}$ minimizes $U_{\bar{\lambda}}$ over K. If not, there would exist $y \in K$ such that

$$U_{\bar{\lambda}}(y) \doteq \max_{i=1,\dots,n} \frac{U_i(y)}{U_i(\bar{x})} < U_{\bar{\lambda}}(\bar{x}) = 1$$

and thus, such that $U_i(y) < U_i(\bar{x})$ for all $i = 1, \dots, n$. This is impossible. □

Therefore, we can associate with any parameter $\lambda \in \mathring{R}^n_+$ the gradient inclusions

$$\begin{aligned} & x'(t) \in -\partial\left(\sum_{i=1}^n \lambda_i U_i + \psi_K\right)(x(t)) \\ & x(0) = x_0 \end{aligned} \tag{42}$$

and

$$\begin{aligned} & x'(t) \in -\partial\left(\max_{i=1,\dots,n} \frac{U_i}{\lambda_i} + \psi_K\right)(x(t)) \\ & x(0) = x_0 \end{aligned} \tag{43}$$

whose trajectories converge to Pareto minima when $t \to \infty$.

For the solutions to the gradient inclusion (42) we know that

$$t \to \sum_{i=1}^n \lambda_i U_i(x(t)) \quad \text{is nonincreasing.} \tag{44}$$

Whereas for the solutions to the gradient inclusion (43), we know that

$$(45) \qquad t \to \max_{i=1,\dots,n} \frac{1}{\lambda_i} U_i(x(t)) \quad \text{is nonincreasing.}$$

We observe that when we assume that

(46) the functions U_i are continuous at each point of K,

then the gradient inclusions (42) and (43) can be written respectively

$$(47) \qquad \begin{aligned} & x'(t) \in -\sum_{i=1}^{n} \lambda_i \partial U_i(x(t)) - N_K(x(t)) \\ & x(0) = x_0 \end{aligned}$$

and

$$(48) \qquad \begin{aligned} & x'(t) \in -\operatorname{co}\left(\bigcup_{i \in I(x(t))} \frac{\partial U_i(x(t))}{\lambda_i} \right) - N_K(x(t)) \\ & x(0) = x_0 \end{aligned}$$

where $I(x) \doteq \left\{ i=1,\dots,n \,\middle|\, \frac{U_i(x)}{\lambda_i} = \max_j \frac{U_j(x)}{\lambda_j} \right\}$.

The trajectories of gradient inclusions (42) and (43) do not necessarily enjoy the monotonicity property

$$(49) \qquad \forall i=1,\dots,n, \quad t \to U_i(x(t)) \quad \text{is nonincreasing.}$$

Example. Let us consider the case when

$$X = R^2, \qquad U_i(x) = \tfrac{1}{2}\|x - u_i\|^2 \qquad (i=1,2)$$

and $\lambda \doteq (2, 1)$.

We set

$$\begin{aligned} \Omega_1 &\doteq \{x \in R^2 \mid U_1(x) > 2U_2(x)\} \\ \Omega_2 &\doteq \{x \in R^2 \mid U_1(x) < 2U_2(x)\} \\ C &\doteq \{x \in R^2 \mid U_1(x) = 2U_2(x)\} \text{ is a circle of center } 2u_2 - u_1. \end{aligned}$$

We check that

$$\partial U_\lambda(x) = \begin{cases} \frac{1}{2}(x - u_1) & \text{when } x \in \Omega_1 \text{ (and } I(x) = \{1\}), \\ (x - u_2) & \text{when } x \in \Omega_2 \text{ (and } I(x) = \{2\}), \\ \left\{ \frac{\lambda}{2}(x - u_1) + (1-\lambda)(x - u_2) \right\}_{\lambda \in [0,1]} & \text{when } x \in C. \end{cases}$$

□

One of the problems arising in planning procedures is to find dynamical systems whose trajectories converge to a Pareto minimum while the n "loss functions" U_i decrease along the trajectories. This is a reason for looking for other differential inclusions (or equations) whose trajectories satisfy the monotonicity property (49) and do converge to Pareto minima. We just have to modify slightly differential inclusion (48) and replace it by

$$\text{(50)}\quad \begin{aligned} &\text{i)}\ x'(t) \in -\operatorname{co}\left(\bigcup_{i=1}^{n} \frac{\partial U_i(x(t))}{\lambda_i}\right) - N_K(x(t)) \\ &\text{ii)}\ x(0) = x_0 \end{aligned}$$

as we shall see later (Chapter 5, §6).

The right-hand side of this differential inclusion is no longer maximal monotone.

Remark. Let us observe that the solutions to the differential inclusions (48) or (50) can be written

$$\text{(51)}\quad \begin{aligned} &\text{i)}\ x'(t) \in -\sum_{i=1}^{n} \mu_i(x(t))\, \partial U_i(x(t)) - N_K(x(t)) \\ &\text{ii)}\ x(0) = x_0 \end{aligned}$$

where the "weights" μ_i depend upon the state $x(t)$ at time t, whereas in the differential inclusion (47), the "weights" λ_i are kept constant. □

Chapter 4. Viability Theory: The Nonconvex Case

Introduction

We devote this chapter to general Viability Theory and we postpone to the next chapter the further results obtained when we assume that the viability subset is convex.

Let K be a closed subset of R^n, the viability subset. A trajectory $t \to x(t)$ is said to be viable when

$$\forall t, \quad x(t) \in K. \tag{1}$$

Let $F: K \to R^n$ be a set-valued map describing the dynamics of our system. Our first problem will be to characterize the relations between the viability subset K and the map F which allow the differential inclusion

$$\begin{aligned} &\text{i)}\ x'(t) \in F(x(t)) \\ &\text{ii)}\ x(0) = x_0,\ x_0 \text{ given in } K \end{aligned} \tag{2}$$

to have a viably trajectory for all $x_0 \in K$.

More generally, we consider a preorder $\succcurlyeq$ (a reflexive and transitive binary relation) defined on K.

We shall solve a stronger viability problem by requiring that the differential inclusion (2) has trajectories *monotone* in the sense that

$$\forall t > s, \quad x(t) \preccurlyeq x(s). \tag{3}$$

We can characterize the preorder $\succcurlyeq$ by the set-valued map P from K to K defined by

$$P(x) = \{y \in K \mid y \preccurlyeq x\}. \tag{4}$$

The set-valued map P satisfies the conditions

$$\begin{aligned} &\text{i)}\ \forall x \in K,\ x \in P(x) \quad \text{(reflexivity)}, \\ &\text{ii)}\ \forall x \in K,\ \forall y \in P(x), \quad \text{then } P(y) \subset P(x)\ \text{(transitivity)}. \end{aligned} \tag{5}$$

Then monotone trajectories are those that satisfy

$$\forall t > s, \quad x(t) \in P(x(s)). \tag{6}$$

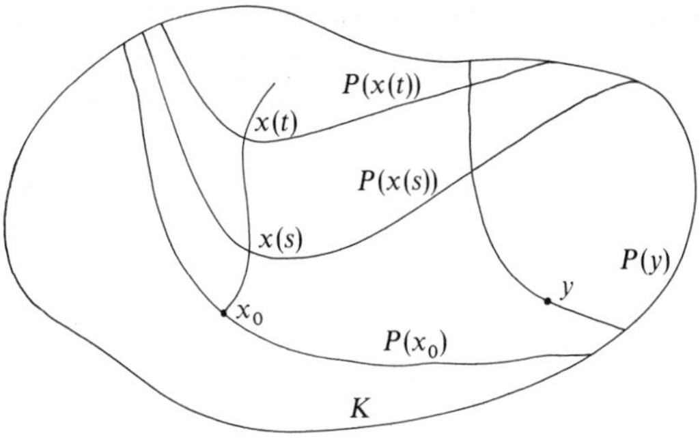

Fig. 13. Monotone trajectory of (2)–(6)

When the preorder $\succcurlyeq$ is regarded as a "preference preorder", monotonicity can be considered as a selection procedure that, among all the trajectories of the differential inclusion, chooses the ones that improve the state of the system as time elapses.

This provides another type of selection procedure of trajectories of differential inclusions than optimal control, where we select trajectories by minimizing a functional defined on a set of trajectories. It is consistent with the behavioral assumption of "limited rationality" proposed by March and Simon, where optimality is replaced by mere satisfaction.

In many instances, the set of constraints K has an explicit form $K = \{x \in L \mid A(x) \in M\}$ where A is a map from R^n to R^p and where $M \subset R^p$. In this case, viability condition (1) is written

$$\forall t \geq 0, \quad x(t) \in L \quad \text{and} \quad A(x(t)) \in M. \tag{7}$$

When $M \subset R^n \times R^n$ is a graph of a set-valued map G and when $A(x) = (x, x)$ for all $x \in L$, then viability condition (7) becomes

$$\forall t \geq 0, \quad x(t) \in L \quad \text{and} \quad x(t) \in G(x(t)). \tag{8}$$

Let us mention also the important case of a preorder defined by *n utility functions* $U_i\colon K \to R$ by $P(x) = \{y \in K \mid \forall i = 1, \ldots, n, \; U_i(y) \geq U_i(x)\}$. In this case condition (3) becomes,

$$\forall s \geq t \geq 0, \quad \forall i = 1, \ldots, n, \quad U_i(x(s)) \geq U_i(x(t)). \tag{9}$$

Naturally it is possible to combine these examples and by doing so, to describe "rich" enough constraints. □

We proceed now to describe the results of this chapter; they show that, independently of the motivations and possible applications, the viability problem is a source of quite interesting results.

It is by no means a new problem. The viability problem was first solved by Nagumo in 1942.

Theorem (Nagumo). *Let K be a closed subset of R^n and $f: K \to R^n$ be a bounded continuous map. The necessary and sufficient condition for the differential equation $x' = f(x)$ to have a viable solution for any initial state $x_0 \in K$ is that*

$$\forall x \in K, \quad \liminf_{h \to 0^+} \frac{d_K(x + hf(x))}{h} = 0. \tag{10}$$ ▲

This theorem suggests to associate with every $x \in K$ the subset

$$T_K(x) \doteq \left\{ v \in R^n \,\middle|\, \liminf_{h \to 0^+} \frac{d_K(x+hv)}{h} = 0 \right\}. \tag{11}$$

We observe that $T_K(x)$ is a *closed cone* which can be written, in a more intrinsic way

$$T_K(x) = \bigcap_{\substack{\varepsilon > 0 \\ \alpha > 0}} \bigcup_{h \in]0, \alpha]} \left[\frac{1}{h} (K - x) + \varepsilon B \right] \tag{12}$$

where B denotes the closed unit ball.

We recognize that $T_K(x)$ is nothing other than the *contingent cone* to K at x, introduced by Bouligand in the 1930's.

It generalizes the concept of tangent space in the following sense: If K is a smooth manifold imbedded in R^n, then $T_K(x)$ is the tangent space to K at x. In this case, the necessary and sufficient conditions states that f is a vector field, a map sending each point x of the manifold K to its tangent space at x.

When X is a closed convex subset, the contingent cone $T_K(x)$ is convex and is equal to the closed convex cone spanned by $K - x$:

$$T_K(x) = \operatorname{cl} \left(\bigcup_{h > 0} \frac{1}{h} (K - x) \right). \tag{13}$$

Note that $T_K(x) = R^n$ whenever x belongs to the interior of K. Then conditions that involve the Bouligand contingent cone are *boundary conditions*.

Hence we can regard Nagumo's Theorem as saying that differential equations $x' = f(x)$ have viable trajectories whenever f is a "generalized vector field" in the sense that

$$\forall x \in K, \quad f(x) \in T_K(x). \tag{14}$$

We observe that we can define a priori many concepts of "tangent cones" to arbitrary subsets of R^n, by replacing in (11) "$\liminf_{h \to 0^+}$" by "$\lim_{h \to 0^+}$", or "$\lim_{\substack{h \to 0^+ \\ y \to x}}$", etc... As far as we are concerned, Nagumo's theorem and its generalizations show that Bouligand contingent cones play a privileged role in our viability problems.

What about differential inclusions?

For the class of differential inclusions $x' \in F(x)$ where F is upper semicontinuous with *compact convex* values, we shall prove that the necessary and sufficient condition for the existence of viable trajectories for each initial state $x_0 \in K$ is that

(15) $$\forall x \in K, \quad F(x) \cap T_K(x) \neq \emptyset.$$

In other words, the convexity of the images $F(x)$ of F allows an absence of regularity in the dependence upon x of the selections $v(x)$ in the intersection $F(x) \cap T_K(x)$.

Let us recall also that this tangential condition is satisfied when $F \doteq -A$, A being maximal monotone, and $K \doteq D(A)$, its domain. (See Chapter 3).

The question arises whether we can drop the assumption of convexity of the images: we shall pay a heavy price for that, since we shall need to suppose that F is continuous and, above all, that

(16) $$\forall x \in K, \quad F(x) \subset T_K(x).$$

We shall distinguish the viability problem and the invariance problem, that we define as follows.

Let G be a set-valued map from an open subset Ω to R^n. We say that a subset K of Ω is invariant by G if, for all $x_0 \in K$, *all* the trajectories of $x' \in G(x)$, $x(0) = x_0$ remain in K.

We shall prove that K is invariant when K is closed and G is Lipschitzean with compact images. The sufficient condition is that

(17) $$\forall x \in K, \quad G(x) \subset T_K(x).$$

As promised, we shall characterize differential inclusions having monotone trajectories $x(\cdot)$, i.e., trajectories satisfying

$$\forall t \geq s, \quad x(t) \in P(x(s)).$$

For an upper semicontinuous set-valued map F with compact convex images and for a preordering $P(\cdot)$ which are lower semicontinuous, the necessary and sufficient condition is, as we expect,

(18) $$\forall x \in K, \quad F(x) \cap T_{P(x)}(x) \neq \emptyset.$$

We shall also consider the time dependent case of differential inclusions

(19) $$x'(t) \in F(t, x(t))$$

having viable trajectories in the sense that

(20) $$\forall t \geq 0, \quad x(t) \in K(t).$$

This problem provides a first reason for introducing the concept of "contingent derivatives" to the set-valued map $K: R_+ \to R^n$.

We proceed by analogy. In the smooth case, we know that the graph of the derivative $\nabla f(x)$ of a single-valued map f at x is the tangent space to the graph of f at $(x, f(x))$.

In the set-valued case, we define the graph of the "contingent derivative" $DF(x, y)$ of a set-valued map F at x and $y \in F(x)$ as the contingent cone to the graph of F at (x, y).

It is convenient to state the analytical characterization of the contingent derivative

(21) $$v \in DF(x, y)(u) \Leftrightarrow \liminf_{\substack{h \to 0^+ \\ u' \to u}} d\left(v, \frac{F(x+hu')-y}{h}\right)=0.$$

The viability criterion for upper semicontinuous maps with compact convex values can be written

(22) $$\forall (t, x) \in \operatorname{graph}(K), \quad F(t, x) \cap DK(t, x)(1) \neq \emptyset.$$

We shall apply the Viability Theorem to prove that a continuous version of Newton's method provides trajectories converging to an equilibrium:

Let f be a continuously differentiable map from an open neighborhood Ω of a compact subset $K \subset R^n$ to R^p. The continuous version of Newton's method is described by the implicit differential equation

(23) $$\begin{aligned} &\text{i)}\ \nabla f(x(t))\, x'(t) = -f(x(t)), \\ &\text{ii)}\ x(0) = x_0, \quad x_0 \text{ given in } K. \end{aligned}$$

The viability condition states that

(24) $$\forall x \in K, \quad \exists u \in T_K(x) \quad \text{such that } \nabla f(x)\, u = -f(x).$$

It is a fact that this viability condition implies the existence of equilibria of f. We shall also prove the convergence of viable trajectories of (23) to the set of equilibria:

(25) $$d(x(t), f^{-1}(0)) \le c e^{-t} \|f(x_0)\|.$$

We conclude this chapter by extending Viability Theory to the case of differential inclusions with memory, also called Functional Differential Inclusions.

1. Bouligand's Contingent Cone

Let K be a nonempty subset of a Hilbert space X. We shall define the Bouligand contingent cone as follows.

Definition 1. We say that the subset

(1) $$T_K(x) \doteq \bigcap_{\varepsilon > 0} \bigcap_{\alpha > 0} \bigcup_{0 < h < \alpha} \left(\frac{1}{h}(K - x) + \varepsilon B\right)$$

is the "*contingent cone*" to K at x. ▲

In other words, $v \in T_K(x)$ if and only if

$$\text{(2)} \qquad \forall \varepsilon > 0, \quad \forall \alpha > 0, \quad \exists u \in v + \varepsilon B, \quad \exists h \in]0, \alpha] \text{ such that } x + hu \in K.$$

It is quite obvious that $T_K(x)$ is a *closed cone.*

We also note that

$$\text{(3)} \qquad \text{if } x \in \operatorname{Int}(K), \quad \text{then } T_K(x) = X.$$

We characterize the contingent cone by using the distance function $d_K(\cdot)$ to K defined by

$$d_K(x) \doteq \inf\{\|x - y\| \mid y \in K\}.$$

Proposition 1. *$v \in T_K(x)$ if and only if* $\liminf_{h \to 0^+} \frac{d_K(x + hv)}{h} = 0.$ ▲

Proof. a) Let $v \in T_K(x)$. For all $\varepsilon > 0$, $\alpha > 0$, there exist $h \in]0, \alpha]$ and $u \in v + \varepsilon B$ such that $x + hu \in K$. Hence

$$\frac{d_K(x + hv)}{h} \leq \frac{1}{h} \|x + hv - (x + hu)\| \leq \|u - v\| \leq \varepsilon.$$

So, $\forall \varepsilon > 0$, $0 \leq \sup_{\alpha} \inf_{h \leqq \alpha} \frac{d_K(x + hv)}{h} \leq \varepsilon$. This proves that

$$\liminf_{h \to 0^+} \frac{d_K(x + hv)}{h} = 0.$$

b) *Conversely,* if $\liminf_{h \to 0^+} \frac{d_K(x + hv)}{h} = \sup_{\alpha > 0} \inf_{h \leqq \alpha} \frac{d_K(x + hv)}{h} = 0$, we deduce that $\forall \varepsilon > 0$, $\forall \alpha > 0$, $\exists h < \alpha$ such that $\frac{d_K(x + hv)}{h} \leq \varepsilon/2$. Thus, there exists $y \in K$ such that $\left\| \frac{x + hv - y}{h} \right\| \leq \frac{d_K(x + hv)}{h} + \varepsilon/2$. Hence $u = \frac{y - u}{h} \in v + \varepsilon B$ and satisfies $x + hu = y \in K$. □

We can also characterize the contingent cone in terms of sequences.

Proposition 2. *$v \in T_K(x)$ if and only if there exist sequences of strictly positive numbers h_n and of elements $u_n \in X$ satisfying*

$$\text{(4)} \qquad \text{i) } \lim_{n \to \infty} u_n = v, \quad \text{ii) } \lim_{n \to \infty} h_n = 0, \quad \text{iii) } \forall n \geq 0, \ x + h_n u_n \in K. \qquad ▲$$

Remark. For all $x \in X$, we have $T_X(x) = X$. We shall set $T_\emptyset(x) \doteq \emptyset$. ▲

Remark. It is easy to see that the contingent cone to K and the contingent cone to the closure $\overline{K}$ of K coincide:

$$\forall x \in K, \quad T_K(x) = T_{\overline{K}}(x).$$

Therefore, there is no danger in speaking of $T_K(x)$ even when $x \in \overline{K}$ and $x \notin K$. ▲

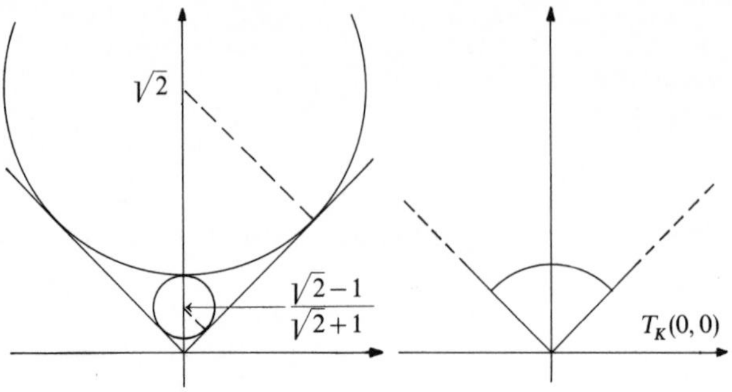

Fig. 14. Example of a contingent cone

Example. Set $\beta \doteq (\sqrt{2}-1)/(\sqrt{2}+1)$. Let $K \subset R^2$ be the union of $(0,0)$ and of the family of circles centered at $(0, \beta^i \sqrt{2})$ with radius β^i. To compute $T_K(0,0)$, observe that a vector $v=(v^1, v^2)$ is such that the half line $x_1(t)=v^1 t$, $x_2(t)=v^2 t$, $t \geq 0$, meets K infinitely often when $|v^1| \leq v^2$ while the distance from $(x_1(t), x_2(t))$ to K is bounded below by some εt when $|v^1| > v^2$. Hence $T_K(0,0)=\{v: |v^1| \leq v^2\}$. Remark that for no v in $T_K(0,0)$ (apart from $v=(0,0)$) the lim inf is a limit. For instance when $v=(0,1)$,

$$\liminf_{h \to 0^+} \frac{1}{h} d_K(hv)=0, \quad \text{while} \quad \limsup_{h \to 0^+} \frac{1}{h} d_K(hv)=(1-\beta)/2.$$

At every other point x of K different from the origin, $T_K(x)$ is the (translate to the origin of the) unique tangent line to K at x.

In particular there is a sequence of points converging to the origin, namely $x_n=(0, \beta^n(\sqrt{2})+\beta^n)$, where $T_K(x_n)$ coincides with the line $v_2=0$. This implies that *the map $x \to T_K(x)$ is neither upper semicontinuous nor lower semicontinuous* at the origin. □

The following are simple technical tools used in the sequel. Recall that by $\pi_K(y)$ we denote the set $\{x \in K: d(y,x)=d(y,K)\}$.

Proposition 3. *Let $K \subset R^n$ be closed. For any $y \in X$, $v \in X$,*

$$\liminf_{h \to 0^+} \frac{1}{h}(d_K(y+hv)-d_K(y)) \leq d(v, T_K(\pi_K(y))). \tag{5}$$ ▲

Proof. a) Assume $y \in K$. For any $v^* \in T_K(y)$, $d_K(y+hv) \leq h d(v, v^*) + d_K(y+hv^*)$, hence $\liminf_{h \to 0^+} \frac{1}{h} d_K(y+hv) \leq d(v, v^*)$ i.e.

$$\liminf_{h \to 0^+} \frac{1}{h} d_K(y+hv) \leq \inf\{d(v, v^*) \mid v^* \in T_K(y)\} = d(v, T_K(y)).$$

b) For $y \in X$, let $x \in \pi_K(y)$. Then

$$\frac{1}{h}(d_K(y+hv)-d_K(y)) \leq \frac{1}{h}(\|y+hv-(x+hv)\|+d_K(x+hv)-d_K(y))$$
$$=\frac{1}{h}d_K(x+hv).$$

By a),

$$\liminf_{h\to 0^+}\frac{1}{h}(d_K(y+hv)-d_K(y)) \leq d(v, T_K(x)).$$

Being x arbitrary in $\pi_K(y)$,

$$\liminf_{h\to 0^+}\frac{1}{h}(d_K(y+hv)-d_K(y)) \leq \inf_{x\in\pi_K(y)} d(v, T_K(x)) = d(v, T_K(\pi_K(y))). \qquad \square$$

Corollary 1. *For $x, v \in R^n$, set $f(t) \doteq d_K(x+tv)$. Then whenever $f'(t)$ exists, $f'(t) \leq d(v, T_K(\pi_K(x+tv)))$.* ▲

2. Viable and Monotone Trajectories

Let K be a subset of a Hilbert space X and $T_K(x)$ denote the contingent cone to K at x. We shall prove that under reasonable assumptions, a necessary and sufficient condition for the differential inclusion

(1) $$x' \in F(x), \quad x(0) = x_0 \quad \text{where } x_0 \text{ is given in } K$$

to have trajectories that remain in K is that the following tangential condition

(2) $$\forall x \in K, \quad F(x) \cap T_K(x) \neq \emptyset$$

holds true.

Note that tangential condition (2) can be written in the form

(3) $$\forall x \in K, \quad \exists v \in F(x) \text{ such that } \liminf_{h\to 0^+}\frac{d_K(x+hv)}{h}=0.$$

We also observed that when A is a maximal monotone map, then $F=-A$ satisfies the tangential condition (2) when K is the domain of A. (See Section 3.1, formula (20)).

Definition 1. We shall say that trajectories $x(\cdot)$ such that

(4) $$\forall t \in [0, T[, \quad x(t) \in K$$

are "viable trajectories" on $[0, T[$. ▲

Let us begin by stating a necessary condition for a differential inclusion to have viable trajectories. We observe that whenever $x(\cdot)$ is a solution to (1) and $x'(t)$ exists, then, $x'(t)\in F(x(t))\cap T_K(x(t))$. However in the case of upper semicontinuous maps with compact convex images we have the following necessary condition:

Proposition 1. *Let F be a strict upper semicontinuous map from a subset $K \subset R^n$ with compact convex values.*

If for all $x_0\in K$, there exist $T>0$ and a viable trajectory on $[0, T[$ of the differential inclusion $x'\in F(x)$ starting at x_0, then the tangential condition

$$\forall x\in K, \quad F(x)\cap T_K(x)\neq\emptyset \tag{2}$$

holds. ▲

Proof. It is a corollary of *Proposition* 2 below. □

The tangential condition is also sufficient:

Theorem 1 (Viability Theorem). *Let K be a subset of X, F be a proper upper hemicontinuous map from K to X with compact convex values. We posit the tangential condition:*

$$\forall x\in K, \quad F(x)\cap T_K(x)\neq\emptyset. \tag{2}$$

a) *Assume that K is locally compact. Then for all $x_0\in K$, there exists $T>0$ such that the differential inclusion $x'\in F(x)$, $x(0)=x_0$ has a viable trajectory defined on $[0, T[$.*

b) *Assume that either K is compact or that X is finite-dimensional and $F(K)$ is bounded. Then for all $x_0\in K$, there exists a viable trajectory of the differential inclusion defined on $[0, \infty[$.* ▲

Proof. It is a corollary of *Theorem* 3 below. □

Remark. When $X=R^n$, any closed subset K is locally compact. When K is an open subset, which is locally compact, we obtain *Theorem* 2.1.3.

Naturally, when F is a single-valued map, we obtain the Nagumo Theorem as a consequence.

Theorem 2 (Nagumo). *Let K be a closed subset of a Hilbert space X and f be a continuous (single-valued) map from K to X satisfying the tangential condition*

$$\forall x\in K, \quad f(x)\in T_K(x). \tag{5}$$

Then for all $x_0\in K$, there exists $T>0$ such that the differential equation $x'(t)=f(x(t))$, $x(0)=x_0$ has a viable trajectory on $[0, T]$. ▲

The problem of finding viable trajectories of a differential inclusion is a particular case of the problem of looking for "monotone trajectories".

Let $\preccurlyeq$ be a preorder defined on K, i.e., a binary relation $x \preccurlyeq y$ which is

(6) $$\begin{cases} \text{i) reflexive (i.e., } \forall x \in K,\ x \preccurlyeq x), \\ \text{ii) transitive (i.e., if } y \preccurlyeq x \text{ and } z \leq y, \text{ then } z \preccurlyeq x). \end{cases}$$

Definition 2. We shall say that a trajectory $x(\cdot)$ defined on $[0, T[$ is "monotone" if and only if

(7) $$\forall t, s \in [0, T[, \quad s \geq t, \quad \text{then } x(s) \preccurlyeq x(t). \qquad \blacktriangle$$

It is convenient to characterize a preorder $\preccurlyeq$ by the set-valued map P defined by

(8) $$y \in P(x) \quad \text{if and only if } y \preccurlyeq x.$$

Conversely, let P be a set-valued map satisfying

(9) $$\begin{cases} \text{i) } \forall x \in K,\ x \in P(x) \quad \text{(reflexivity)}, \\ \text{ii) } \forall x \in K,\ \forall y \in P(x), \quad \text{we have } P(y) \subset P(x) \text{ (transitivity)}. \end{cases}$$

Then the relation (8) defines a preorder $\preccurlyeq$ by $y \preccurlyeq x$ if y belongs to $P(x)$. So, a trajectory $x(\cdot)$ is monotone if and only if

(10) $$\forall t, s \in [0, T[, \quad s \geq t, \quad \text{then } x(s) \in P(x(t)).$$

The typical example of a preorder is the one defined by m real-valued functions $V_j\colon K \to R (j=1, \ldots, m)$:

$$\forall x \in K, \quad P(x) \doteq \{y \in K \mid \forall j=1, \ldots, m,\ V_j(y) \leq V_j(x)\}.$$

For this preorder, a trajectory $x(\cdot)$ is monotone if and only if

$$\forall j=1, \ldots, m, \quad \forall s, t \in [0, T[, \quad s \geq t, \quad \text{then } V_j(x(s)) \leq V_j(x(t)).$$

We study this specific example in Chapter 6.

We shall prove that a necessary and sufficient condition for a differential inclusion to have monotone trajectories is that the following tangential condition

(11) $$\forall x \in K, \quad F(x) \cap T_{P(x)}(x) \neq \emptyset$$

holds true. Note that when P is the constant map defined by $P(x)=K$, monotone trajectories are viable trajectories.

Let us begin proving the necessity.

Proposition 2. *Let F be an upper semicontinuous map from* $\operatorname{Dom}(F) \subset R^n$ *into the compact convex subsets of R^n. Let P be a map from* $\operatorname{Dom}(F)$ *into its subsets satisfying $x \in P(x)$ for every x in* $\operatorname{Dom}(F)$. *Assume that for every x_0 in* $\operatorname{Dom}(F)$, *there exists $T>0$ and a solution $x(\cdot)$ of* (1) *satisfying*

(12) $$\liminf_{t\to 0+} \frac{1}{t} d_{P(x_0)}(x(t))=0$$

(in particular: $x(t)\in\overline{P(x_0)}$*). Then for every* $x_0\in\mathrm{Dom}(F)$

$$F(x_0)\cap T_{P(x_0)}(x_0)\neq\emptyset.$$ ▲

Proof. Let $\varepsilon_i\to 0$. By the upper semicontinuity of F and the lipschitzeanity of x there exist $\eta_n>0$ such that $|t|\le\eta_n$ implies $F(x(t))\subset F(x^0)+1/2n\mathring{B}$. By (12) there exist $t_n<\eta_n$ such that

$$\frac{1}{2n}>\frac{1}{t_n} d_{P(x_0)}(x(t_n))=\frac{1}{t_n} d_{P(x_0)}\left(x_0+\int_0^{t_n} x'(s)\,ds\right).$$

Let w_n in $F(x_0)+\frac{1}{2n}\mathring{B}$ be such that $x_0+\int_0^{t_n} x'(s)\,ds=x_0+t_n w_n$, and let v_n in $F(x_0)$ satisfy $\|v_n-w_n\|<1/2n$. Then

$$\begin{aligned}\frac{1}{t_n} d_{P(x_0)}(x_0+t_n v_n)&\le\|v_n-w_n\|+\frac{1}{t_n} d_{P(x_0)}(x_0+t_n w_n)\\ &\le\frac{1}{2n}+\frac{1}{2n}=\frac{1}{n}.\end{aligned}$$

By the compactness of $F(x_0)$, we can assume that $v_n\to v_*\in F(x_0)$. Then

$$\frac{1}{t_n} d_{P(x_0)}(x_0+t_n v_*)\le\|v_*-v_n\|+\frac{1}{t_n} d_{P(x^0)}(x_0+t_n v_n)\to 0$$

i.e. (11) holds. □

The tangential condition is also sufficient.

Theorem 3. *Let K be a subset of X, F be a strict upper hemicontinuous map from K to X with compact convex values and $P\colon K\to K$ be a lower semicontinuous map with closed graph satisfying*

(9) $$\begin{cases}\text{i) } \forall x\in K,\ x\in P(x) & (\textit{reflexivity}),\\ \text{ii) } \forall x\in K,\ \forall y\in P(x), \text{ then } P(y)\subset P(x) & (\textit{transitivity}).\end{cases}$$

We posit the following tangential condition:

(11) $$\forall x\in K,\quad F(x)\cap T_{P(x)}(x)\neq\emptyset.$$

a) *Assume that K is locally compact. Then for all $x_0\in K$, there exists $T>0$ such that the differential inclusion*

(1) $$x'\in F(x),\quad x(0)=x_0$$

has a solution on $[0,T[$ that satisfies

(10) $$\forall t,\quad \forall s\ge t,\quad x(s)\in P(x(t)).$$

b) *Assume that either K is compact or that X is finite-dimensional and $F(K)$ is bounded. Then there exists a solution to the problem* (1), (10) *which is defined on* $[0, \infty[$. ▲

Proof. a) Let K and $x_0 \in K$ be given.

α) If K is locally compact, there exists $r>0$ such that $K_0 \doteq K \cap (x_0 + rB)$ is compact. We set $T = r/(\|F(K_0)\| + 1)$.

β) If K is compact, we set $K_0 \doteq K$ and $T = \infty$.

γ) If $F(K)$ is bounded and X is finite dimensional we take $T>0$ arbitrary and we set $K_0 \doteq K \cap \mathrm{cl}(x_0 + B + TF(K))$.

b) We set $C \doteq F(K_0) + B$, which is a bounded set. For all $y \in K$, we can find $h_y < 1/k$ and $v_y \in F(y)$ satisfying, by the tangential assumption (11)

$$d_{P(y)}(y + h_y v_y) \le h_y / 2k.$$

We introduce the subsets

$$(13) \qquad N(y) \doteq \{x \in X \mid d_{P(x)}(x + h_y v_y) < h_y / 2k\}.$$

Since P is lower semicontinuous, Corollary 1.2.1 shows that the function $x \to d_{P(x)}(x + h_y v_y)$ is upper semicontinuous. Hence the subsets $N(y)$ are open. Since y belongs to $N(y)$, there exists a ball $B(y, \eta_y)$ of radius $\eta_y < 1/k$ contained in $N(y)$. Therefore, the compact subset K_0 can be covered by q such balls $B(y_j, \eta_{y_j})$. For simplicity, we set

$$\eta_j \doteq \eta_{y_j}, \quad h_j \doteq h_{y_j}, \quad v_j \doteq v_{y_j} (j = 1, \dots, q) \quad \text{and} \quad h_0(k) \doteq \min_{j=1,\dots,q} h_j > 0.$$

c) Now, let $x \in K_0$ be fixed. It belongs to $B(y_j, \eta_j) \subset N(y_j)$ for some index j. Consequently, we can find $x_j \in P(x)$ such that

$$\|v_j - (x_j - x)/h_j\| \le \frac{1}{h_j} d_{P(x)}(x + h_j v_j) + \frac{1}{2k} \le \frac{1}{k}.$$

We set $u_j = (x_j - x)/h_j$. So, cancelling the index j, we have proved that for all $x \in K_0$, there exist $h \in \left[h_0(k), \frac{1}{k}\right]$ and $u \in X$ such that

$$(14) \qquad \begin{cases} \text{i) } x + hu \in P(x), \quad u \in C \doteq F(K_0) + B, \\ \text{ii) } \exists\, y \in K \text{ and } v \in F(y) \quad \text{such that } \|x - y\| \le \frac{1}{k}, \ \|u - v\| \le \frac{1}{k}. \end{cases}$$

d) Let $x_0 \in K$ be fixed. Then we can find $h_0 \in \left[h_0(k), \frac{1}{k}\right]$ and $u_0 \in C$ such that $x_1 \doteq x_0 + h_0 u_0 \in P(x_0)$ and such that $(x_0, u_0) \in \mathrm{graph}\, F + \frac{1}{k}(B \times B)$. Furthermore, $x_1 - x_0 = h_0 u_0 \in h_0 C$ and thus, $\|x_1 - x_0\| \le h_0(\|F(K_0)\| + 1)$. Therefore, in the cases α) and γ), $\|x_1 - x_0\| \le r$ when $h_0 \le T$ and thus, $x_1 \in K_0$. In the case β), x_1 belongs also to K_0.

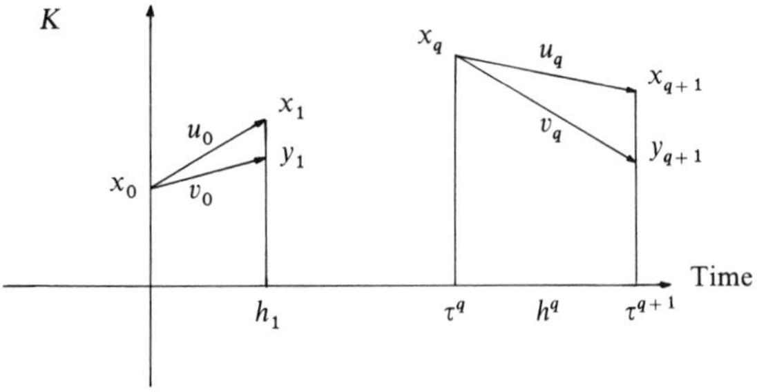

Fig. 15. Modified Euler method

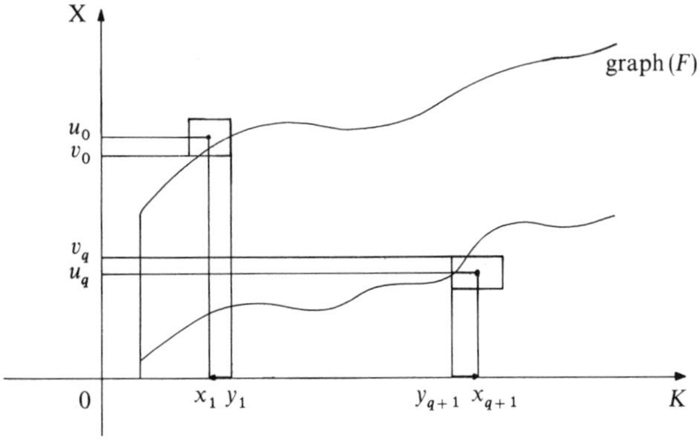

Fig. 16. Construction of pairs (x_q, u_q)

Hence, there exist $h_1 \in \left[h_0(k), \frac{1}{k}\right]$ and $u_1 \in C$ such that $x_2 \doteq x_1 + h_1 u_1 \in P(x_1)$ and such that $(x_1, u_1) \in \operatorname{graph} F + \frac{1}{k}(B \times B)$. Furthermore, $x_2 - x_0 \in (h_0 + h_1)\, C$ and thus, $\|x_2 - x_0\| \leq (h_0 + h_1)(\|F(K_0)\| + 1)$. Therefore, in the cases α) and γ), $\|x_2 - x_0\| \leq r$ when $h_0 + h_1 \leq T$ and thus, $x_2 \in K_0$. In the case β), x_2 belongs to K_0. We reiterate this process for constructing the sequence of elements x_p.

Since the h_j's belong to $\left[h_0(k), \frac{1}{k}\right]$, we are sure that there exists an integer m such that

$$h_0 + h_1 + \ldots + h_m \leq T < h_1 + \ldots + h_m + h_{m+1}.$$

In summary, we have constructed sequences of $h_p \in \left[h_0(k), \frac{1}{k}\right]$, of x_p's$\in K_0$ and u_p's$\in C$ such that

$$(15) \quad \begin{cases} \text{i)}\ x_{p+1} \doteq x_p + h_p u_p \in P(x_p), \quad u_p \in C, \\ \text{ii)}\ (x_p, u_p) \in \operatorname{graph}(F) + \frac{1}{k}(B \times B). \end{cases}$$

This sequence is finite in the cases α) and γ) and infinite in the case β).

e) Let us set $\tau_k^q \doteq h_0 + \ldots + h_{q-1}$, $\tau_k^0 \doteq 0$. We interpolate this sequence by the piecewise linear function $x_k(\cdot)$ defined on each interval $]\tau_k^{p-1}, \tau_k^p]$ by

(16)
$$x_k(t) \doteq x_{p-1} + (t - \tau_k^{p-1})\, u_{p-1}.$$

Let t be fixed in $]\tau_k^{p-1}, \tau_k^p]$. We have $|t-\tau_k^p| \le \frac{1}{k}$ and there exists $(y, v) \in \operatorname{graph}(F)$ such that

$$\|x_k'(t) - v\| = \|u_{p-1} - v\| \le \frac{1}{k}$$

and

$$\|x_k(t) - y\| \le \|x_k(t) - x_{p-1}\| + \|x_{p-1} - y\|$$
$$\le |t - \tau_k^p|\, \|u_{p-1}\| + \|x_{p-1} - y\| < \frac{1}{k}\,(\|F(K_0)\| + 2).$$

We have proved that $\forall\, t \ge 0$,

(17)
$$(x_k(t), x_k'(t)) \in \operatorname{graph}(F) + \varepsilon(k)(B \times B)$$

where $\varepsilon(k) \to 0$ when $k \to \infty$. We also know that

(18)
$$\|x_k'(t)\| \le \|F(K_0)\| + 1$$

and that

(19)
$$x_k(t) \in \operatorname{co}(K_0), \quad \text{which is compact.}$$

Then the assumptions of Theorem 0.3.4 and the Convergence Theorem 1.4.1 are satisfied. A subsequence (again denoted by) $x_k(\cdot)$ converges uniformly to a solution $x(\cdot)$ to the differential inclusion $x' \in F(x)$ and the derivatives $x_k'(\cdot)$ weakly to $x'(\cdot)$ in $L^\infty(0, T; X)$.

f) It remains to check that the trajectory is monotone. Let us choose $s > t$. For k large enough, we can find $p > q$ such that $\tau_k^p > \tau_k^q$ converge to s and t respectively. Then, thanks to the transitivity of P, inclusions $x_k(\tau_k^j) \in P(x_k(\tau_k^{j-1}))$ imply that $x_k(\tau_k^p) \in P(x_k(\tau_k^q))$, i.e., that $(x_k(\tau_k^q)), x_k(\tau_k^p)) \in \operatorname{graph}(P)$. Since the latter is closed and since $x_k(\tau_k^p)$ and $x_k(\tau_k^q)$ converge to $x(s)$ and $x(t)$ respectively, we deduce that $(x(t), x(s)) \in \operatorname{graph}(P)$, i.e., that $x(s) \in P(x(t))$.

The solution $x(\cdot)$ belongs to $\mathscr{C}(0, T; X)$ in the cases α) and γ) and to $\mathscr{C}(0, \infty; X)$ in the case β). Since T is chosen independently of x_0 in the case γ), we can extend the monotone trajectory $x(\cdot)$ defined on $[0, T]$ to a trajectory defined on $[0, 2T]$, $[0, 3T]$, and so on. Hence, there exists a monotone trajectory $x(\cdot) \in \mathscr{C}(0, \infty; X)$ in the case γ). □

Viable Trajectories on Sets Defined by Constraints

We consider the case when the subset K is defined by constraints. For instance, if $A \in \mathscr{L}(X, Y)$ is a continuous linear operator, if $L \subset X$ and $M \subset Y$

are nonempty subsets, we define K by

$$K \doteq \{x \in L \mid Ax \in M\} = L \cap A^{-1}(M). \tag{20}$$

We would like to express the tangential condition

$$\forall x \in K, \quad F(x) \cap T_K(x) \neq \emptyset$$

in terms of L, M and A.

If we are in the favorable situation when

$$\forall x \in K, \quad T_K(x) = T_L(x) \cap A^{-1} T_M(Ax), \tag{21}$$

the tangential condition is indeed couched in terms involving L, M and A:

$$\forall x \in K, \quad F(x) \cap T_L(x) \cap A^{-1} T_M(Ax) \neq \emptyset. \tag{22}$$

We shall see that property (21) holds when, for instance, L and M are closed convex subsets satisfying $0 \in \mathrm{Int}(A(L) - M)$. (See Theorem 5.1.2.)

Otherwise, we devise other sufficient conditions for the existence of viable trajectories. For instance, in the case when $K \doteq L \cap A^{-1}(M)$, we may assume that

$$\begin{aligned} &\forall x \in L, \ \exists v \in F(x) \text{ such that} \\ &\liminf_{h \to 0^+} \frac{1}{h} \max\{d_L(x+hv), d_M(Ax+hAv) - d_M(Ax)\} = 0. \end{aligned} \tag{23}$$

Theorem 4. *Let $L \subset R^n$ and $M \subset R^m$ be two closed subsets and $A \in \mathscr{L}(R^n, R^m)$. Let F be a proper upper hemicontinuous map from L to the compact convex subsets of R^n. We posit the above assumption* (23).

Then for all $x_0 \in L$ satisfying $A(x_0) \in M$, there exists $T > 0$ such that the differential inclusion $x' \in F(x)$, $x(0) = x_0$ has a solution on $[0, T[$ satisfying

$$\forall t \in [0, T[, \quad x(t) \in L \quad \textit{and} \quad Ax(t) \in M. \tag{24}$$

If $F(L)$ is bounded, we can take $T = +\infty$. ▲

Proof. It is largely similar to the proof of Theorem 2.

a) We set $L_0 \doteq L \cap (x_0 + rB)$ and $T \doteq r/(\|F(L_0)\| + 1)$ or T arbitrary and $L_0 \doteq L \cap (cl(x_0 + TF(L) + B))$ when $F(L)$ is bounded.

b) We associate with any $y \in L$ an element $v_y \in X$ and $h_y < \frac{1}{k}$ satisfying

$$\max\{d_L(y + h_y v_y), d_M(Ay + h_y Av_y) - d_M(Ay)\} < h_y/2k \tag{25}$$

and we introduce the open subset

$$N(y) \doteq \{x \in X \mid \max\{d_L(x + h_y v_y), d_M(Ax + h_y v_y) - d_M(Ax)\} < h_y/2k\}. \tag{26}$$

Since $y \in N(y)$, we can find a ball $B(y, \eta_y)$ with radius $\eta_y < 1/k$ contained in $N(y)$. Therefore, the compact subset L_0 can be covered by q such balls $B(y_j, \eta_{y_j})$. We set

$$\eta_j \doteq \eta_{y_j}, \quad h_j \doteq h_{y_j}, \quad v_j \doteq v_{y_j} \quad \text{and} \quad h_0(k) \doteq \min_{j=1,\dots,q} h_j > 0.$$

c) Now, let $x \in L_0$ be given. It belongs to some $B(y_j, \eta_j) \subset N(y_j)$. Consequently, we can find $x_j \in L_0$ such that

$$\left\| v_j - \frac{x_j - x}{h_j} \right\| \leq \frac{1}{h_j} d_L(x + h_j v_j) + \frac{1}{2k} \leq \frac{1}{k}.$$

Hence, if we set $u_j \doteq \frac{x_j - x}{h_j}$, we obtain

$$d_M(Ax + h_j A u_j) \leq d_M(Ax + h_j A v_j) + h_j \|A\| \, \|u_j - v_j\|$$

and since $x \in N(y_j)$,

$$\begin{aligned} d_M(Ax + h_j A u_j) &\leq d_M(Ax) + h_j/2k + h_j \|A\| \, \|u_j - v_j\| \\ &\leq d_M(Ax) + (\|A\| + \tfrac{1}{2}) \, h_j/k. \end{aligned}$$

So, cancelling the index j, we have proved that for all $x \in L$, $\exists\, h \in \left[h_0(k), \frac{1}{k} \right]$ and $u \in C$ such that

$$\text{(27)} \quad \begin{cases} \text{i)} \; x + hu \in L, \\ \text{ii)} \; d_M(Ax + hAu) - d_M(Ax) \leq (\|A\| + \tfrac{1}{2}) \, h/k, \\ \text{iii)} \; (x, u) \in \operatorname{graph}(F) + \frac{1}{k} (B \times B). \end{cases}$$

d) Therefore, we can construct sequences of scalars $h_p \in \left[h_0(k), \frac{1}{k} \right]$, of elements $x_p \in L_0$ and of elements $u_p \in X$ such that

$$\text{(28)} \quad \begin{cases} \text{i)} \; x_{p+1} \doteq x_p + h_p u_p \in L \quad \text{for all } p = 0, 1, \ldots, u_p \in C, \\ \text{ii)} \; d_M(Ax_{p+1}) - d_M(Ax_p) \leq (\|A\| + \tfrac{1}{2}) \, h_p/k, \\ \text{iii)} \; (x_p, u_p) \in \operatorname{graph}(F) + \frac{1}{k} (B \times B). \end{cases}$$

e) We set $\tau_k^q \doteq h_0 + \ldots + h_{q-1}$. We interpolate this sequence by the piecewise linear functions $x_k(\cdot)$ defined on each interval $]\tau_k^{p-1}, \tau_k^p]$ by $x_k(t) \doteq x_{p-1} + (t - \tau_k^{p-1}) u_{p-1}$. As in the proof of Theorem 3 we prove that a subsequence $x_k(\cdot)$ converges to a solution $x(\cdot)$ of the differential inclusion satisfying

$$\text{(29)} \qquad \forall\, t \in [0, T[, \quad x(t) \in L.$$

Since the approximate solutions $x_k(\cdot)$ satisfy, thanks to (28) ii),

$$\text{(30)} \qquad \forall\, p = 0, \ldots \; d_M(Ax_k(\tau_k^{p+1})) - d_M(Ax_k(\tau_k^p)) \leq (\|A\| + \tfrac{1}{2}) \, h_p/k,$$

we deduce, by summing up these inequalities from $p = 0$ to $p = q - 1$ and using the fact that $d_M(Ax(0)) = d_M(Ax_0) = 0$ for $x_0 \in K \doteq L \cap A^{-1}(M)$, that

$$\text{(31)} \qquad d_M(Ax_k(\tau_k^q)) \leq (\|A\| + \tfrac{1}{2}) \, \tau_k^q/k.$$

Since any $t \in [0, T[$ can be approximated by suitable τ_k^q and since $x_k(\tau_k^q)$ converges to $x(t)$, it follows that $d_M(Ax(t)) = 0$, i.e., that $Ax(t) \in M$. Hence, $\forall\, t \in [0, T[$, $Ax(t) \in M$. □

3. Contingent Derivative of a Set-Valued Map

We adapt to the case of a set-valued map the intuitive definition of a derivative of a function in terms of the tangent to its graph.

Let F be a strict set-valued map from $K \subset X$ to Y and (x_0, y_0) belong to the graph of F.

We denote by $DF(x_0, y_0)$ the set-valued map from X to Y whose graph is the contingent cone $T_{\text{graph}(F)}(x_0, y_0)$ to the graph of F at (x_0, y_0).

In other words

$$(1) \qquad v_0 \in DF(x_0, y_0)(u_0) \quad \text{if and only if } (u_0, v_0) \in T_{\text{graph}(F)}(x_0, y_0).$$

Definition 1. We shall say that the set-valued map $DF(x_0, y_0)$ from X to Y is the "contingent derivative" of F at $x_0 \in K$ and $y_0 \in F(x_0)$. ▲

It is a closed process (since its graph is a closed cone). Also, we note that

$$(2) \qquad \operatorname{Dom} DF(x_0, y_0) \subset T_K(x_0)$$

i.e., that the domain of $DF(x_0, y_0)$ is contained in the contingent cone to K at x_0.

We point out that

$$(3) \qquad \forall x_0 \in K,\ \forall y_0 \in F(x_0), \quad DF(x_0, y_0)^{-1} = D(F^{-1})(y_0, x_0).$$

Indeed, to say $(u_0, v_0) \in T_{\text{graph}(F)}(x_0, y_0)$ amounts to saying that $(v_0, u_0) \in T_{\text{graph}(F^{-1})}(y_0, x_0)$.

The contingent derivative allows the derivative of restrictions of functions to subsets with empty interior. If F is a map from X to Y, we denote

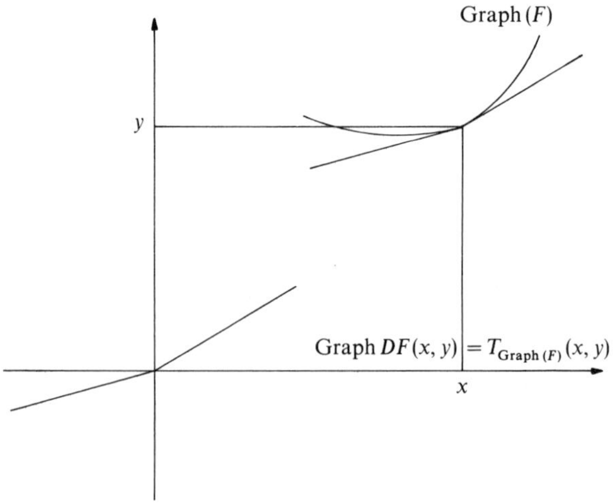

Fig. 17. Graph of the contingent derivative

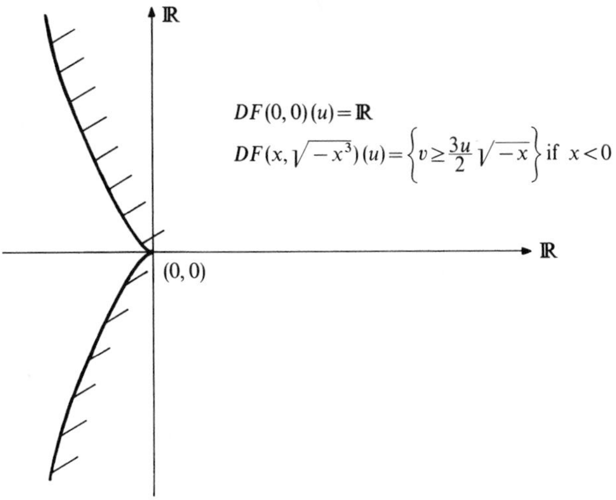

Fig. 18. Graph of the contingent derivative of the map F defined by

$$F(x) = \begin{cases} \mathbb{R} & \text{if } x \geqq 0 \\ \{y \geq \sqrt{-x^3}\} \cup \{y \leq -\sqrt{-x^3}\} & \text{if } x \leq 0 \end{cases}$$

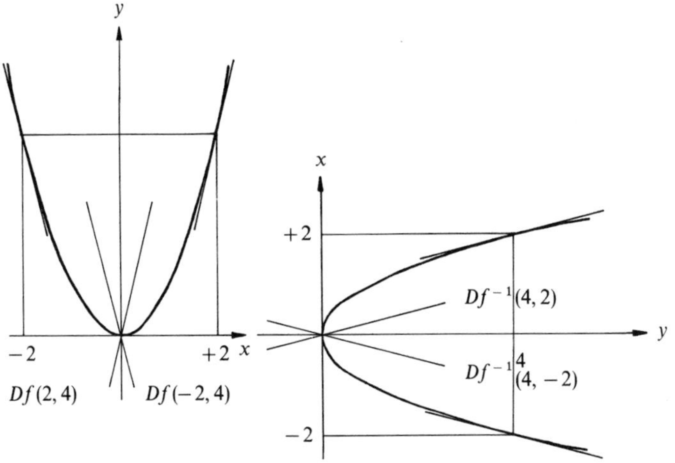

Fig. 19. Derivative of the inverse of the map $x \to x^2$

by $F|_K$ its restriction to K defined by

$$F|_K \doteq \begin{cases} F(x) & \text{when } x \in K \\ \emptyset & \text{when } x \notin K. \end{cases}$$

Proposition 1. *Let F be a continuously differentiable single-valued map on a neighborhood of K and $F|_K$ be its restriction to K. Then*

$$(4)\qquad \forall x_0 \in K, \quad DF|_K(x_0)(u_0) = \begin{cases} \{\nabla F(x_0)\cdot u_0\} & \text{when } u_0 \in T_K(x_0) \\ \emptyset & \text{when } u_0 \notin T_K(x_0). \end{cases} \qquad \blacktriangle$$

Characterization of the Contingent Derivative

We shall give an analytical characterization of $DF(x_0, y_0)$, which justifies that the above definition is a reasonable candidate for capturing the idea of a derivative as a (suitable) limit of differential quotients. We extend F to X by setting $F(x) = \emptyset$ when $x \notin K$.

Proposition 2. *Let F be a set-valued map from $K \subset X$ to Y and $(x_0, y_0) \in \operatorname{graph}(F)$. Then $v_0 \in DF(x_0, y_0)(u_0)$ if and only if*

$$(5)\qquad \liminf_{\substack{h \to 0+ \\ u \to u_0}} d\left(v_0, \frac{F(x_0 + hu) - y_0}{h}\right) = 0. \qquad \blacktriangle$$

Proof. To say that $v_0 \in DF(x_0, y_0)(u_0)$, i.e., that $(u_0, v_0) \in T_{\operatorname{graph}(F)}(x_0, y_0)$, amounts to saying that for all $\varepsilon_1, \varepsilon_2 > 0$ and $\alpha > 0$, there exist $u \in u_0 + \varepsilon_1 B$ and $v \in v_0 + \varepsilon_2 B$ such that $v \in \dfrac{F(x_0 + hu) - y_0}{h}$.

This is equivalent to say $\forall \varepsilon_1 > 0$, $\varepsilon_2 > 0$, $\alpha > 0$, we have

$$\inf_{h < \alpha} \inf_{\|u - u_0\| < \varepsilon_1} d\left(v_0, \frac{F(x_0 + hu) - y_0}{h}\right) < \varepsilon_2.$$

The last statement is equivalent to (5). □

When F is a single-valued map, we set

$$(6)\qquad DF(x_0, y_0) = DF(x_0)$$

since $y_0 = F(x_0)$. The above formula shows that in this case, $v_0 \in DF(x_0)(u_0)$ if and only if

$$(7)\qquad \liminf_{\substack{h \to 0+ \\ u \to u_0}} \frac{\|F(x_0 + hu) - F(x_0) - h v_0\|}{h} = 0.$$

If F is a single-valued map which is *regularly Gâteaux differentiable*, in the sense that there exists $\nabla F(x) \in \mathscr{L}(X, Y)$ satisfying:

$$(8)\qquad \forall u_0 \in X, \quad \lim_{\substack{h \to 0+ \\ u \to u_0}} \frac{F(x_0 + hu) - F(x_0)}{h} = \nabla F(x_0)\, u_0$$

then the contingent derivative coincides with the Gâteaux derivative:

$$(9)\qquad \forall u_0 \in X, \quad DF(x_0)(u_0) = \nabla F(x_0)\, u_0.$$

When F is locally Lipschitzean, formula (5) becomes

(10) $$v_0 \in DF(x_0, y_0)(u_0) \Leftrightarrow \liminf_{h \to 0^+} d\left(v_0, \frac{F(x_0 + h u_0) - y_0}{h}\right) = 0.$$

When the graph of F is convex, we can show that

(11) $$v_0 \in DF(x_0, y_0)(u_0) \Leftrightarrow \liminf_{u \to u_0} \left(\inf_{h > 0} d\left(v_0, \frac{F(x_0 + h u) - y_0}{h}\right)\right) = 0.$$

4. The Time Dependent Case

We shall study now time dependent differential inclusions

(1) $$x' \in F(t, x); \quad x(0) = x_0.$$

We shall look for time-dependent viable trajectories, i.e., trajectories $x(\cdot)$ defined on $[0, T[$ and satisfying:

(2) $$\forall t \in [0, T[, \quad x(t) \in K(t)$$

where $t \to K(t)$ is a strict set-valued map from $[0, \infty[$ to X.

Let us consider the contingent derivative $DK(t, x)$ of the set-valued map K.

Theorem 1. *Let K be a strict set-valued map from $[0, \infty[$ to a Hilbert space X with closed graph and F be a strict upper hemicontinuous map from graph(K) to $R \times X$ with compact convex values. We posit the condition:*

(3) $$\forall t \geq 0, \ \forall x \in K(t), \quad F(t, x) \cap DK(t, x)(1) \neq \emptyset.$$

a) *Assume that* graph(K) *is locally compact. Then for all $t_0 > 0$, for all $x_0 \in K(t_0)$, there exists $T > 0$ such that the differential inclusion* (1) *has a viable trajectory on $[t_0, t_0 + T[$.*

b) *Assume that X is finite dimensional and that F(graph(K)) is bounded. Then for all $t_0 > 0$ and for all $x_0 \in K(t_0)$, there exists a viable trajectory on $[t_0, \infty[$ of the differential inclusion* (1). ▲

Proof. As usual, statements concerning the time dependent case follow from analogous statements on the time independent case by using the following transformation.

We consider the graph $\hat{K} \subset R \times X$ of the set-valued map $K(\cdot)$ and the set-valued map $\hat{F}$ from $\hat{K}$ to $R \times X$ defined by

(4) $$\hat{F}(t, x) \doteq (1, F(t, x)).$$

So the viable trajectories $\tau \to (t(\tau), x(\tau))$ of the time-independent differential inclusion $(t, x)' \in \hat{F}(t, x)$, i.e.,

(5)
$$\text{i)}\ \frac{dt(\tau)}{d\tau}=1,$$
$$\text{ii)}\ \frac{dx(\tau)}{d\tau}\in F(t(\tau),x(\tau)) \quad \text{for almost all } \tau\in[0,T],$$

starting at $(t_0, x_0)\in\hat{K}$ yield the trajectories of the time dependent differential inclusion

(6) $$x'(t)\in F(t,x), \quad x(t_0)=x_0 \text{ given in } K(t_0)$$

satisfying

(7) $$\forall t\in[t_0, t_0+T[, \quad x(t)\in K(t).$$

Also, we note that the tangential condition for the differential inclusion $(t,x)'\in\hat{F}(t,x)$ is

(8) $$\forall(t,x)\in\hat{K}=\operatorname{graph}(K), \quad (1,F(t,x))\in T_{\operatorname{graph}(K)}(t,x).$$

By the very definition of the contingent derivative of K, this is equivalent to the condition (3). □

Remark. Condition (3) is also necessary for viable trajectories to exist for all $(t_0, x_0)\in\operatorname{graph}(F)$. The above transformation and *Proposition 1.1* imply the following statement:

Proposition 1. *Let K be a strict set-valued map from $[0,\infty[$ to R^n and F be a strict bounded upper semicontinuous map from* graph(K) *to R^n with compact convex values. If for all $t_0\geq 0$ and $x_0\in K(t_0)$, there exist $T>0$ and a viable trajectory on $[t_0, t_0+T[$ of the differential inclusion*

(1) $$x'\in F(t,x), \quad x(t_0)=x_0$$

then the condition

(3) $$\forall t\geq 0,\ \forall x\in K(t), \quad F(t,x)\cap DK(t,x)(1)\neq\emptyset$$

holds true. ▲

Application: Implicit Differential Inclusions

Let f be a single-valued map from $R_+\times R^n\times R^n\to R^m$. We shall solve the implicit differential inclusion.

(9) $$f(t,x(t),x'(t))\in F(t,x(t)); \quad x(0)=x_0 \text{ given in } K.$$

Corollary 1. *We assume that the graph $\mathscr{K}$ of the set-valued map $t\in R_+\to K(t)\subset R^n$ is closed, that the map f: $(t,x,v)\in\mathscr{K}\times R^n\to f(t,x,v)\in R^m$ is continuous and affine with respect to v and that the proper set-valued map F: $\mathscr{K}\to R^m$ is upper hemicontinuous with closed convex images. We posit the following* *tangential condition: There exists a constant $c>0$ such that*

$$(10)\qquad \begin{aligned}&\forall t\geq 0,\ \forall x\in K(t), \quad \exists v\in DK(t,x)(1)\cap cB\\ &\text{such that } f(t,x,v)\in F(t,x).\end{aligned}$$

Then, for all $x_0\in K(0)$, there exists a viable solution $x(\cdot)$ to the implicit differential inclusion (9). ▲

Proof. We set $G(t,x)\doteq\{v\in R^n$ such that $f(t,x,v)\in F(t,x)\}$. Since f is continuous and since the graph of F is closed, then, obviously, the graph of G is also closed. Therefore, the set-valued map H: $(t,x)\to G(t,x)\cap cB$ has nonempty values (by assumption (10)), is upper semicontinuous (having a closed graph and taking its values in the compact set cB), is bounded and has compact convex values. It also satisfies the condition:

$$\forall (t,x)\in\operatorname{graph}(K), \quad \exists v\in H(t,x)\cap DK(t,x)(1).$$

Hence Theorem 1 implies the existence of viable solutions to the differential inclusion $x'(t)\in H(t,x(t))$, $x(0)=x_0$, which are obviously viable solutions to the implicit differential inclusion (9). □

Example. Let $X=R^n$ be the space of states of the system we wish to describe and $Y=R^m$ be the space of "observations". We denote by B: $X\to Y$ the "observation map" of the system and by C: $Y\to X$ the "feedback map".

In this model, we assume that the evolution law is

$$(11)\qquad x'(t)\in F(x(t))+C\frac{d}{dt}B(x(t)); \quad x(0)=x_0.$$

In other words, we assume that the velocity depends not only upon the state of the system but also upon the variations of observations of the state.

We assume that

(12)
- i) $C\in\mathscr{L}(Y,X)$ is linear,
- ii) B: $X\to Y$ is continuously differentiable on an open subset Ω containing K.

We set

$$(13)\qquad A(x)\doteq I-C\nabla B(x).$$

So, the system can be written

$$(14)\qquad A(x(t))\,x'(t)\in F(x(t)).$$

Corollary 2. *Assume that $K\subset X=R^n$ is a closed subset, F is an upper hemicontinuous set-valued map from K into X with nonempty closed convex values. Let $C\in\mathscr{L}(Y,X)$ and $B\in\mathscr{C}^1(\Omega,Y)$. We suppose that there exists $c>0$ such that,*

$$(15)\qquad \forall x\in K, \quad \exists v\in F(x)+C\nabla B(x)\,v \text{ such that } v\in T_K(x)\cap cB.$$

Then, for any initial state $x_0 \in K$, *there exists a viable solution to the inclusion* (11). ▲

We can also, by the same device, characterize the existence of time-dependent monotone trajectories. We define the time-dependent preorder by a set-valued map from graph(K) to graph(K) satisfying:

$$(16) \quad \begin{array}{lll} \text{i) } \forall t \geq 0, \ \forall x \in K(t), & (t,x) \in P(t,x) & \text{(reflexivity)}, \\ \text{ii) } \forall (s,y) \in P(t,x), & \text{we have } P(s,y) \subset P(t,x) & \text{(transitivity)}. \end{array}$$

So, for each $(t,x) \in \text{graph}(K)$, $P(t,x)$ can be regarded as the graph of a set-valued map $Q_{t,x}(\cdot)$ from $[t, \infty[$ to X defined by

$$(17) \quad \forall s \geq t, \quad y \in Q_{t,x}(s) \quad \text{if} \quad (s,y) \in P(t,x).$$

In other words if $P(t,x)$ is regarded as defining a "preference preorder", then $Q_{t,x}(s)$ is the subset of elements y which are "preferred at time s to an element x given at time t". Therefore, if F is a set-valued map defined on graph(K), to say that

$$(18) \quad \forall t \geq 0, \ \forall x \in K(t), \quad (1, F(t,x)) \cap T_{P(t,x)}(t,x) \neq \emptyset$$

amounts to saying that

$$(19) \quad \forall t \geq 0, \ \forall x \in K(t), \quad F(t,x) \cap DQ_{t,x}(t,x)(1) \neq \emptyset.$$

This condition is necessary and sufficient for the existence of time dependent monotone trajectories. We only state the following consequence of Theorem 1.2.

Corollary 3. *Let K be a strict set-valued map from $[0, \infty[$ to a Hilbert space X with closed graph, P:* graph$(K) \to$ graph(K) *be a strict lower semicontinuous map with closed graph satisfying reflexivity and transitivity conditions* (16) *and F:* graph$(K) \to R \times X$ *be a strict upper hemicontinuous map with compact convex values. We posit the following condition:*

$$(20) \quad \forall t \geq 0, \ \forall x \in K(t), \quad F(t,x) \cap DQ_{t,x}(t,x)(1) \neq \emptyset.$$

a) *Assume that* graph(K) *is locally compact. Then for all $t_0 \geq 0$ and $x_0 \in K(t_0)$, there exists $T > 0$ such that the differential inclusion*

$$(1) \quad x' \in F(t,x), \quad x(0) = x_0$$

has a solution on $[t_0, t_0 + T[$ that is monotone in the sense that

$$(21) \quad \forall s, t \in [t_0, t_0 + T[, \quad s \geq t, \quad x(s) \in Q_{t,x(t)}(s).$$

b) *Assume that X is finite dimensional and that F is bounded. The above holds for $T = \infty$.* □

5. A Continuous Version of Newton's Method

We prove a version of the Inverse Function Theorem that we state in the following form.

Theorem 1. *Let f be a continuously differentiable map from a neighborhood of a closed subset K of a Hilbert space X to a Hilbert space Y. We assume that*

$$\text{(1)} \qquad \begin{cases} \exists\, c>0 \text{ such that, } \forall x\in K,\ \exists\, u\in T_K(x) \text{ satisfying} \\ \nabla f(x)\,u = -f(x) \text{ and } \|u\| \le c\,\|f(x)\|. \end{cases}$$

Then there exists an equilibrium of f;

$$\text{(2)} \qquad \exists\, \bar{x}\in K \quad \text{such that } f(\bar{x})=0.$$

Furthermore, for all $x\in K$,

$$\text{(3)} \qquad d(x, f^{-1}(0)) \le c\,\|f(x)\|. \qquad \blacktriangle$$

Actually, this result is a consequence of its set-valued version.

Theorem 2. *Let F be a proper set-valued map with closed graph from a Banach space X to a Banach space Y. We assume that*

$$\text{(4)} \qquad \begin{cases} \exists\, c>0 \text{ such that } \forall x\in \operatorname{Dom} F,\ \forall y\in F(x),\ \exists\, u\in X \text{ satisfying} \\ -y\in DF(x,y)(u) \text{ and } \|u\| \le c\,\|y\|. \end{cases}$$

Then there exists an equilibrium of F:

$$\text{(5)} \qquad \exists\, \bar{x}\in X \text{ such that } 0\in F(\bar{x}).$$

Furthermore, for all $x\in K$, $y\in F(x)$,

$$\text{(6)} \qquad d(x, F^{-1}(0)) \le c\,\|y\|. \qquad \blacktriangle$$

These results follow from Ekeland's Theorem (see for instance Aubin [1979a], Theorem 4.9.6, p. 174).

Theorem 3 (Ekeland). *Let V be a proper lower semicontinuous function from a Banach space Z to $R\cup\{+\infty\}$. Let $z_0\in \operatorname{Dom} V$ be fixed and $\varepsilon>0$ chosen. Then there exists $\bar{z}\in \operatorname{Dom} V$ satisfying*

$$\text{(7)} \qquad \begin{aligned} &\text{i) } V(\bar{z})+\varepsilon\,\|\bar{z}-z_0\| \le V(z_0), \\ &\text{ii) } \forall z\in Z, \quad V(\bar{z}) \le V(z)+\varepsilon\,\|z-\bar{z}\|. \end{aligned} \qquad \blacktriangle$$

Proof of Theorem 2. We apply Ekeland's Theorem to the case when $Z \doteq X \times Y$, $V(x,y)=\|y\|$ when $(x,y)\in \operatorname{graph} F$ and $V(x,y)=+\infty$ otherwise. Let $z_0 \doteq (x_0, y_0)$ belong to graph(F) and $\varepsilon<(1+c)^{-1}$ be fixed. By Ekeland's Theorem, there exists $\bar{z}\doteq(\bar{x},\bar{y})\in\operatorname{graph}(F)$ such that

$$\text{(8)} \qquad \begin{aligned} &\text{i) } \|\bar{y}\| + \varepsilon(\|\bar{x}-x_0\| + \|\bar{y}-y_0\|) \le \|y_0\|, \\ &\text{ii) } \forall (x,y)\in\operatorname{graph}(F), \quad \|\bar{y}\| \le \|y\| + \varepsilon(\|x-\bar{x}\| + \|y-\bar{y}\|). \end{aligned}$$

We now use assumption (4) for deriving that $\bar{y}=0$ and consequently, that $\bar{x}$ is an equilibrium of F. Indeed, we take $\bar{u}\in X$, solution to

$$(9)\qquad -\bar{y}\in DF(\bar{x},\bar{y})(\bar{u}),\qquad \|\bar{u}\|\le c\|\bar{y}\|.$$

By the very definition of contingent derivative, we can associate to any α and β converging to 0 elements u_α and v_α and $h\in]0,\beta[$ satisfying $\|u_\alpha\|\le\alpha$, $\|v_\alpha\|\le\alpha$ and

$$(10)\qquad (\bar{x}+h\bar{u}+hu_\alpha,\bar{y}-h\bar{y}+hv_\alpha)\in\operatorname{graph}(F).$$

We plug this pair in formula (8) ii): we get

$$\begin{aligned}\|\bar{y}\|&\le\|(1-h)\bar{y}+hv_\alpha\|+\varepsilon h(\|\bar{u}+u_\alpha\|+\|\bar{y}+v_\alpha\|)\\&\le(1-h)\|\bar{y}\|+\varepsilon h(1+c)\|\bar{y}\|+\alpha h(1+2\varepsilon).\end{aligned}$$

In summary, we obtain after division by $h>0$

$$\|\bar{y}\|\le\varepsilon(1+c)\|\bar{y}\|+\alpha(1+2\varepsilon).$$

By letting α converge to 0, we deduce that

$$(1-\varepsilon(1+c))\|\bar{y}\|\le 0.$$

We infer that $\bar{y}$ is equal to 0. By taking $\bar{y}=0$ in inequality (8) i), we deduce that

$$d(x_0,F^{-1}(0))\le\|x_0-\bar{x}\|\le\left(\frac{1}{\varepsilon}-1\right)\|y_0\|.$$

By letting ε converge to $(1+c)^{-1}$, we deduce that $d(x_0,F^{-1}(0))\le c\|y_0\|$. □

It is remarkable that assumption (1) is the necessary and sufficient condition for the implicit differential equation

$$(11)\qquad \begin{aligned}&\text{i)}\ \nabla f(x(t))\,x'(t)=-f(x(t)),\\&\text{ii)}\ x(0)=x_0\in K,\end{aligned}$$

to have *viable trajectories*. We observe that such trajectories satisfy the equation

$$(12)\qquad \frac{d}{dt}f(x(t))=\nabla f(x(t))\,x'(t))=-f(x(t))$$

and thus, verify

$$(13)\qquad f(x(t))=e^{-t}f(x_0).$$

Consequently, $f(x(t))$ converges to 0 and, by formula (3), we have

$$(14)\qquad d(x(t),f^{-1}(0))\le c\,e^{-t}\|f(x_0)\|.$$

Hence any cluster point $\bar{x}\in K$ of such trajectory, limit of a subsequence $x(t_n)$ when $t_n\to\infty$, is an equilibrium of F. □

We recognize that trajectories of the implicit differential equation (11) are the continuous analogs of the classical *Newton method*, which yields the discrete trajectories defined recursively by

(15) $$\nabla f(x_n)(x_{n+1}-x_n)=-f(x_n); \quad x_0 \text{ given}.$$

We summarize the above remarks in the following statement.

Theorem 4. *Let f be a bounded continuous differentiable map from a neighborhood of a closed subset $K\subset R^n$ to R^m. Then condition (1) is necessary and sufficient for the implicit differential equation (11) to have, for all $x_0\in K$, viable trajectories $x(\cdot)$ on $[0,\infty[$ satisfying $\|x'(t)\|\le c\|f(x(t))\|$. Furthermore, for all $t\ge 0$, we have*

(16) $$\begin{cases} \text{i)}\ f(x(t))=e^{-t}f(x_0), \\ \text{ii)}\ d(x(t),f^{-1}(0))\le c\,e^{-t}\|f(x(0))\|. \end{cases}$$

Every cluster point of such trajectories is an equilibrium of F. ▲

Proof. We apply the Viability Theorem to the set-valued map G defined by

$$G(x)\doteq -\nabla f(x)^{-1}f(x)\cap c\|f(x)\|\,B$$

which is a bounded upper semicontinuous map with compact convex values. We observe that condition (1) is equivalent to the tangential condition

(17) $$\forall x\in K, \quad G(x)\cap T_K(x)\neq\emptyset.$$ □

We can adapt Newton's Method to the case of set-valued map.

Theorem 5. *Let F be a proper bounded set-valued map with closed graph from R^n to R^m. We assume that there exist a constant $c>0$ and a continuous map g from the graph of F to R^n satisfying*

(18) $$\forall (x,y)\in\text{graph}(F), \quad -y\in DF(x,y)(g(x,y)) \quad \textit{and} \quad \|g(x,y)\|\le c\|y\|.$$

Then, for all $(x_0,y_0)\in\text{graph}(F)$, there exists a solution to the differential equation

(19) $$x'(t)=g(x(t),e^{-t}y_0), \quad x(0)=x_0$$

that satisfies, for all $t>0$,

(20) $$\begin{cases} \text{i)}\ x(t)\in\text{Dom}(F), \\ \text{ii)}\ e^{-t}y_0\in F(x(t)), \\ \text{iii)}\ d(x(t),F^{-1}(0))\le c\,e^{-t}\|y_0\|. \end{cases}$$

Every cluster point of the trajectory is an equilibrium of F. ▲

Proof. We consider the differential equation

(21) $$x'=g(x,y), \quad y'=-y$$

with the initial condition $x(0)=x_0$, $y(0)=y_0$.

Condition (18) implies that

(22) $$\forall (x, y) \in \operatorname{graph}(F), \quad (g(x, y), -y) \in T_{\operatorname{graph}(F)}(x, y) \neq \emptyset.$$

Then the Viability Theorem implies that there exists a trajectory $(x(t), y(t))$ of the differential equation (21) which remains in graph(F). Furthermore, $y(t) = e^{-t} y_0$. Hence $e^{-t} y_0 \in F(x(t))$. The rest of the theorem ensues. □

Remark. We note that we can devise a whole family of algorithms that converge to an equilibrium of F. Let h be any map from R^n to itself such that

(23) $$\begin{array}{l}\text{the solution of } y' = h(y),\ y(0) = y_0 \text{ is unique} \\ \text{and converges to 0 when } t \to \infty.\end{array}$$

We associate with such a map h a bounded continuous map g (single-valued for the sake of simplicity) such that

(24) $$\forall (x, y) \in \operatorname{graph}(F), \quad h(y) \in DF(x, y)(g(x, y)).$$

Then there exists a solution to the differential equation

(25) $$x'(t) = g(x(t), y(t)), \quad y'(t) = h(y(t)); \quad x(0) = x_0, \ y(0) = y_0$$

such that

(26) $$\forall t \geq 0, \quad y(t) \in F(x(t)).$$

Since $\lim_{t \to \infty} y(t) = 0$ by assumption, the cluster points of $x(t)$ (if any) are equilibria of F. □

6. A Viability Theorem for Continuous Maps with Nonconvex Images

We relax in this section the assumption that the images of the map F are convex. As we saw in proving Filippov's Theorem 2.3.1, we impose a stronger regularity assumption, namely, that F is continuous. For proving the existence of a viable trajectory, we shall be led to require also a stronger condition, namely

(1) $$\forall x \in K, \quad F(x) \subset T_K(x).$$

Theorem 1. *Let $K \subset R^n$ be closed, F from K into the compact subsets of R^n be continuous and such that $F(x) \subset T_K(x)$, $x \in K$. Then given $x_0 \in K$, there exist $T > 0$ and a viable trajectory on $[0, T]$ of the differential inclusion*

(2) $$x'(t) \in F(x(t)), \quad x(0) = x_0.$$

Moreover $x'(\cdot)$ is a regulated function. ▲

Before proving this theorem, we need a lemma showing how the strong tangential condition (1) and the regularity of F strengthen the regularity of the function $h \to \frac{d_K(x+hv)}{h}$ when v belongs to $F(x)$. In the following lemma, two sets $K_0 \subset K$ and K are used. For instance K_0 could be $B_K(x_0, r)$. Observe also that a map F satisfying $F(x) \subset T_K(x)$ need not satisfy $F(x) \subset T_{K_0}(x)$.

Lemma 1. *Let $K_0 \subset K$ be closed, F a map from K into the closed subsets of R^n and $r>0$ such that on $B_K(K_0, r)$, F is uniformly continuous and bounded. Moreover, assume that for x in $B_K(K_0, r)$, $F(x) \subset T_K(x)$.*

Then for every $\varepsilon>0$ there exists $h>0$ such that,

(3) *for every $t \in]0, h]$, $x \in K_0$ and $v \in F(x)$,* $\quad \frac{d_K(x+tv)}{t} \le \varepsilon.$ ▲

Proof. Let M be such that $\|F(x)\| \le M$ on $B_K(K_0, r)$. There exists h (we assume $0<h \le r/2M$) such that $x, x' \in B_K(K_0, r)$ and $d(x, x')<2Mh$ imply $\mathfrak{d}(F(x), F(x')) \le \varepsilon$. Fix $x \in K_0$, $v \in F(x)$. We have to show that $f(t) \doteq d_K(x+tv) \le \varepsilon t$ for $t \in]0, h]$. Since f is Lipschitzean, hence absolutely continuous, it is enough to show that whenever $f'(t)$ exists, its norm is bounded by ε. A fortiori, from *Corollary 1.1* it is enough to show that $d(v, T_K(\pi_K(x+tv))) \le \varepsilon$. Since for $t \in [0, h]$, $\pi_K(x+tv) \subset B_K(x, 2Mh)$, at once we have that $\pi_K(x+tv) \subset B(K_0, r)$ and that $d(v, T_K(\pi_K(x+tv))) \le d(v, F(x+tv)) \le \varepsilon$. □

Proof of Theorem 1. The proof follows the same pattern as the proof of Theorem 2.3.1. Consider x_0: since F is continuous with compact values, there exists $M>1$ such that $\|F(x)\| \le M-1$ for $x \in B_K[x_0, r]$. Set $T \doteq r/M$, $K_0 \doteq B_K[x_0, r]$, $\eta_k \doteq 2^{-k}$, let δ_k and h_k be such that $x' \in B_K(x, \delta_k)$ implies $\mathfrak{d}(F(x), F(x')) < \frac{1}{3}\eta_k$; $t \in [0, h_k]$ implies $d_K(x+tv) < \frac{1}{3}\eta_k t$, any v in $F(x)$, any $x \in K_0$. Choose numbers $l_k \le \inf\{\delta_k, h_k\}$ such that T/l_1 is an integer, l_{k-1}/l_k is an integer, and define a sequence of partitions of $[0, T]$ into right open intervals $[i\,l_k, (i+1)\,l_k[$; by construction each partition is a refinement of the preceding. The points $i\,l_k$ are the nodal points of the k-th partition.

Let us define the first polygonal approximate solution, x_1, as follows. At each nodal point t^* of the first partition, starting with $t^*=0$, fix arbitrarily a $v \in F(x_1(t^*))$. Choose $y \in \pi_K(x_1(t^*)+l_1 v)$ and call $x_1'(t^*)=(y-x_1(t^*))/l_1$. Set on $]t^*, t^*+l_1]$

$$x_1(t)=x_1(t^*)+x_1'(t^*)(t-t^*).$$

Then x_1 has a derivative constant on $[t^*, t^*+l_1[$ (a right derivative at t^*) and at the nodal point t^*+l_1, $x_1(t^*+l_1)=y \in K$. Moreover

$$\begin{aligned} d(x_1'(t^*), F(x_1(t^*))) &\le \|x_1'(t^*)-v\| = \frac{1}{l_1}\|y-x_1(t^*)-l_1 v\| \\ &= \frac{1}{l_1} d_K(x_1(t^*)+l_1 v) \le \tfrac{1}{3}\eta_1 \end{aligned}$$

because of our choice of l_1. In particular this last estimate yields that $\|x_1'(t^*)\| \leq \|F\| + \frac{1}{3}\eta_1 \leq M$ hence that x_1 is Lipschitzean of constant M on $]t^*, t^*+l_1]$. By defining recursively x_1 on each $]t^*, t^*+l_1]$ for every nodal point t^*, starting with $t^*=0$, in a finite number of steps x_1 is defined on $[0, T]$.

The construction of x_m will now be presented, by a recurrence argument on the nodal points of the m-th partition. We use the same maps $0(\tau)$ and $s(\tau)$, defined on the nodal points, as presented in the proof of Theorem 2.3.1.

We wish to construct a function x_m satisfying at each nodal point t of the m-th partition

(4)
- i) x_m is a continuous map from $[0, t]$ with $x_m(0)=x^0$; its derivative x_m' is constant on each interval of the m-th partition (it is only a right derivative at the nodal points);
- ii) at each nodal point $\tau \in [0, t]$, $x_m(\tau) \in K$ and (for $\tau < t$),
  $x_m'(\tau) \in F(x_m(\tau)) + \frac{1}{3}\eta_m B$;
- iii) at each nodal point $\tau \in [0, t]$ such that $O(\tau) > 1$,
  $d(x_m'(\tau), x'\ (s(\tau))) \leq \eta_{O(\tau)-1}$.

Remark that x_1 satisfies (4) i) and ii), while (4) iii) is empty since there are no points in the first partition with $O(\tau) > 1$. We claim that x_m satisfying (4) i), ii) and iii) above can be defined on $[0, T]$.

When $t=0$, the function $x_m = x_0$ defined on $[0, 0]$ satisfies (4) i), ii) and iii). Assume that x_m has been constructed up to the nodal point t^*.

Consider $O(t^*)$ and $s \doteq s(t^*)$. Since $|t^* - s| < l_{O(\tau)-1}$, we have

$$d(x_m(t^*), x_m(s)) \leq M\, l_{O(\tau)-1},$$

$$\text{hence } \mathfrak{d}(F(x_m(t^*)), F(x_m(s))) \leq \tfrac{1}{3}\eta_{O(\tau)-1}.$$

By assumption (4) ii) $d(x_m'(s), F(x_m(s))) \leq \frac{1}{3}\eta_m$; hence there exists $v \in F(x_m(t^*))$ such that

$$d(v, x_m'(s)) \leq \tfrac{1}{3}\eta_{O(\tau)-1} + \tfrac{1}{3}\eta_m.$$

Choose $y \in \pi_K(x_m(t^*) + l_m v)$, call $x_m'(t^*) \doteq (y - x_m(t^*))/l_1$ and set

$$x_m(t) \doteq x_m(t^*) + x_m'(t^*)\,(t - t^*) \quad \forall t \in]t^*, t^* + l_m]. \tag{5}$$

At the new nodal point $t^* + l_1$, $x_m(t^* + l_1) = y$ belongs to K; also

$$d(x_m'(t^*), F(x_1(t^*)) \leq \|x_m'(t^*) - v\| = \frac{1}{l_m}\|y - x_m(t^*) - l_m v\|$$

$$= \frac{1}{l_m} d_K(x_m(t^*) + l_m v) \leq \eta_m/3$$

so that (4) ii) holds. About (4) iii),

$$d(x'_m(t^*), x'_m(s(t^*))) \leq d(x'_m(t^*), v) + d(v, x'_m(s(t^*)))$$
$$\leq \tfrac{1}{3}\eta_m + \tfrac{1}{3}\eta_{O(\tau)-1} + \tfrac{1}{3}\eta_m < \eta_{O(\tau)-1}.$$

Hence there exists a sequence $\{x_m(\cdot)\}$ with each x_m satisfying (4) i), ii) and iii) on $[0, T]$.

Since the oscillation of x'_m on each interval is estimated by using (4) iii) that is identical with point (2) iii) of the proof of Theorem 2.3.1 the sequence $\{x'_m\}$ is equi-oscillating. Hence the Compactness Theorem 0.3.5 in $\mathscr{B}(0, T; R^n)$ implies that there is a subsequence, that we keep with the same indexes, converging to some $v_*(\cdot) \in \mathscr{B}(0, T; R^n)$. Then $\{x_m(\cdot)\}$ converges uniformly to $x(\cdot)$ satisfying

$$x(t) = x^0 + \int_0^t v_*(s)\, ds.$$

The function $x(\cdot)$ satisfies $x(0) = x_0$; let t be a nodal point of one of the partitions, say the q-th. Then $x_m(t)$ belongs to K for $m > q$ and, K being closed, $x(t) \in K$. The set of such points, where $x(t) \in K$, is dense. Since K is closed and $x(\cdot)$ is continuous, $x(t)$ belongs to K for every $t \in [0, T]$.

Let t be a point such that $d(x'_m(t), v_*(t)) \to 0$ and $x'(t) = v_*(t)$ (almost every t in $[0, T]$ has the above properties); call t_m the initial point of the interval of the m-th partition to which t belongs. Then

$$d(v_*(t), F(x(t)) \leq d(v_*(t), x'_m(t)) + d(x'_m(t), x'_m(t_m))$$
$$+ d(x'_m(t_m), F(x_m(t_m))) + \mathfrak{d}(F(x_m(t_m), F(x_m(t)))$$
$$+ \mathfrak{d}(F(x_m(t)), F(x(t))).$$

To show that $d(v_*(t), F(x(t))) = 0$ the same reasoning of the proof of Theorem 2.3.1 applies. The only modification required is that $d(x'_m(t_m), F(x_m(t_m))$ this time is not zero, but by (4) ii) is bounded by $\frac{1}{3}\eta_m$, converging to 0. □

Corollary 1. *To the assumptions of* Theorem 1 *add that F is bounded. Then for every $x_0 \in K$ there exists a viable trajectory of the differential inclusion defined on $[0, \infty)$.* ▲

Proof. There exists $M > 1$ such that $M - 1$ bounds $\|F\|$ on the whole of K. Then we can take T arbitrary in the proof of the above Theorem, independently of the initial point t_0. It is therefore sufficient to piece together the viable trajectories of the differential inclusions

$$\forall t \in [jT, (j+1)T], \quad x'_j(t) \in F(x_j(t)), \quad x_j(jT) = x_{j-1}(jT). \qquad \square$$

The viability theorem of Section 2 requires the intersection of the velocity field and of the contingent cone of K to be non empty in order to provide existence of viable trajectories while Theorem 1 of this section demands that F be contained in the contingent cone. The lack of convexity of the images $F(x)$ makes the difference. One can wonder whether by imposing more conditions on K (say: K is convex) or on the regularity of F

(say: F is Lipschitzean) it would be possible to obtain the existence of viable trajectories by imposing only the non-empty intersection condition 4.6.1, even in the case when F is not convex valued.

This is not so, as the following examples shows.

Example. Let $X = R^2$, K the closed unit ball, $\bar{B} \doteq \{x \mid \|x\| \leq 1\}$. Define F on K to be the constant map:

$$F(x) \equiv \{(-1, 0), (1, 0)\}.$$

Then the problem

$$x' \in F(x), \qquad x(0) = x^0$$

has no solution when x^0 is either $(0, 1)$ or $(0, -1)$.

Invariant Subsets

Theorem 1 and its Corollary 1 guarantee the existence of at least a viable trajectory to the differential inclusion (2), where F is defined on K only. Taking this point of view, we are not allowed to ask whether there exist solutions that do leave K, since F is not defined outside K! Assuming that F has been, or that it can be, extended to an open neighborhood of K, the possibility of having solutions to (2) that do leave K *depends entirely* on *what happens outside* K, i.e. is not a problem that can be solved by imposing additional regularity for F on K. For instance, for $K = (-\infty, 0] \subset R$ and $F = \{0\}$ on K, $x(t) \equiv 0$ is a solution to $x' \in F(x)$, $x(0) = 0$. Being F constant, it has all the regularity we wish. However F can be extended to R as a continuous map by setting $F(x) = \{\sqrt{x}\}$ for $x > 0$. The same problem $x' \in F(x)$, $x(0) = 0 \in K$, admits solutions that do leave K.

The following is a result along these lines.

Theorem 2. *Let K be closed, F from an open neighborhood $\mathcal{N}$ of K into the compact subsets of R^n be Lipschitzean of constant L and such that for $x \in K$, $F(x) \subset T_K(x)$. Then any solution to*

$$x' \in F(x), \qquad x(0) = x_0 \in K \tag{2}$$

is viable. ▲

The same proof as for Proposition 1.3 yields the following slightly more general statement.

Proposition 1. *Let $y \in X$, $\varepsilon(\cdot)$ a map with values in R^n such that $\varepsilon(h) \to 0$ as $h \to 0$. Then*

$$\liminf_{h \to 0^+} \frac{1}{h} \left(d_K(y + h(v + \varepsilon(h))) - d_K(y) \right) \leq d(v, T_K(\pi_K(y))). \tag{6}$$ ▲

Proof of Theorem 2. Let $x(\cdot)$ be a solution, defined on $[0, T]$ and call $g(t)$ the non negative, Lipschitzean map $g(t) \doteq d_K(x(t))$.

Let t be a point where both $g'(t)$ and $x'(t)$ exist. Then there exists $\varepsilon(h)$, $\varepsilon(h) \to 0$ as $h \to 0$, such that $x(t+h) = x(t) + h(x'(t) + \varepsilon(h))$ and

$$(7) \qquad g'(t) = \lim_{h \to 0} \frac{1}{h}(d_K(x(t) + h(x'(t) + \varepsilon(h))) - d_K(x(t))).$$

By *Proposition 1*, $g'(t) \leq d_K(x'(t), T_K(\pi_K(x(t))))$. Let $y \in \pi_K(x(t))$. Then

$$\begin{aligned} d(x'(t), T_K(\pi_K(x(t)))) &\leq d(x'(t), T_K(y)) \leq d(x'(t), F(y)) \\ &\leq \mathfrak{d}(F(x(t)), F(y)) \leq L\, d_K(x(t)) = L\, g(t). \end{aligned}$$

Then g satisfies a.e.

$$(8) \qquad g'(t) \leq L\, g(t), \quad g(0) = 0$$

i.e. $g \equiv 0$ on $[0, T]$. □

Remark that in above Theorem, existence of a solution having values in K was obtained by taking any solution of $x' \in F(x)$ with F defined on an open set, and then by proving that such a solution remains in K. No use was made of the viability Theorem. One should not be led to believe that this can be a general way to prove existence of solutions remaining in a set K.

For the above method to be feasible, one has to prove that, given: K, F on K and the viability conditions on F, there exists an extension $\tilde{F}$ of F to an open neighborhood of K such that all solutions to $x' \in \tilde{F}(x)$ issuing from K do not leave K.

Definition 1. Let $\Omega \subset X$ be an open neighborhood of K, $\tilde{F}$ a map from Ω into the subsets of X. We say that K is *invariant* under $\tilde{F}$ if for every $x_0 \in K$ all solutions to $x' \in \tilde{F}(x)$, $x(0) = x_0$ remain in K. ▲

In general *a viability problem cannot be imbedded as an invariance problem.* In other words, given K and F on K so that for any $x^0 \in K$ there are solutions to $x' \in F(x)$, $x(0) = x^0$, in general there are no extension $\tilde{F}$ of F to an open $\Omega \subset K$ so that K is invariant under F. The following is a single-valued example.

Example. Let $K \subset R^2$ be the union of the two circles C_1 and C_2, where $C_i = \{(x_1, y_2) \colon (x_1 - (-1)^i)^2 + x_2^2 = 1\}$ and let $f(x)$ be the vector of unit length tangent to K at x in the direction of counterclockwise rotation for C_1 and of clockwise rotation for C_2 (at their intersection, the origin $(0, 0)$, f is the vector $(0, 1)$). The map f is continuous on K and satisfies the viability condition. There are two solutions to $x' = f(x)$, $x(0) = 0$, namely the two solutions with angular velocity 1, counterclockwise on C_1 and clockwise on C_2. Let $\tilde{f}$ be any continuous extension of f to a neighborhood of K.

Consider a neighborhood of (0, 0) where $\tilde{f}$ is defined and such that the second component of $\tilde{f}$ is at least $\frac{1}{2}$ on it (it is 1 at (0, 0)). Let δ be so small that no solution to $x' = \tilde{f}(x)$, $x(0) = 0$ can leave this neighborhood on $[0, \delta]$. If K is invariant under $\tilde{f}$, in particular $\mathscr{A}_0(\delta) \subset K$. Since $\mathscr{A}_0(\delta)$ is connected and contains points both of C_1 and of C_2 (belonging to the two solutions mentioned above), $\mathscr{A}_0(\delta)$ must contain the origin (0, 0). However the mean value theorem implies that for every solution $x(\cdot)$, $\|x(\delta)\| \geq \frac{1}{2}\delta$, a contradiction. Hence K cannot be invariant under $\tilde{f}$. □

We postpone to Section 5.2 the study of invariance of closed convex subsets (Theorem 5.2.7).

7. Differential Inclusions with Memory

Differential inclusions express that at every instant the velocity of the system depends upon its state at this very instant. *Differential inclusions with memory*, or, as they are also called, *functional differential inclusions*, express that the velocity depends not only of the state of the system at this instant, but depends upon the history of the trajectory until this instant. To formalize this concept, we introduce the Fréchet space $\mathscr{C}(-\infty, 0; X)$ of continuous functions from $]-\infty, 0[$ to the Hilbert space X supplied with the topology of uniform convergence on compact intervals.

We "embed" the "past history" of a trajectory $x(\cdot)$ of $\mathscr{C}(-\infty, +\infty; X)$ in this space $\mathscr{C}(-\infty, 0; X)$ by associating with it the function $T(t)x$ of $\mathscr{C}(-\infty, 0; X)$ defined by

(1) $$\forall \tau \in]-\infty, 0[, \quad T(t)x(\tau) \doteq x(t+\tau).$$

Hence a differential inclusion with memory describes the dependence of the velocity $x'(t)$ upon the history $T(t)x$ of $x(\cdot)$ up to time t through a set-valued map F from a subset $\Omega \subset R \times \mathscr{C}(-\infty, 0; X)$ to X.

Solving a differential inclusion with memory is the problem of finding an absolutely continuous function $x(\cdot) \in \mathscr{C}(-\infty, T; X)$ satisfying

(2) $$\forall t \geq 0, \quad x'(t) \in F(t, T(t)x).$$

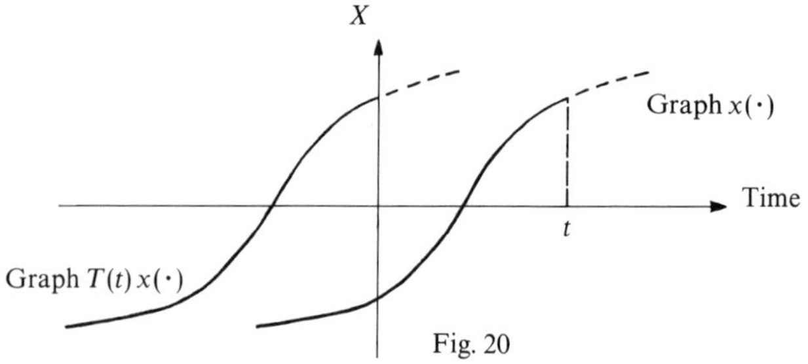

Fig. 20

This class of problems covers many examples:

a) Differential inclusions $x'(t) \in G(t, x(t))$, where G maps a subset of $R \times X$ to X, by setting

$$F(t, \varphi) \doteq G(t, \varphi(0))$$

because $F(t, T(t)x) = G(t, (T(t)x)(0)) = G(t, x(t))$.

b Differential-difference inclusions, associated to a set-valued map G from a subset of $R \times X^p$ to X, defined by

$$x'(t) \in G(t, x(t - r_1(t)), \ldots, x(t - r_p(t))) \tag{3}$$

belong to this class since we can define the set-valued map F by

$$F(t, \varphi) \doteq G(t, \varphi(-r_1(t)), \ldots, \varphi(-r_p(t))).$$

The functions $r_i(t)$ $(1 \leq i \leq p)$ are called the *delay functions.*

c) Volterra inclusions, which are inclusions of the form

$$x'(t) \in G\left(t, \int_{-\infty}^{t} k(t, s, x(s))\, ds\right) \tag{4}$$

where k maps $R \times R \times X$ to X and where G is a set-valued map from $R \times X$ to X are also differential inclusions with memory. Indeed, we define F from $R \times \mathscr{C}(-\infty, 0; R)$ by

$$F(t, \varphi) \doteq G\left(t, \int_{-\infty}^{0} k(t, t+\tau, \varphi(\tau))\, d\tau\right).$$

d) Differential trajectory processing inclusions. A "trajectory-processor" is a family of maps $P(t)$ from $\mathscr{C}(-\infty, +\infty; X)$ to a Hilbert space Y satisfying the property

$$\text{if } \varphi(s) = \psi(s) \text{ for all } s \leq t, \text{ then } P(t)\varphi = P(t)\psi. \tag{5}$$

Differential Trajectory processing inclusions are problems of the form

$$x'(t) \in G(t, P(t)x) \tag{6}$$

where G maps $R \times Y$ to X.

This is a particular differential inclusion with memory.

Indeed, we introduce the left-inverse $R(t)$ of $T(t)$, which maps any $\varphi \in \mathscr{C}(-\infty, 0; X)$ to $R(t)\varphi \in \mathscr{C}(-\infty, +\infty; X)$ defined by

$$R(t)\varphi(s) = \begin{cases} \varphi(s-t) & \text{when } s \leq t \\ \varphi(0) & \text{when } s \geq t. \end{cases} \tag{7}$$

We set

$$F(t, \varphi) \doteq G(t, P(t)R(t)\varphi).$$

Any solution to the differential trajectory processing inclusion (6) is a solution to the differential inclusion with memory

$$x'(t) \in F(t, T(t)x)$$

and conversely. ☐

Initial-value problems for differential inclusions with memory are problems of the form

(8)
$$\begin{aligned} &\text{i) for almost all } t \geq 0, \quad x'(t) \in F(t, T(t)\,x),\\ &\text{ii) } T(0)\,x = \varphi_0 \quad \text{where } \varphi_0 \text{ is given in } \mathscr{C}(-\infty, 0: X). \end{aligned}$$

Theorems about differential inclusions whose right hand-side are upper semicontinuous with compact convex images can be extended to differential inclusions with memory.

We choose, for instance, to state and prove the time dependent Viability Theorem.

Theorem 1. *Let X be a Hilbert space, K be a set-valued map with closed graph from $[0, \infty[$ to X. We set*

(9)
$$\forall t \geq 0, \quad \mathscr{K}(t) \doteq \{\varphi \in \mathscr{C}(-\infty, 0; X) \mid \varphi(0) \in K(t)\}.$$

Let F be an upper semicontinuous map from graph $(\mathscr{K})$ to the compact convex subsets of X, whose image is relatively compact.

We assume that

(10)
$$\begin{aligned} &\forall t \geq 0, \ \forall \varphi \text{ such that } \varphi(0) \in K(t), \ \forall x \in K(t),\\ &\qquad F(t, \varphi) \cap DK(t, \varphi(0))(1) \neq \emptyset. \end{aligned}$$

Then, for all $\varphi_0 \in \mathscr{K}(0)$, there exists a solution to the differential inclusion with memory

(11)
$$\begin{aligned} &\text{for almost all } t \geq 0, \quad x'(t) \in F(t, T(t)\,x)\\ &T(0)\,x = \varphi_0 \end{aligned}$$

which is viable in the sense that

(12)
$$\forall t \geq 0, \quad x(t) \in K(t).$$
▲

Remark. As in the case of differential inclusion, we can prove that when the dimension of X is finite, condition (10) is necessary.

Proof. We shall construct a sequence of approximate solution $x_n(\cdot)$ to the differential inclusion with memory in the following way.

First, we divide the interval $[0, \infty[$ into subintervals $[jh, (j+1)h]$ where $h = 1/n$.

We associate with any $j \in N$, $x \in X$ and $\varphi \in \mathscr{C}(-\infty, jh; X)$ the function $\phi_j^\varphi(x)(\cdot) \in \mathscr{C}(-\infty, 0; X)$ defined by

(13)
$$\phi_j^\varphi(x)(\tau) \doteq \begin{cases} \varphi((j+1)\,h + \tau) & \text{if } \tau \leq -h \\ \varphi(jh) + \left(\dfrac{\tau}{h} + 1\right)(x - \varphi(jh)) & \text{if } -h \leq \tau \leq 0. \end{cases}$$

It satisfies obviously

(14)
$$\phi_j^\varphi(x)(-h) = \varphi(jh), \quad \phi_j^\varphi(x)(0) = x.$$

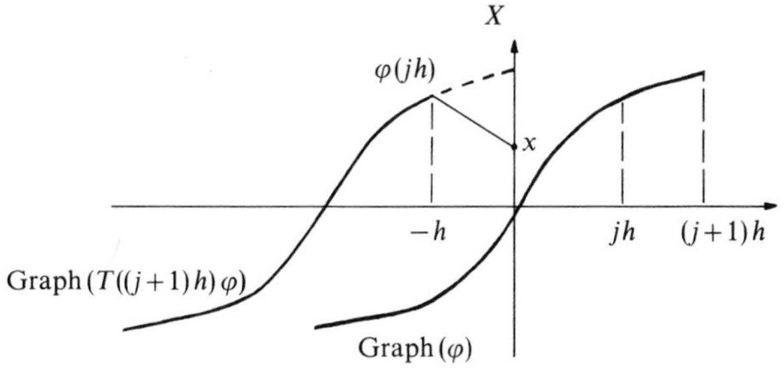

Fig. 21

We define set-valued maps G_j^φ from graph(K) to X by setting

(15) $$G_j^\varphi(t, x) \doteq F(t, \phi_j^\varphi(x)).$$

(This has a meaning because $\phi_j^\varphi(x) \in \mathscr{K}(t)$ whenever $x \in K(t)$.)

We observe that the set-valued maps G_j^φ are upper semicontinuous with compact convex values and that they satisfy

(16) $$\forall (t, x) \in \text{graph}(K), \qquad G_j^\varphi(t, x) \cap DK(t, x)(1) \neq \emptyset$$

by assumption (10), because

$$DK(t, x)(1) \doteq DK(t, \phi_j^\varphi(x)(0))(1).$$

Consequently, the assumptions of Theorem 2.1 are satisfied and we know that there exists a viable solution $x_j^\varphi(\cdot)$ on $[jh, (j+1)h]$ to the differential inclusion

(17) $$\begin{cases} \text{i) for almost all } t \in [jh, (j+1)h], \quad (x_j^\varphi)'(t) \in G_j^\varphi(t, x(t)), \\ \text{ii) } x_j^\varphi(jh) = \varphi(jh) \in K(jh), \\ \text{iii) } \forall t \in [jh, (j+1)h], \quad x_j^\varphi(t) \in K(t). \end{cases}$$

This allows the construction by induction of an approximate solution $x_n(\cdot)$ to the differential inclusion with memory.

a) For $t \leq 0$, we naturally set $x_n(t) \doteq \varphi(t)$.

b) Assume that $x_n(\cdot)$ is defined on $[-\infty, jh]$. We take, for $t \in [jh, (j+1)h]$, the solution

(18) $$x_n(t) \doteq x_j^{x_n}(t)$$

to the differential inclusion (17) with $\varphi \doteq x_n(\cdot)$.

They are approximate solutions in the sense that for almost all $t \in R$,

(19) $$(t, T(t)x_n, x_n'(t)) \in \text{graph}(F) + \{0\} \times 2h \|\text{Im} F\| B \times \{0\}$$

where B denotes the unit ball of $\mathscr{C}(-1, 0, X)$.

Indeed, when $t \in [jh, (j+1)h]$, $(t, \phi_j^{x_n}(x_n(t)), x_n'(t))$ belongs to the graph of F by (15) and (17). On the other hand,

$$T(t)\,x_n(\tau)-\phi_j^{x_n}(x_n(t))\,(\tau) = \begin{cases} 0 & \text{if } \tau\le -h \\ \left(1+\frac{\tau}{h}\right)(x_n(t+\tau)-x_n(t))-\frac{\tau}{h}(x_n(t+\tau)-x_n(jh)) \end{cases}$$

and thus, it belongs to $2h\,\overline{\mathrm{co}}(\mathrm{Im}(F))$, since

$$x_n(t+\tau)-x_n(t)=\int_t^{t+\tau} x_n'(\tau)\,d\tau$$

and

$$x_n(t+\tau)-x_n(jh)=\int_{jh}^{t+\tau} x_n'(\tau)\,d\tau$$

belong to $h\,\overline{\mathrm{co}}\,F(K)$ and $h\,\overline{\mathrm{co}}\,F(K)$. Hence

$$\|T(t)\,x_n-\phi_j^{x_n}(x_n(t))\|_{\mathscr{C}(-1,0;X)}\le 2h\,\|\overline{\mathrm{co}}\,F(K)\|.$$

Assumptions of the Compactness Theorem are satisfied because

(20)
i) $\|x_n'(t)\|\le\|\mathrm{co}(F(K))\|$,
ii) $x_n(t)\in\varphi(0)+t\,\mathrm{co}\,F(K)$, which is relatively compact.

Hence $x_n(\cdot)$ converges uniformly over compact intervals to some function $x(\cdot)$ in $\mathscr{C}(R,X)$, so that, for all $t\ge 0$, $T(t)\,x_n$ converges to $T(t)\,x$ in $\mathscr{C}(-\infty,0;X)$, and $x_n'(\cdot)$ converges weakly to $x'(\cdot)$ in $L^\infty(0,\infty;X)$, so that, for all $T>0$, $x_n'(\cdot)$ converges weakly to $x'(\cdot)$ in $L^1(0,T;X)$.

Therefore, the assumptions of the Convergence Theorem 1.4.1 are satisfied with $R\times\mathscr{C}(-\infty,0;X)$ playing the role of X, Y the role of X, $T(t)\,x_n$ the role of $x_k(t)$ and $x_n'(\cdot)$ the role of $y_k(\cdot)$. We deduce that for almost all t, $x'(t)$ belongs to $F(t,T(t)\,x)$. We know that $T(0)\,x=\varphi$, $(t,x_n(t))$ belongs to the graph of K, we infer that $x(t)\in K(t)$.

Corollary 1. *Let K be a closed convex subset of a Hilbert space X. We set $\mathscr{K}\doteq\{\varphi\in\mathscr{C}(-\infty,0;X)\,|\,\varphi(0)\in K\}$.*

Let F be an upper semicontinuous map from $\mathscr{K}$ to the compact convex subsets of X, whose image is relatively compact.

We assume that:

(21) $$\forall\varphi\in\mathscr{K},\quad F(\varphi)\cap T_K(\varphi(0))\neq\emptyset.$$

Then, for all $\varphi\in\mathscr{K}$, there exists a solution to the differential inclusion with memory

(22)
i) *for almost all $t\ge 0$,* $x'(t)\in F(T(t)\,x)$,
ii) $T(0)\,x=\varphi$,

which is viable in the sense that

(23) $$\forall t\ge 0,\quad x(t)\in K.$$ ▲

We can also deduce by the same techniques the existence of monotone trajectories to a differential inclusion with memory.

Theorem 2. *Let K be a closed subset of a Hilbert space X, $P\colon K \to K$ be a lower semicontinuous map with closed graph satisfying*

$$(24)\qquad \begin{cases} \text{i)}\ \forall x \in X, \quad x \in P(x) \quad (\textit{reflexivity}), \\ \text{ii)}\ \forall x \in K,\ \forall y \in P(x), \quad P(y) \subset P(x) \quad (\textit{transitivity}), \end{cases}$$

and F be a set-valued map from $\mathscr{K}$ to the compact convex subsets of X, whose image is relatively compact.

We posit the following tangential condition

$$(25)\qquad \forall \varphi \in \mathscr{K}, \quad F(\varphi) \cap T_{P(\varphi(0))}(\varphi(0)) \neq \emptyset.$$

Then, for all $\varphi \in \mathscr{K}$, there exists a solution to the differential inclusion with memory

$$(26)\qquad \begin{cases} \text{i)}\ x'(t) \in F(T(t)\,x), \\ \text{ii)}\ T(0)\,x = \varphi, \end{cases}$$

which is monotone in the sense that

$$(27)\qquad \forall s \geq t \geq 0, \quad x(s) \in P(x(t)).$$

▲

Properties of the Set of Trajectories

Theorems 2.2.1 and Corollary 2.2.4 stating properties of the set of trajectories to a differential inclusion remain true for differential inclusions with memory.

To provide some change, we shall make assumptions that guarantee global solutions instead of studying local solutions, by taking $K(t) \doteq X$ for all t in Theorem 1.

Theorem 3. *Let F be a bounded upper semicontinuous map from $R_+ \times \mathscr{C}(-\infty, 0; R^n)$ to the compact convex subsets of R^n.*

Let $\mathscr{T}_\infty(\varphi)$ denote the set of solutions to the differential inclusion

$$(28)\qquad \begin{cases} \text{i)}\ x'(t) \in F(t, T(t)\,x), \\ \text{ii)}\ T(0)\,x = \varphi. \end{cases}$$

Then $\mathscr{T}_\infty$ is an upper semicontinuous map from $\mathscr{C}(-\infty, 0; R^n)$ to the compact connected subsets of $\mathscr{C}(-\infty, +\infty; R^n)$ and is σ-selectionable. ▲

Actually the map $\mathscr{T}_\infty$ is even more regular. Theorem 3 follows from the following one.

Theorem 4. *We posit the assumptions of* Theorem 3. *There exists a sequence of convex compact metrizable subsets U^m and of continuous maps S_m from*

$\mathscr{C}(-\infty,0;R^n)\times U^m$ to $\mathscr{C}(-\infty,+\infty;R^n)$ satisfying

$$(29)\qquad \begin{cases} \text{i) } \forall\varphi,\ \forall n\geq 0, \quad \mathscr{T}_\infty(\varphi)\subset\ldots\subset S_{m+1}(\varphi,U^{m+1})\subset S_m(\varphi,U^m)\subset\ldots, \\ \text{ii) } \forall\varphi,\ \forall\mathscr{N},\exists\ N_{\varphi,\mathscr{N}}|\forall m\geq N_{\varphi,\varepsilon}, \quad S_m(\varphi,U^m)\subset\mathscr{T}_\infty(\varphi)+\mathscr{N}, \end{cases}$$

where $\mathscr{N}$ denotes a neighborhood of 0 in $\mathscr{C}(-\infty,+\infty;R^n)$ and

$$(30)\qquad \forall\varphi,\ \forall x\in\mathscr{T}_\infty(\varphi), \quad \{u\in U^m|S_m(\varphi,u)=x\} \text{ is convex.}$$ ▲

Proof. The proof is not as simple as the one of *Theorem 2.2.3* for differential inclusions.

Indeed, in the above case, the set-valued map F was defined on a compact subset Q and thus, could be approximated by maps F_m which are *finite* combinations of constant convex subsets.

In our case, F is defined on the space $R_+\times\mathscr{C}(-\infty,0;R^n)$. Hence the Approximate Theorem 1.13.1 states that F can be approximated by a decreasing sequence of set-valued maps F_m defined on graph ($\mathscr{K}$) by

$$(31)\qquad F_m(t,\varphi)=\sum_{i\in I_m}\psi_i^m(t,\varphi)\,C_i^m\subset\overline{\text{co}}\,F(K)$$

where $\psi_i^m(\cdot,\cdot)$ is a locally finite locally Lipschitzean partition of the unity and where C_i^m are convex compact subsets of R^n contained in $\overline{\text{co}}\,F(K)$, which is compact. Since $R\times\mathscr{C}(-\infty,0;R^n)$ is separable, the set I_m of indexes is countable.

We denote by $\mathscr{T}_\infty^m(\varphi)$ the set of trajectories of the differential inclusion with memory:

$$(32)\qquad \begin{cases} \text{i) } x'(t)\in F_m(t,T(t)x), \\ \text{ii) } T(0)x=\varphi, \end{cases}$$

and we deduce easily from the Approximation Theorem that

$$(33)\qquad \begin{cases} \text{i) } \mathscr{T}_\infty(\varphi)\subset\ldots\subset\mathscr{T}_\infty^{m+1}(\varphi)\subset\mathscr{T}_\infty^m(\varphi)\subset\ldots, \\ \text{ii) for all neighborhood } \mathscr{N} \text{ of 0 in } \mathscr{C}(0,\infty;R^n), \text{ for all } \varphi\in\mathscr{K}, \\ \quad \text{there exists } N_{\varphi,\mathscr{N}} \text{ such that for all } m\geq N_{\varphi,\mathscr{N}}, \\ \quad \mathscr{T}_\infty^m(\varphi)\subset\mathscr{T}_\infty(\varphi)+\mathscr{N}. \end{cases}$$

We use the Convergence Theorem 1.4.1 and the Compactness Theorem 0.3.4 for proving the later statement, as in the proof of Theorem 2.2.3.

It remains to prove that

$$(34)\qquad \begin{cases} \forall m\geq 0, \text{ there exist a convex compact subset } U^m \text{ and a} \\ \text{continuous map } S_m \text{ from } \mathscr{K}\times U^m \text{ to } \mathscr{C}(0,\infty;R^n) \text{ such that} \\ \mathscr{T}_\infty^m(\varphi)=S_m(\varphi,U^m). \end{cases}$$

At this point we need to be careful.

We introduce the space

$$(35)\qquad \mathscr{F}_\infty^m\doteq\{u^m\doteq(u_i)_{i\in I_m}\in L^\infty(R,R^n)^{I_m}|\sup_{i\in I_m}\|u_i^m\|_{L^\infty}<+\infty\}.$$

The set U^m of controls is defined by

(36) $$U^m \doteq \{u^m \in \mathscr{F}^m \mid \text{for almost all } t \in R,\ \forall i \in I_m,\ u_i^m(t) \in C_i^m\}.$$

Since C_i^m is convex for every i, U^m is obviously a convex set.

Let u^m belong to U^m and φ belong to $\mathscr{K}$. Then there exists a unique solution to the differential equation with memory

(37) $$\begin{cases} \text{i)}\ x'(t) = \sum_{i \in I_m} \psi_i^m(t, T(t)\,x)\,u_i^m(t), \\ \text{ii)}\ T(0)\,x = \varphi. \end{cases}$$

(See Hale [1977].) We shall denote it by $S_m(\varphi, u^m)$. Hence

(38) $$S_m(\varphi, U^m) \subset \mathscr{T}_\infty^m(\varphi).$$

To go further, we need to supply $\mathscr{F}_\infty^m$ with a topology for which U^m is compact and S_m is continuous.

For that purpose, we observe that $\mathscr{F}_\infty^m$ is in duality with the vector space $\mathscr{F}_1^m$, defined as

(39) $$\mathscr{F}_1^m \doteq \{v^m = (v_i^m)_{i \in I_m} \in L^1(R, R^n)^{I_m} \mid v_i^m = 0 \text{ except for a finite number of indexes}\}.$$

The duality pairing is

(40) $$\langle u^m, v^m \rangle \doteq \sum_{i \in I_m} \int_R u_i^m(t) \cdot v_i^m(t)\,dt.$$

We supply $\mathscr{F}_1^m$ and $\mathscr{F}_\infty^m$ with the associated weak topologies $\sigma(\mathscr{F}_1^m, \mathscr{F}_\infty^m)$ and $\sigma(\mathscr{F}_\infty^m, \mathscr{F}_1^m)$.

On the other hand, $\mathscr{F}_1^m$ is a normed space for the norm

$$\|v^m\|_{\mathscr{F}_1^m} = \sum_{i \in I_m} \|v_i^m\|_{L^1}$$

for which $\mathscr{F}_y^m$ is separable, whose dual is $\mathscr{F}_\infty^m$, supplied with the norm

$$\|u^m\|_{\mathscr{F}_\infty^m} = \sup_{i \in I_m} \|u_i^m\|_{L^\infty}.$$

Since the subset U^m is bounded in $\mathscr{F}_\infty^m$ because all the subsets C_i^m are contained in a bounded subset of R^n, we deduce that U^m is bounded, and thus, *relatively compact* for the weak-topology. It is weakly closed: if a sequence of elements u_p^m of U^m converges weakly to some u^m, each component $u_{p_i}^m$ converges weakly to u_i^m in $L^\infty(R, R^n)$. Since $L^\infty(R, C_i^m)$ is weakly compact, we deduce that $u_i^m(t)$ belongs to C_i^m for almost all t. Hence the subset U^m is weakly compact in $\mathscr{F}_m^\infty$ and, since $\mathscr{F}_1^m$ is separable, U^m is metrizable.

The continuity of $S_m(\cdot,\cdot)$ follows from the fact that the solution $x_m = S_m(\varphi, u^m)$ of the differential equation (37) with memory do satisfy

(41) $$x_m(t) = \varphi(0) + \sum_{i \in I_m} \int_0^t \psi_i(\tau, T(\tau)\,x_m)\,u_i^m(\tau)\,d\tau.$$

Indeed, let us consider a sequence of elements $u_p^m \in U^m$ and $\varphi_p \in \mathscr{C}(-\infty, 0; R^n)$ converging to u^m and φ respectively. It is clear that the derivatives $x'_{m,p}$ of the solutions $x_{m,p} = S_m(\varphi_p, u_p^m)$ remain in a bounded set and thus, that the assumptions of the Compactness Theorem 0.3.4 are satisfied. Hence a subsequence converges uniformly over compact intervals to a function $x_m(\cdot)$.

Since the functions $x_{m,p}$ are in a compact subsets of $\mathscr{C}(R, R^n)$, then the pairs $(\tau, T(\tau)\, x_m)$ when τ ranges over $[0, T]$ and $m \in N$ remain in a compact set of $R \times \mathscr{C}(-\infty, 0; R^n)$.

The partition being locally finite, there exists a finite set of indexes J such that

$$(42) \qquad x_{m,p}(t) = \varphi_p(0) + \sum_{i \in J} \int_0^t \psi_i(\tau, T(\tau)\, x_{m,p})\, u_{i,p}^m(\tau)\, d\tau.$$

Since $\tau \to (\tau, T(\tau)\, x_{m,p})$ converges uniformly to $\tau \to (\tau, T(\tau)\, x_m)$ on $[0, t]$, it follows that $\tau \to \psi_i(\tau, T(\tau)\, x_{m,p})$ converges uniformly to $\tau \to \psi_i(\tau, T(\tau)\, x_m)$. The functions $u_{i,p}^m(\cdot)$ converging weakly to $u_i^m(\cdot)$, we infer that equations (42) imply Equation (41) by letting p go to ∞.

Finally, we have to prove that S_m maps U^m onto $\mathscr{T}_\infty(\varphi)$ for all $\varphi \in \mathscr{C}(-\infty, 0; R^n)$. But this is a consequence of Corollary 1.14.1 to the Measurable Selection Theorem: if x is an absolutely continuous function, there exists a measurable function u^m from R_+ to $\prod_{i \in I_m} C_i^m$ satisfying, for almost all $t \geq 0$

$$(43) \qquad x'(t) = \sum_{i \in I_m} \psi_i(t, T(t)\, x_m)\, u_i^m(t).$$

Such a measurable function u^m belongs to the set U^m of parameters and the subset of such functions is obviously convex. □

Theorem 4 is naturally as useful as its analogue for differential inclusion. For instance, we can prove the existence of periodic solutions under assumptions analogous to the assumptions of Theorem 5.3.4 in the next chapter.

Chapter 5. Viability Theory and Regulation of Controled Systems: The Convex Case

Introduction

When we assume that the viability subset K is convex and compact, we obtain many more properties. The most striking one is that the *tangential condition*

(1) $$\forall x \in K, \quad F(x) \cap T_K(x) \neq \emptyset$$

which is necessary and sufficient when F has convex values for the differential inclusion

(2) $$\begin{aligned} &\text{i)}\ x'(t) \in F(x(t)),\\ &\text{ii)}\ x(0) = x_0,\ x_0 \text{ given in } K, \end{aligned}$$

to have viable trajectories for all initial states x_0 in K, is also a sufficient condition for F to have an equilibrium state $\bar{x}$ in K.

This is one of the most powerful theorems of nonlinear analysis, which is equivalent to the Brouwer fixed point Theorem:

Theorem. *Let K be convex compact and F be an upper semicontinuous map with closed convex images from K to R^n. If the tangential condition* (1) *holds true, then there exists an equilibrium $\bar{x} \in K$:*

(3) $$\bar{x} \in K \quad \text{is a solution to } 0 \in F(\bar{x}).$$ ▲

So, under the above assumptions, viability implies the existence of an equilibrium.

This theorem is itself equivalent to the Kakutani fixed point Theorem in locally convex spaces, which we already proved in Chapter 1 (Corollary 1.12.1) and extended in Chapter 2 (Corollary 2.2.2) to maps defined by the solution sets to differential inclusions.

Actually, we extend Kakutani's Theorem to the larger class of σ-selectionable maps. For instance, this allows us to prove the existence of periodic solutions to differential inclusions.

When K is convex, the contingent cone $T_K(x)$, also called the tangent cone to K at x, is a *closed convex cone*, which enjoys a rather rich calculus.

For instance, consider the frequent case when the viability subset is defined by

$$K \doteq L \cap A^{-1}(M).$$

So, we need to couch the tangential condition in terms of contingent cones to L and M, since K is not explicitely known. We shall prove for instance that when L and M are closed and convex, when A is linear and continuous, and when $0 \in \operatorname{Int}(A(L)-M)$, then tangential condition (1) can be written

$$\forall x \in K, \quad F(x) \cap T_L(x) \cap A^{-1} T_M(Ax) \neq \emptyset.$$

Let us say a few words for the particular case of controled systems

(4) $$x'(t) = f(x(t), u(t)).$$

When we assume that the images $F(x) \doteq f(x, U)$ are compact convex, the tangential condition (1) can be written

(5) $$\forall x \in K, \quad C(x) \doteq \{u \in U \mid f(x,u) \in T_K(x)\} \neq \emptyset.$$

Since it is a necessary and sufficient condition, we see that the differential equation (4) has viable trajectories if and only if

(6) $$\text{for almost all } t > 0, \quad u(t) \in C(x(t)).$$

This is the reason why C is called the *feedback map*. Nagumo's Theorem shows that any continuous selection $\underset{\sim}{u}$ of the feedback map C is a *closed loop control*, because the differential equations $x'(t) = f(x(t), \underset{\sim}{u}(x(t)))$ have viable trajectories.

When $f(x, U)$ is no longer convex, we are forced to require not only that the images $C(x)$ of the feedback map C are non empty, but that

$$\forall x \in K, \quad C(x) = U.$$

Hence, the absence of convexity of the images $f(x, U)$ leaves no interest in the realm of regulation of controled systems.

Since the tangent cones to closed convex subsets are closed convex cones, we may think to exploit duality relations. We remark that the negative polar cone of the tangent cone is simply the normal cone $N_K(x)$ to K at x. We shall regard elements of the dual space as "co-states p".

Therefore, a separation argument shows that the tangential condition (1) can be written in the form

(7) $$\forall x \in K, \quad \forall p \in N_K(x), \quad \inf_{v \in F(x)} \langle p, v \rangle \leq 0.$$

This is the reason why, in the case of controled systems, the existence of a map $\underset{\sim}{w}$ from graph $N_K(\cdot)$ to U satisfying

(8) $$\forall x \in K, \quad \forall p \in N_K(x), \quad \langle p, f(x, \underset{\sim}{w}(x,p)) \rangle \leq 0$$

implies the viability of the controled system: the values of the feedback set-valued map C are nonempty and so, both open-loop and closed-loop controls do exist. Such dual conditions are useful when we use regulation models in economic theory, because co-states have a natural interpretation as price systems.

We mention that equilibria of controled systems are pairs $(\bar{x}, \bar{u}) \in K \times U$ satisfying $f(\bar{x}, \bar{u}) = 0$.

Assumption (7) implies the existence of such a pair $(\bar{x}, \bar{u})$. We shall use this regulation point of view for proposing a decentralized dynamical model for allocating scarce resources. We shall explain the justifications for this model in due time. Now, we briefly describe the model. Let R^l be a commodity space, $M \subset R^l$ be the set of resources (or available commodities, supplies, etc....). Let n *consumers* i choose commodities $x_i \in L_i$ (the commodity set of consumer i) under the scarcity constraint $\sum_{i=1}^{n} x_i \in M$. This amounts to finding *allocations* of M, i.e., elements of the subset

(9) $$K \doteq \left\{ \vec{x} = (x_1, \ldots, x_n) \in \prod_{i=1}^{n} L_i \,\middle|\, \sum_{i=1}^{n} x_i \in M \right\}.$$

We describe the behavior of consumers i by n "controled" differential equations, where the controls are price systems $p \in R^{l*}$.

(10) $$\forall i = 1, \ldots, n, \quad x_i'(t) = d_i(x_i(t), p(t)).$$

This is a *decentralized system*, because the action of consumer i, which is a rate of change of his consumption, does not depend upon the actions of the other consumers: it depends only upon his own consumption and a control – the price system – which is known to every consumer. The price system must give enough information to the consumers for the trajectories to be viable in the sense that

(11) $$\forall t > 0,\ \forall i = 1, \ldots, n, \quad x_i(t) \in L_i \quad \text{and} \quad \sum_{i=1}^{n} x_i(t) \in M$$

and even, to be monotone in the sense that they also satisfy

(12) $$\forall i = 1, \ldots, n,\ \forall t \geq s, \quad U_i(x_i(t)) \geq U_i(x_i(s))$$

where U_i are utility functions.

The dual approach shall be used, since it states that financial constraints of the type

(13) $$\left\langle p, \sum_{i=1}^{n} d_i(x_i, p) \right\rangle \leq 0$$

imply the possibility of allocating commodities of M in a decentralized and dynamical way. Such financial constraints play the role of a Walras law, because they *forbid consumers to spend more than they earn.*

We shall argue that this model retains the main ideas underlying the Walrasian model of an economy while taking in account dynamical features which are absent of the Walrasian world.

Any closed loop control $\underset{\sim}{p}(\vec{x})$ can be regarded as an example of a planning procedure: The planning bureau associates to each allocation $\vec{x}$ a price system $\underset{\sim}{p}(\vec{x})$ such that, by solving their differential equations

$$x_i'(t) = d_i(x_i(t), \underset{\sim}{p}(\vec{x}(t))) \tag{14}$$

the consumers implement viable allocations.

More general *planning procedures* can be regarded as procedures yielding viable trajectories (satisfying (11)), or, more generally, monotone trajectories (satisfying (11) and (12)). The problem amounts to constructing dynamical systems defined by n set-valued maps $F_i\colon \prod_{i=1}^{n} L_i \to R^l$ such that the system of differential inclusions

$$x_i'(t) \in F_i(x_1(t), \ldots, x_n(t)) \tag{15}$$

yields trajectories satisfying the viability conditions (11) and the monotonicity conditions (12). In most cases, the set-valued maps must obey several additional constraints. For instance, the planning bureau sends to the consumers "signals" which can be regarded as controls which guide the dynamical behavior of the consumers through differential equations of the form

$$x_i'(t) = f_i(x_1(t), \ldots, x_n(t), u(t)). \tag{16}$$

The problem of the planning bureau is to find closed loop controls $\underset{\sim}{u}$ that associate to allocations $\vec{x} = (x_1, \ldots, x_n)$ a convenient signal $\underset{\sim}{u}(x_1, \ldots, x_n)$ such that the differential equations

$$x_i'(t) = f_i(x_1(t), \ldots, x_n(t), \underset{\sim}{u}(x_1(t), \ldots, x_n(t))) \tag{17}$$

yield viable or monotone trajectories.

For illustrating the above results, we propose a model in game theory whose purpose is to explain the formation of coalitions. We choose the framework (familiar in Game Theory) of players forming a "continuum" of players Ω: this is a set on which we define a σ-algebra $\mathscr{A}$ and a *nonatomic* measure μ. We regard elements $\omega \in \Omega$ as *players* and subsets $A \in \mathscr{A}$ as *coalitions* of players. Players $\omega \in \Omega$, by acting, modify the state $x \in R^n$ of the environment described by a finite dimensional vector space. Let $f(x, \omega)$ denote the rate of change caused by player ω when the state of the environment is x. We assume that the rate of change produced by a coalition $A \in \mathscr{A}$ of players ω is the aggregation $\int_A f(x, \omega)\, d\mu(\omega)$ of the rates of each players. Hence the state evolves according to the controled differential equation

$$x'(t) = \int_{A(t)} f(x(t), \omega)\, d\mu(\omega) \tag{18}$$

where the *role of the control is played by a coalition* A.

If we require viability conditions on the state of the environment, open loop controls $t \to A(t) \in \mathscr{A}$ or closed loop controls $x \in K \to \underset{\sim}{A}(x) \in \mathscr{A}$ yielding viable trajectories provide an explanation of the formation of coalitions.

The question arises whether it is possible to construct dynamical systems having viable trajectories, or, more precisely, to modify a given system in order to insure the existence of viable trajectories. By the viability Theorem, we have to transform a given set-valued map $G: K \to R^n$ to a map $F: K \to R^n$ that satisfies condition (1). Since the latter is a boundary condition, the modification of G should occur only at the boundary of K. A solution to the problem is to take:

$$F(x) \doteq G(x) - N_K(x). \tag{19}$$

Let $\pi_{T_K(x)}(u)$ denote the projection of u onto the closed convex cone $T_K(x)$. It is easy to check that in this case

$$\pi_{T_K(x)}[G(x)] \subset F(x). \tag{20}$$

So, the set-valued map F coincides with G on the interior of K and, by (20), satisfies the tangential condition (1) on the boundary of K.

We deduce from the viability Theorem the following result proved by Cl. Henry for constructing planning procedures.

Theorem. *Let K be a closed convex subset of R^n and $G: K \to R^n$ be a set-valued map satisfying*

$$\begin{cases} \text{i) the graph of } G \text{ is closed,} \\ \text{ii) } G(K) \text{ is bounded,} \\ \text{iii) } \forall x \in K,\ G(x) \text{ is convex.} \end{cases} \tag{21}$$

Then, for all $x_0 \in K$, there exists a viable trajectory of the differential inclusion

$$x'(t) \in G(x(t)) - N_K(x(t)), \qquad x(0) = x_0. \tag{22}$$

Furthermore, the set of viable trajectories of the differential inclusion (22) *and of the differential inclusion*

$$x'(t) \in \pi_{T_K(x(t))}(G(x(t))), \qquad x(0) = x_0 \ \text{coincide.} \tag{23}$$

▲

Note that the values $\pi_{T_K(x)}$ are not necessarily convex and that the set-valued map $x \to \pi_{T_K(x)}(G(x))$ has no continuity properties. Still, this theorem states the existence of trajectories of the differential inclusion (23).

Note also that solutions to the differential inclusion (22) can be written

$$\begin{cases} \text{i) } \langle x'(t) - g(x(t)), x(t) - y \rangle \le 0, \quad \forall y \in K \text{ where } g(x(t)) \in G(x(t)), \\ \text{ii) } x(0) = x_0. \end{cases} \tag{24}$$

Problem (24) is often called *differential variational inequalities.*

Existence of equilibria for such systems was proved in the 1960's by Browder, Lions, Stampacchia, etc.

Theorem (Browder-Lions-Stampacchia). *Let $K \subset R^n$ be convex and compact and $G: K \to R^n$* satisfy assumption (21). There exists a solution to

$$\bar{x} \in K \quad \text{and} \quad 0 \in G(\bar{x}) - N_K(\bar{x})). \tag{25}$$ ▲

Note that when $\bar{x} \in \mathrm{Int}(K)$, a solution of (25) is an equilibrium of G. Solutions of (25) are solutions to

$$\exists \bar{x} \in K \quad \text{and} \quad \exists \bar{g} \in G(\bar{x}) | \langle \bar{x} - \bar{g}, \bar{x} - y \rangle \leq 0 \quad \forall y \in K. \tag{26}$$

Problem (26) is often called *variational inequalities.*

Still, we are not satisfied since, in the framework of planning procedures, deterministic procedures are needed: it is possible to satisfy this request by proving that *slow solutions* to differential inclusions (22) and (23) coincide and exist. Slow solutions are defined in the following way. If $m(A)$ denotes the element of minimal norm of a closed convex subset A, a slow solution to the differential inclusion $x' \in F(x)$, $x(0) = x_0$ is a solution to the differential equation

$$x'(t) \in m(F(x(t))), \quad x(0) = x_0. \tag{27}$$

Theorem. *Let K be a closed convex subset of R^n and $G: K \to R^n$ be a set-valued map satisfying*

$$\begin{cases} \text{i) } \textit{the graph of } G \textit{ is closed and } G \textit{ is continuous,} \\ \text{ii) } G(K) \text{ is bounded,} \\ \text{iii) } \forall x \in K, \ G(x) \textit{ is convex.} \end{cases} \tag{28}$$

Then the slow solution to the differential inclusions (22) *and* (23) *coincide and exist for all $x_0 \in K$.* ▲

1. Tangent Cones and Normal Cones to Convex Sets

Let X be a Hilbert space and $K \subset X$ be a convex subset. We associate to any $x \in K$ the cone

$$S_K(x) \doteq \bigcup_{h>0} \frac{1}{h}(K - x) \tag{1}$$

spanned by $K - x$. We note that v belongs to $S_K(x)$ if and only if $x + hv$ belongs to K for some $h > 0$.

We observe that

$$\forall v \in S_K(x), \ \exists h > 0 \text{ such that, } \forall t \in [0, h], \quad x + tv \in K \tag{2}$$

Indeed, $x+tv=\left(1-\frac{t}{h}\right)x+\frac{t}{h}(x+hv)$ is a convex combination of elements belonging to the convex set K. We now prove that the contingent cone $T_K(x)$ to K at x is the closure of the cone $S_K(x)$.

Proposition 1. *Let K be a convex subset and x belong to K. Then*

$$T_K(x)=cl\left(\bigcup_{h>0}\frac{1}{h}(K-x)\right) \tag{3}$$

and thus, is a closed convex cone. ▲

Proof. a) It is obvious that $T_K(x)$ is contained in the closure of $S_K(x)$.

Let v belong to $cl(S_K(x))$. Then, for all $\varepsilon>0$, there exist $y_\varepsilon\in K$ and $h_\varepsilon>0$ such that

$$v-\frac{1}{h_\varepsilon}(y_\varepsilon-x)\in\varepsilon B$$

i.e., such that $x+h_\varepsilon v_\varepsilon\in K$ where $v_\varepsilon=\frac{1}{h_\varepsilon}(y_\varepsilon-x)$.

Let us associate with any $\alpha>0$ the positive number $h\doteq\min(\alpha,h_\varepsilon)>0$. *Since K is convex*, $x+h\left(\frac{y_\varepsilon-x}{h_\varepsilon}\right)=\left(1-\frac{h}{h_\varepsilon}\right)x+\frac{h}{h_\varepsilon}(x+h_\varepsilon v_\varepsilon)$ belongs to K. This proves that for any $\varepsilon>0$ and $\alpha>0$, there exist $h\le\alpha$ and $v_\varepsilon\in v+\varepsilon B$ such that $x+hv_\varepsilon\in K$, i.e. that $v\in T_K(x)$.

b) $S_K(x)$ is a convex cone: Indeed, if v_1 and v_2 belong to $S_K(x)$, then $x+h_iv_i\in K$ for $i=1,2$; let $h\doteq\min(h_1,h_2)$. Then, by the preceding remark, $x+hv_i\in K$ for $i=1, 2$. Hence $x+h(\alpha v_1+(1-\alpha)v_2)\in K$ when $\alpha\in[0,1]$. Since the closure of a convex cone is still a convex cone, we have proved the proposition. □

Therefore, v belongs to $T_K(x)$ if and only if there exist a sequence of elements v_n converging to v and positive h_n such that $x+h_nv_n\in K$ for all $n>0$.

Definition 1. When K is convex, the contingent cone $T_K(x)$ is simply called the tangent cone to K at x. ▲

We observe that

$$\forall x\in K,\quad K\subset x+S_K(x)\subset x+T_K(x) \tag{4}$$

and that $T_K(x)$ is contained in the closed vector space $M(K)\doteq\overline{\{\alpha K+\beta K\}}_{\alpha,\beta\in R}$ spanned by K.

A normal to a smooth manifold at a point x is any vector orthogonal to the tangent space. In the case of tangent cones to convex subsets, the orthogonal subspace to the tangent space will be replaced by the concept of negative polar cone to the tangent cone, which will be called the normal cone. The introduction of these two concepts, tangent and normal cones, allows the use of duality relations.

Definition 2. *Let K be a nonempty closed convex subset of X. The "normal cone" $N_K(x)$ to K at x is defined by*

$$(5) \qquad N_K(x) \doteq \{p \in X^* \text{ such that } \langle p, x\rangle = \max\{\langle p, y\rangle \mid y \in K\} \doteq \sigma_K(p)\}. \qquad \blacktriangle$$

Proposition 2. *The normal cone $N_K(x)$ is the (negative) polar cone $T_K(x)^-$ of the tangent cone $T_K(x)$.* ▲

Therefore, $T_K(x) = N_K(x)^-$ since $T_K(x)$ is a closed convex cone.

Proof. a) If $p \in T_K(x)^-$, then $\langle p, y-x\rangle \le 0$ for all $y \in K$ since $v = y - x \in T_K(x)$ when $y \in K$. Hence $p \in N_K(x)$.

b) *Conversely*, let $p \in N_K(x)$ and $v = \lim_{n\to\infty} \lambda_n(y_n - x) \in T_K(x)$, where $\lambda_n \ge 0$ and $y_n \in K$. Hence $\langle p, v\rangle \le 0$ since $\langle p, \lambda_n(y_n - x)\rangle = \lambda_n \langle p, y_n - x\rangle \le 0$ for all $n > 0$. □

Proposition 3. *Let π_K be the projection of best approximation onto the closed convex subset K. Then $\pi_K^{-1}(x) = x + N_K(x)$ and v belongs to $T_K(x)$ if and only if $\langle y - x, v\rangle \le 0 \ \forall y \in \pi_K^{-1}(x)$.* ▲

Proof. a) If $p \in N_K(x)$, then $x = \pi_K(x+p)$ since $\langle (x+p) - x, y - x\rangle = \langle p, y - x\rangle \le 0$ for all $y \in K$.

b) *Conversely*, let $y \in \pi_K^{-1}(x)$; then

$$\langle y - x, z - x\rangle \le 0 \quad \text{for all } z \in K.$$

Hence $p \doteq y - x \in N_K(x)$.

c) The last statement follows from the first and from the fact that $T_K(x) = N_K(x)^-$. ■

Continuity Properties of the Tangent and Normal Cones

Theorem 1. *Let K be a closed convex subset of a finite dimensional space. Then*

$$(6) \qquad \begin{cases} \text{i) } x \to N_K(x) \text{ has a closed graph,} \\ \text{ii) } x \to T_K(x) \text{ is lower semicontinuous.} \end{cases} \qquad \blacktriangle$$

Proof. a) Let (x_n, p_n) be a sequence of elements of the graph of $N(\cdot)$ converging to (x, p). For all $y \in K$, we have $\langle p_n, y\rangle \le \langle p_n, x_n\rangle$. Hence, letting $n \to \infty$, we deduce that $\langle p, y\rangle \le \langle p, x\rangle$. Thus (x, p) belongs to the graph of $N(\cdot)$.

b) The second statement is equivalent to the first by a result of Proposition 1.2.2.

Proposition 4. *If K has a nonempty interior, then, for all $x \in K$, the interior of the tangent cone is nonempty and spanned by* $\operatorname{Int} K - x$:

(7) $$\operatorname{Int} T_K(x) = \bigcup_{h>0} \frac{1}{h}(\operatorname{Int} K - x)$$

Furthermore, the set-valued map $x \to \operatorname{Int} T_K(x)$ has an open graph. ▲

Proof. The cone $\bigcup_{h>0} \frac{1}{h}(\operatorname{Int} K - x)$ is open, being an union of open subsets. Hence it is contained in $\operatorname{Int} T_K(x)$. Since $\operatorname{Int} T_K(x) = \operatorname{Int} S_K(x)$, it suffices to prove that any $v \in \operatorname{Int} S_K(x)$ belongs to some $\frac{1}{h}(\operatorname{Int} K - x)$. Let $\eta > 0$ be such that $v + \eta B \subset S_K(x)$.

If $x + v$ belongs to $\operatorname{Int} K$, the proof is finished. If not, let x_0 belong to $\operatorname{Int} K$ and let us set $v_0 \doteq x_0 - x$. Hence $v - \frac{\eta}{\|v_0\|} v_0$ belongs to $S_K(x)$ and consequently, there exists $h > 0$ such that $x + h\left(v - \frac{\eta}{\|v_0\|} v_0\right)$ belongs to K. Let us set $\alpha \doteq \frac{h\eta}{h\eta + \|v_0\|}$.

We thus observe that

$$x + (1-\alpha) h v = \alpha x_0 + (1-\alpha)\left(x + h\left(v - \frac{\eta}{\|v_0\|} v_0\right)\right).$$

Since x_0 belongs to $\operatorname{Int} K$, $x + h\left(v - \frac{\eta}{\|v_0\|} v_0\right)$ belongs to K and since α belongs to $]0,1[$, then $x + (1-\alpha) h v$ belongs to the interior of K (see Theorem 0.5.1). This proves that v belongs to $\frac{1}{(1-\alpha)h}(\operatorname{Int} K - x)$.

b) Let $v_0 \in \operatorname{Int} T_K(x_0)$. Therefore $v_0 \in \frac{1}{h_0}(\operatorname{Int} K - x_0)$ for some $h_0 > 0$; hence there exists $\varepsilon > 0$ such that $x_0 + h_0 v_0 + \varepsilon B = x_0 + h_0\left(v_0 + \frac{\varepsilon}{h_0} B\right) \subset \operatorname{Int} K$. Take $x \in x_0 + \varepsilon/2B$ and $v \in v_0 + \varepsilon/2h_0 B$. Then $x + h_0 v \in h_0 v_0 + \varepsilon B \subset \operatorname{Int} K$ and therefore, $v \in \operatorname{Int} T_K(x)$. Hence the graph of $x \to \operatorname{Int} T_K(x)$ is open. □

We proceed by presenting several "simple" examples of tangent cones.

a) Let B be the unit ball of a Hilbert space and $x \in B$. Then

(8) $$\begin{cases} \text{i)} \ T_B(x) = X \text{ if } x \in \operatorname{Int} B \text{ and } T_B(x) = \{x\}^- \text{ if } \|x\| = 1, \\ \text{ii)} \ N_B(x) = \{0\} \text{ if } x \in \operatorname{Int} B \text{ and } N_B(x) = R_+ x \text{ if } \|x\| = 1. \end{cases}$$ ▲

Proof. We take $\|x\| = 1$.

Then $p \in N_B(x)$ if and only if $\|p\|_* = \sup_{y \in B} \langle p, y \rangle = \langle p, x \rangle$. By the Cauchy-Schwarz inequality, this is equivalent to $p = \lambda x$ with $\lambda \geq 0$.

By polarity, we deduce the formula for the tangent cone. □

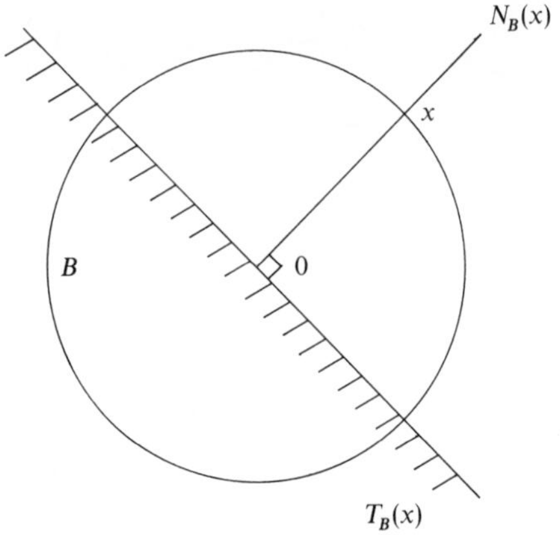

Fig. 22. Tangent core to the ball

b) **Proposition 5.** *Let $K \subset X$ be a closed convex cone. Then $N_K(x) = K^- \cap \{x\}^\perp$ and thus:*

(9) $$v \in T_K(x) \quad \text{if and only if } \langle p, v\rangle \le 0 \text{ for all } p \in K^- \text{ satisfying } \langle p, x\rangle = 0.$$

If K is a closed subspace, then $T_K(x) = K$ and $N_K(x) = K^\perp$. ▲

Proof. It is clear that $K^- \cap \{x\}^\perp$ is contained in $N_K(x)$. Conversely, if $p \in N_K(x)$, then $\langle p, x\rangle = \max_{y \in K} \langle p, y\rangle$. Since K is a cone, we deduce that $\langle p, x\rangle = 0$ and that $p \in K^-$. □

c) **Proposition 6.** *Let $A \in \mathscr{L}(X, Y)$ and $K = A^{-1}(y)$ be an affine subspace. Then, if $Ax = y$,*

(10) $$T_{A^{-1}(y)}(x) = \operatorname{Ker} A.$$ ▲

Proof. a) If $v \in \operatorname{Ker} A$, then $v + x \in A^{-1}(y) = K$ and thus, $v = v + x - x$ belongs to $T_K(x)$.

b) Conversely, if $v = \lim_{n \to \infty} v_n \in T_K(x)$, where $v_n = \lambda_n(x_n - x)$ with $x_n \in K$ and $\lambda_n \ge 0$, then $v_n \in \operatorname{Ker} A$ and thus $v \in \operatorname{Ker} A$.

d) **Proposition 7.** *Let $S^n \doteq \left\{x \in R^n_+ \,\middle|\, \sum_{i=1}^n x_i = 1\right\}$ and $I(x) \doteq \{i = 1, \dots, n \mid x_i = 0\}$. Then*

(11) $$v \in T_{R^n_+}(x) \quad \text{if and only if } v_i \ge 0 \text{ for all } i \in I(x)$$

and

(12) $$v \in T_{S^n}(x) \quad \text{if and only if } v_i \ge 0 \text{ for all } i \in I(x) \text{ and } \sum_{i=1}^n v_i = 0.$$ ▲

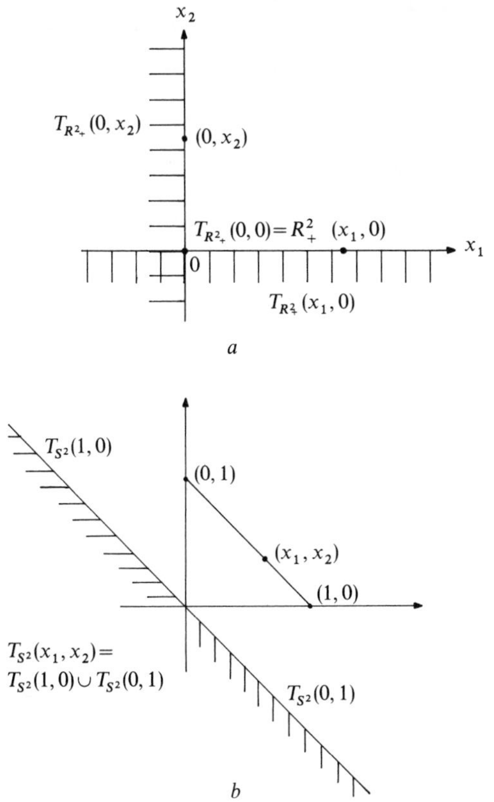

Fig. 23. *a* Tangent cone to R^2_+; *b* Tangent cone to the simplex

Proof. a) If $K=R^n_+$, the first statement follows from Proposition 3: If $p \in R^n_-$ satisfies $\sum_{i=1}^{n} p_i x_i = \sum_{i \notin I(x)} p_i x_i = 0$, then $p_i = 0$ whenever $i \notin I(x)$; hence $v \in T_{R^n_+}(x)$ if $\sum_{i \in I(x)} p_i v_i \leq 0$ for all $p \in R^n_-$, i.e., if and only if (4) holds.

b) Let v satisfy $v_i \geq 0$ if $i \in I(x)$ and $\sum_{i=1}^{n} v_i = 0$. If $v_i = 0$ for all $i \notin I(x)$, then $v = 0$. If not, let $\lambda = \min_{\substack{i \notin I(x) \\ v_i \neq 0}} \frac{x_i}{|v_i|} > 0$. Therefore $x + \lambda v \in S^n$ since $x_i + \lambda v_i = \lambda_i v_i \geq 0$ if $i \in I(x)$, $x_i + \lambda v_i \geq x_i - \lambda |v_i| \geq x_i - x_i = 0$ if $i \notin I(x)$ and $\sum_{i=1}^{n} (x_i + \lambda v_i) = 1 + 0 = 1$. Hence $v \in \frac{1}{\lambda}(S^n - x) \in T_{S^n}(x)$.

c) If $v = \lambda(y - x)$ where $y \in S^n$ and $\lambda > 0$, then we deduce that $v_i = \lambda(y_i - x_i) = \lambda y_i \geq 0$ when $i \in I(x)$ and that $\sum_{i=1}^{n} v_i = 0$. Therefore

$$\bigcup_{\lambda>0} \lambda(S^n - x) \subset \left\{ v \in T_{R^n_+}(x) \,\middle|\, \sum_{i=1}^n v_i \right\} = 0.$$

Since the latter subset is closed, we deduce that

$$T_{S^n}(x) = \left\{ v \in T_{R^n_+}(x) \,\middle|\, \sum_{i=1}^n v_i = 0 \right\}. \qquad \square$$

Calculus on Tangent and Contingent Cones

Since the tangent cones to a closed convex subset $K \subset X$ coincide with the contingent cones, the tangent cones inherit the properties of the contingent cones. They do enjoy other properties that are quite useful. We begin by the obvious properties.

We assume once and for all that all the subsets that are involved in the following statements are nonempty.

Proposition 8. a) *If* $x \in K \subset L$, *then*

(13) $\quad T_K(x) \subset T_L(x)$ *and when the sets are convex,* $N_L(x) \subset N_K(x)$.

b) *Let* $K \doteq \bigcap_{i \in J} K_i$, *and* $J(x) = \{i \mid x_i \notin \operatorname{Int} K_i\}$. *Then*

(14) $$T_K(x) \subset \bigcap_{i \in J(x)} T_{K_i}(x).$$

c) *Let* $K \doteq \bigcup_{i \in I} K_i$, *and* $I(x) = \{i \in I \mid x \in K_i\}$. *Then*

(15) $$\bigcup_{i \in I(x)} T_{K_i}(x) \subset T_K(x)$$

Equality holds when I is finite or, more generally, locally finite. ▲

Proposition 9. *Let* $\vec{K} \doteq \prod_{i=1}^n K_i$ *and* $\vec{x} = (x_1, \ldots, x_n) \in \vec{K}$. *Then* $T_K(\vec{x}) \subset \prod_{i=1}^n T_{K_i}(x_i)$ *and, when the subsets* K_i *are convex, we have*

(16) $$T_K(\vec{x}) = \prod_{i=1}^n T_{K_i}(x_i) \quad \text{and} \quad N_K(\vec{x}) = \prod_{i=1}^n N_{K_i}(x_i). \qquad ▲$$

Proof. It is obvious that $T_{\vec{K}}(\vec{x}) \subset \prod_{i=1}^n T_{K_i}(x_i)$. Conversely assume that the subsets K_i are convex; let $v_i \in T_{K_i}(x_i)$ for $i = 1, \ldots, n$. Then there exist sequences of elements v_i^k converging to v_i and of $h_i^k > 0$ such that $x_i + h_i^k v_i^k \in K_i$ for all $i = 1, \ldots, n$. We set $h^k = \min_{i=1,\ldots,n} h_i^k > 0$. Since the subsets K_i are convex, $x_i + h^k v_i^k \in K_i$ for all i, i.e., $\vec{x} + h\vec{v}^k \in \vec{K}$. Hence $\vec{v} \in T_{\vec{K}}(\vec{x})$. We deduce the formula on normal cones by polarity. $\square$

Proposition 10. *Let $A \in \mathscr{L}(X, Y)$ and $K \subset X$. Then, $\forall x \in K$, $A T_K(x) \subset T_{A(K)}(Ax)$. When K is convex, these two cones are equal:*

(17) $$\forall x \in K, \quad T_{A(K)}(Ax) = cl(A T_K(x))$$

and, by polarity,

(18) $$N_{\overline{A(K)}}(x) = A^{*-1}(N_K(x)). \qquad \blacktriangle$$

Proof. Since $\langle p, Ax \rangle = \max_{y \in K} \langle p, Ay \rangle = \max_{y \in K} \langle A^* p, y \rangle = \langle A^* p, x \rangle$, we obtain the formula for the normal cones and deduce it by polarity for tangent cones. □

Corollary 1. *Let K and L be two closed convex subsets, $x \in K$ and $y \in L$. Then*

(19) $$T_{K+L}(x+y) = cl(T_K(x) + T_L(y)) \quad \text{and} \quad N_{K+L}(x+y) = N_K(x) \cap N_L(y). \qquad \blacktriangle$$

Theorem 2. *Let $A \in \mathscr{L}(X, Y)$, $L \subset X$ and $M \subset Y$. We set*

(20) $$K \doteq \{x \in L \mid Ax \in M\} = L \cap A^{-1}(M).$$

Assume that $K \neq \emptyset$, i.e., that $0 \in A(L) - M$. The inclusion

(21) $$T_K(x) \subset T_L(x) \cap A^{-1}(T_M(Ax))$$

is always true. If we assume that L and M are closed and convex and that

(22) $$0 \in \operatorname{Int}(A(L) - M),$$

then equalities

(23) $$T_K(x) = T_L(x) \cap A^{-1}(T_M(Ax)) \quad \text{and} \quad N_K(x) = N_L(x) + A^* N_M(Ax)$$

hold true. ▲

Proof. a) The first inclusion is obvious. The equality shall follow from the Robinson-Ursescu theorem. (See Corollary 1.3.2.) Let $x_0 \in K$ and $v_0 \in T_L(x_0) \cap A^{-1} T_M(Ax_0)$. There exist sequences of elements $v_n \in X$ and $u_n \in Y$ converging to v_0 and Av_0 respectively such that, for all n, $x_0 + h_n^1 v_n \in L$ and $Ax_0 + h_n^2 u_n \in M$. We set $h_n \doteq \min(h_n^1, h_n^2, 1) > 0$. Since L and M are convex, we deduce that

(24) $$\text{for all } n, \quad x_n \doteq x_0 + h_n v_n \in L \quad \text{and} \quad y_n \doteq Ax_0 + h_n u_n \in M.$$

The theorem is proved if $u_n = Av_n$ for an infinite subset of indexes. If not, we apply the Robinson-Ursescu theorem to the set-valued map F defined by L to Y by

(25) $$F(x) \doteq Ax - M \quad \text{if } x \in L, \qquad F(x) \doteq \emptyset \quad \text{if } x \notin L.$$

We take $y_0 = 0$ and $x_0 \in F^{-1}(0) = K$. By assumption (22), $y_0 \in \operatorname{Int} F(L) = \operatorname{Int}(A(L) - K)$. Hence there exists $\gamma > 0$ such that $\forall y \in y_0 + \gamma B$,

(26) $$\forall x \in L, \quad d(x, F^{-1}(y)) \leq \frac{1}{\gamma} d(y, F(x)) (\|x_0 - x\| + 1).$$

We take $y=0$ and $x=x_0+h_n v_0$. So $\|x_0-x\|=h_n\|v_0\|$,

$$d(0,F(x_0+h_n v_0))=d(Ax_0+h_n A v_0, M)$$
$$\leq d(Ax_0+h_n u_n, M)+h_n\|Av_0-u_n\|=h_n\|Av_0-u_n\|.$$

Therefore,

$$(27)\qquad \frac{d(x_0+h_n v_0,K)}{h_n}=\frac{1}{h_n}d(x_0+h_n v_0,F^{-1}(0))$$
$$\leq\frac{1}{\gamma h_n}d(0,F(x_0+h_n v_0))(\|x-x_0\|+1)$$
$$\leq\frac{1}{\gamma}\|Av_0-u_n\|(h_n\|v_0\|+1).$$

Hence

$$(28)\qquad \inf_{h_n>0}\frac{d(x_0+h_n v_0,K)}{h_n}=0.$$

This means that $v_0\in T_K(x_0)$.

b) By polarity, we deduce that

$$(29)\qquad N_K(x)=cl(N_L(x)+A^*N_M(Ax)).$$

Assumption (22) implies that $N_L(x)+A^*N_M(Ax)$ is *closed*.

For that purpose, let $r_n\doteq p_n+A^*q_n$ be a sequence converging to r, where $p_n\in N_L(x)$ and $q_n\in N_M(Ax)$. We prove that

$$(30)\qquad \forall v\in Y,\quad \sup_n\langle q_n,v\rangle<+\infty.$$

Indeed, there exist $\lambda>0$, $y\in L$ and $z\in K$ such that $v=\lambda(z-Ay)$ by assumption (20). Hence

$$\langle q_n,v\rangle=\lambda\langle q_n,z-Ay\rangle=\lambda(\langle q_n,z\rangle-\langle A^*q_n,y\rangle)$$
$$=\lambda(\langle p_n,y\rangle+\langle q_n,z\rangle-\langle r_n,y\rangle)$$
$$\leq(\langle p_n,x\rangle+\langle q_n,Ax\rangle-\langle r_n,y\rangle)\quad(\text{since } p_n\in N_L(x) \text{ and } q_n\in N_M(Ax))$$
$$=\lambda\langle r_n,x-y\rangle\leq\lambda\|r_n\|\,\|x-y\|<+\infty$$

since the converging sequence r_n is bounded. Therefore, the sequence of elements q_n is weakly bounded and thus, weakly relatively compact; some subsequence (again denoted by) q_n converges to $q\in N_M(Ax)$. Hence $p_n=r_n-A^*q_n$ converges to $p=r-A^*q\in N_L(x)$ and thus, $r=p+A^*q$ belongs to $N_L(x)+A^*N_M(x)$. □

Corollary 2. *If L is a closed convex subset of X and if $A\in\mathscr{L}(X,Y)$ then, for any $y\in\operatorname{Int}A(L)$ and $x\in L\cap A^{-1}(y)$, we have*

$$(31)\qquad T_{L\cap A^{-1}(y)}(x)=T_L(x)\cap\operatorname{Ker}A. \qquad ▲$$

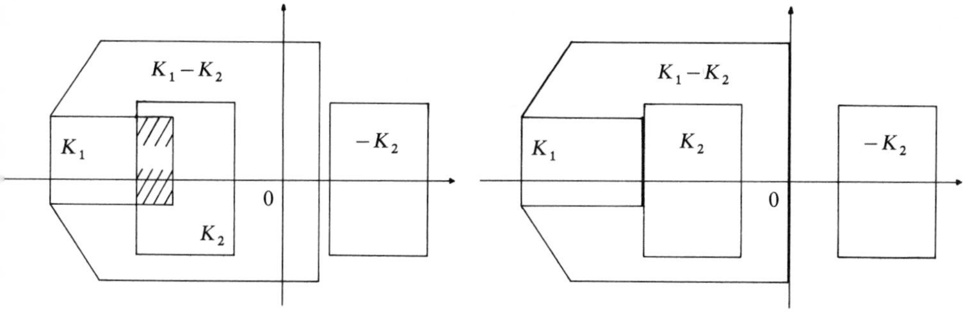

a $0 \in \operatorname{Int}(K_1 - K_2)$ *b* $0 \notin \operatorname{Int}(K_1 - K_2)$

Fig. 24. Tangent cone to $K_1 \cap K_2$

$$\forall x \in K_1 \cap K_2, \quad T_{K_1 \cap K_2}(x) = T_{K_1}(x) \cap T_{K_2}(x)$$

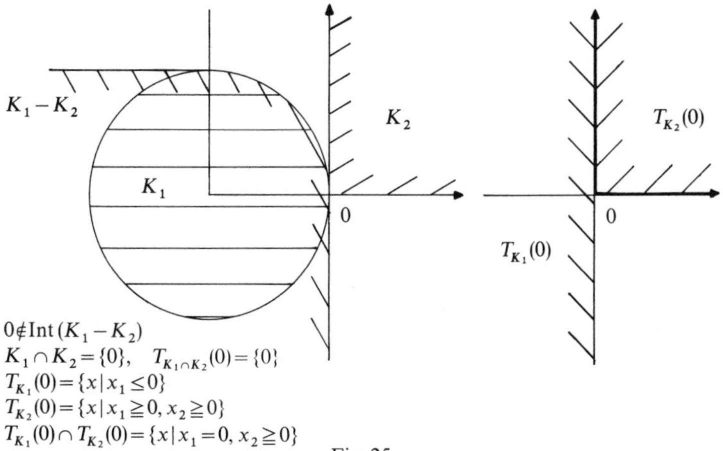
$K_1 - K_2$
K_1
K_2
0
$T_{K_2}(0)$
0
$T_{K_1}(0)$
$0 \notin \operatorname{Int}(K_1 - K_2)$
$K_1 \cap K_2 = \{0\}, \quad T_{K_1 \cap K_2}(0) = \{0\}$
$T_{K_1}(0) = \{x \mid x_1 \leq 0\}$
$T_{K_2}(0) = \{x \mid x_1 \geqq 0, x_2 \geqq 0\}$
$T_{K_1}(0) \cap T_{K_2}(0) = \{x \mid x_1 = 0, x_2 \geqq 0\}$

Fig. 25

Corollary 3. *Let $P \subset X$ be a closed convex cone and $p_0 \in P^-$ such that $K = \{x \in P$ such that $\langle p_0, x \rangle = -1\}$ is not empty. Then*

$$v \in T_K(x) \quad \text{if and only if } \langle p_0, v \rangle = 0 \text{ and } \langle p, v \rangle \leq 0 \text{ for all } p \in P^- \text{ satisfying } \langle p, x \rangle = 0, \tag{32}$$

and

$$N_K(x) = p_0 R + (P^- \cap \{x\}^\perp). \tag{33}$$

▲

By taking $X = Y$ and A to be the identity, we obtain:

Corollary 4. *Let K and L be two closed convex subsets of X. If*

$$0 \in \operatorname{Int}(K - L) \tag{34}$$

then

(35) $$\forall x \in K \cap L, \quad T_{K \cap L}(x) = T_K(x) \cap T_L(x). \quad \blacktriangle$$

2. Viability Implies the Existence of an Equilibrium

The Viability Theorem states that when F is a strict upper semicontinuous map from a closed subset $K \subset R^n$ to the compact convex subsets of R^n, then the tangential condition

(1) $$\forall x \in K, \quad F(x) \cap T_K(x) \neq \emptyset$$

is necessary and sufficient for the existence of viable trajectories for every initial state x_0 in K. It is remarkable that when K is convex and compact, this viability property implies the existence of an *equilibrium* (or a *stationary* solution or *rest* point) of the dynamical system, i.e., a solutions $\bar{x} \in K$ to the inclusion

(2) $$0 \in F(\bar{x}).$$

Theorem 1. *Let $K \subset X$ be compact convex and F be an upper hemicontinuous set-valued map from K into X with nonempty closed convex images. We posit the following tangential condition:*

(3) $$\forall x \in K, \quad F(x) \cap T_K(x) \neq \emptyset.$$

Then

a) *There exists an equilibrium $\bar{x} \in K$ of F.*

b) $\forall h > 0$, $\forall x_0 \in K$, *there exists a sequence of elements $x^n \in K$ such that $x^0 = x_0$ and*

(4) $$\forall n \in N, \quad \frac{(x^n - x^{n-1})}{h} \in F(x^n). \quad \blacktriangle$$

Remark. We can regard the sequence of elements $x^n \in K$ as a viable *discrete* trajectory of the dynamical system. The finite-difference system (4) is called the "implicit" finite difference scheme of the differential inclusion $x' \in F(x)$, by opposition to the *explicit* scheme

$$\forall n \in N, \quad \frac{y^n - y^{n-1}}{h} \in F(y^{n-1}).$$

We can write that $y^n \in (1 + hF)(y^{n-1})$ whereas the implicit scheme yields $x^n \in (1 - hF)^{-1}(x^{n-1})$. Usualy, $1 + hF$ does not map K into itself. But Theorem 1 states that $(1 - hF)^{-1}$, a formal approximant of $1 + hF$, does map F into itself in the sense that:

(5) $$\forall x \in K, \quad (1 - hF)^{-1}(x) \cap K \neq \emptyset. \quad \square$$

Remark. Sum of two maps satisfying the tangential condition.

We single out a simple remark, although quite important, due to the *convexity* of the tangent cone. If *two set-valued maps* F_1 *and* F_2 *satisfy the tangential condition, so does the set-valued map* $\alpha_1 F_1 + \alpha_2 F_2$ *when* $\alpha_1, \alpha_2 \geq 0$.

Corollary 1. *Let* $K \subset X$ *be convex compact and* F_i $(i=1,2)$ *be upper hemicontinuous maps from* K *to* X *with compact convex values. If*

(6) $$\forall i=1,2,\ \forall x \in K, \quad F_i(x) \cap T_K(x) \neq \emptyset,$$

then the differential inclusions

$$x' \in F_i(x) \quad (i=1,2) \quad \text{and} \quad x' \in \alpha_1 F_1(x) + \alpha_2 F_2(x) \quad (\alpha_1, \alpha_2 > 0)$$

have both equilibria and viable trajectories. ▲

Remark. Stability under perturbations.

We assume now that $X = R^n$ and that K is a compact convex subset whose interior is not empty. Hence we know that the graph of the set-valued map $y \to \operatorname{Int} T_K(y)$ is open (see Proposition 1.4).

We prove now the following stability result:

Lemma 1. *Let* F *be an upper semicontinuous compact valued map from* K *to* X *satisfying the "strong internal tangential condition":*

$$\forall x \in K, \quad F(x) \subset \operatorname{Int} T_K(x).$$

Then there exists $\alpha > 0$ *such that any set-valued map* G *"close" to* F *in the sense that*

(7) $$\operatorname{graph}(G) \subset \operatorname{graph}(F) + \alpha B$$

satisfies also the strong internal condition. ▲

Proof. Since the graph of F is compact and contained in the open set $\operatorname{Int} T_K(\cdot)$, there exists $\alpha > 0$ such that

$$\operatorname{graph}(F) + \alpha B \subset \operatorname{graph}(\operatorname{Int} T_K).$$

Then the graph of any G satisfying (7) is contained in the graph of $\operatorname{Int} T_K$. □

This yields a very important property of *stability:*

Corollary 2. *Let* $K \subset R^n$ *be a compact convex subset with nonempty interior and* F *be an upper semicontinuous map from* K *to* R^n *with convex compact values. Assume that it satisfies the strong internal tangential condition*

(8) $$\forall x \in K, \quad F(x) \subset \operatorname{Int} T_K(x).$$

Then there exists $\alpha>0$ such that every upper semicontinuous map G from K to R^n with compact convex values satisfying

$$\operatorname{graph}(G)\subset\operatorname{graph}(F)+\alpha B$$

has an equilibrium in K and viable trajectories. ▲

Ky Fan's Inequality

Instead of using the Brouwer fixed point theorem, which states that any continuous map from a compact convex subset of R^n to itself has a fixed point, we shall use an equivalent statement, Ky Fan's inequality, equivalent but still much more powerful.

Theorem 2 (Ky Fan). *Let K be a convex compact subset and $\varphi\colon K\times K\to R$ be a function satisfying*

$$(1)\qquad\begin{cases}\text{i)}\ \forall y\in K,\ x\to\varphi(x,y)\ \text{is lower semicontinuous,}\\ \text{ii)}\ \forall x\in K,\ y\to\varphi(x,y)\ \text{is concave,}\\ \text{iii)}\ \forall y\in K,\ \varphi(y,y)\le 0.\end{cases}$$

Then there exists $\bar{x}\in K$ such that

$$(9)\qquad \forall y\in K,\quad \varphi(\bar{x},y)\le 0.$$ ▲

Proof. a) Ky Fan's inequality implies Brouwer's fixed point theorem: Let K be a convex compact subset of R^n and f be a continuous map from K to K. Hence the function φ defined by $\varphi(x,y)=\langle f(x)-x,\ y-x\rangle$ satisfies Assumption (8) of Theorem 2: Hence there exists $\bar{x}\in X$ such that $\varphi(\bar{x},y)\le 0$ for all $y\in K$ and, in particular, such that $\|f(\bar{x})-\bar{x}\|=\varphi(\bar{x},f(\bar{x}))\le 0$. Therefore $\bar{x}$ is a fixed point of f.

b) Brouwer's fixed point theorem implies Ky Fan's inequality: See for instance, Aubin [1979b], Chapter 5, § 6, p. 199–203 for a proof. For more sophisticated results, see Aubin [1979a], Chapter 7, § 1 and Chapter 13, § 2.

Proof of Theorem 1. a) The second conclusion follows from the first: Since the set-valued map G defined by $G(x)\doteq hF(x)+x^n-x$ is the sum of the two maps hF and $x\to x^n-x$ that satisfy the tangential condition, then G satisfies it and consequently, has an equilibrium x^{n+1} that is a solution to the inclusion $x^{n+1}-x^n\in hF(x^{n+1})$.

b) We denote by $\sigma(F(x),q)\doteq\sup_{v\in F(x)}\langle q,v\rangle$ the support function of the closed convex subset $F(x)$.

For proving the existence of an equilibrium, we assume the contrary: $\forall x\in K,\ 0\notin F(x)$ and we derive a contradiction. Since the subsets $F(x)$ are closed and convex, the separation theorem implies that:

$$(10)\qquad \forall x\in K,\quad \exists p\in X^*\quad \text{such that } \sigma(F(x),-p)<0.$$

We set

$$\Delta_p \doteq \{x \in K \mid \sigma(F(x), -p) < 0\}. \tag{11}$$

So, the statement that no equilibrium exists takes the form

$$K \subset \bigcup_{p \in X^*} \Delta_p. \tag{12}$$

c) Since F is upper hemicontinuous, the subsets Δ_p are open. Hence the compact subset K can be covered by n open subsets Δ_{p_i}. Let $\{a_i\}_{i=1,\ldots,n}$ be a continuous partition of unity associated to this covering.

d) We introduce the function φ defined on $K \times K$ by

$$\varphi(x, y) = -\sum_{i=1}^{n} a_i(x) \langle p_i, x - y \rangle. \tag{13}$$

It is continuous with respect to x, affine with respect to y and satisfies $\varphi(y, y) = 0$ for all $y \in K$.

So the assumptions of the Ky Fan inequality hold and, consequently, there exists $x_* \in K$ such that

$$\forall y \in K, \quad \varphi(x_*, y) = \langle -p_*, x_* - y \rangle \leq 0 \tag{14}$$

where we set $p_* = \sum_{i=1}^{n} a_i(x_*) p_i$.

In other words, p_* belongs to $N_K(x_*)$. The tangential condition (3) implies the existence of $v \in F(x_*) \cap T_K(x_*)$. Hence, since $N_K(x) = T_K(x)^-$,

$$\sigma(F(x_*), -p_*) \geq \langle -p_*, v \rangle \geq 0. \tag{15}$$

e) The latter inequality is impossible: Let I be the set of indices i such that $a_i(x_*) > 0$. It is nonempty since $\sum_{i=1}^{n} a_i(x_*) = 1$. If $i \in I$, then $x_* \in \Delta_{p_i}$ and thus, $\sigma(F(x_*), -p_i) < 0$. Therefore:

$$\begin{aligned} \sigma(F(x_*), -p_*) &= \sigma(F(x_*), -\sum_{i \in I} a_i(x_*) p_i) \\ &\leq \sum_{i \in I} a_i(x_*) \sigma(F(x_*), -p_i) \text{ (by the convexity of support functions).} \end{aligned}$$

We have proved that $\sigma(F(x_*), -p_*) < 0$, which is the contradiction we were looking for. □

Remark. We used only the weaker tangential condition

$$\forall x \in K, \ \forall p \in N_K, \quad \sigma(F(x), -p) \geq 0. \tag{16}$$

Remark that condition (16) is equivalent to the tangential condition (1) when the images $F(x)$ are convex compact: If $F(x) \cap T_K(x) = \emptyset$, then $0 \notin F(x) - T_K(x)$, which is a closed convex set. The separation theorem implies the existence of $p \in X^*$ such that

$$\sigma(F(x), -p) + \sup_{v \in T_K(x)} \langle -p, -v \rangle \leq -\varepsilon < 0.$$

Hence $\sigma(T_K(x), p)$ is bounded above and thus, is equal to 0 and $p \in T_K(x)^- = N_K(x)$. This is a contradiction of condition (1). □

Remark. The Kakutani fixed point theorem.

We deduce from Theorem 1 another proof of the Kakutani fixed point theorem, (Corollary 1.12.1), which holds in any Hausdorff locally convex space.

Theorem 3 Kakutani. *Let K be a compact convex subset and G be an upper semicontinuous map from K to K with compact convex images. Then there exists a fixed point $x_* \in K$ of G.* ▲

Proof. We set $F(x) \doteq G(x) - x \subset K - x \subset T_K(x)$. Hence $F(\cdot)$ is an upper hemicontinuous map from K to X that satisfies the tangential condition (1). By Theorem 1, it is an equilibrium $x_* \in K$, which is a fixed point of G. □

Actually, the same proof implies the following statement. We say that G is *inward* if

(17) $$\forall x \in K, \quad G(x) \cap (x + T_K(x)) \neq \emptyset.$$

Theorem 4 Browder-Ky Fan. *Let G be an upper hemicontinuous map from a compact convex subset $K \subset X$ to the closed convex subsets of X. If G is inward, then it has a fixed point $x_* \in K$.* ▲

We also mention the following result.

We say that G is *outward* if

(18) $$\forall x \in K, \quad G(x) \cap (x - T_K(x)) \neq \emptyset.$$

Theorem 5 (Ky Fan-Rogalski). *Let G be an upper hemicontinuous map from a compact convex subset $K \subset X$ to the closed convex subsets of X. If G is outward, then*

a) *it has a fixed point $x_* \in K$.*

b) $K \subset G(K)$. (*i.e., for all $y \in K$, $\exists x \in K$ such that $y \in G(x)$.*) ▲

Proof. It follows from Theorem 1 applied to the map F defined on K by $F(x) \doteq x - G(x)$. □

Remark. It is not more difficult to prove a slightly more general theorem. Let X and Y be two Hilbert spaces (or, more generally, Hausdorff locally convex vector spaces).

Theorem 6. *We introduce*

(19) $$\begin{cases} \text{i)} & K, \text{ a compact convex subset of } X, \\ \text{ii)} & F, \text{ an upper hemicontinuous map from } K \text{ to } Y \text{ with closed convex values}, \\ \text{iii)} & A: K \to \mathscr{L}(X, Y), \text{ a continuous map associating with} \\ & \text{each } x \in K \text{ a continuous linear operator from } X \text{ to } Y. \end{cases}$$

We posit the tangential condition.

(20) $$\forall x \in K, \quad F(x) \cap cl(A(x)\, T_K(x)) \neq \emptyset$$

Then

a) *there exists an equilibrium* $x_* \in K$ *of* F: $0 \in F(x_*)$,

b) $\forall y \in K$, *there exists* $\hat{x} \in K$ *satisfying,*

(21) $$A(\hat{x})(\hat{x} - y) \in F(\hat{x}).$$ ▲

Note that the second statement allows the construction of a solution to the implicit finite difference scheme:

$$A(x^n)(x^n - x^{n-1}) \in F(x^n); \qquad x^0 = x_0.$$

It also amounts to saying that the set-valued map $x \to (x - A(x)^{-1} F(x)) \cap K$ from K to K has nonempty values.

When $A \in \mathscr{L}(X, Y)$ does not depend on x, the second statement can be stated as follows.

(22) Whenever there exists a solution x_0 in K to the linear equation $y = A x_0$, then there exists also a solution $\hat{x}$ in K to the perturbed inclusion $y \in A(\hat{x}) + F(\hat{x})$.

Proof. a) The second statement follows from the first applied to the map G defined by $G(x) = F(x) + A(x)(y - x)$.

b) The proof of the first statement is the same as the proof of Theorem 1, where the function φ defined by (13) is replaced by the function φ defined by

(23) $$\varphi(x, y) = -\sum_{i=1}^{n} a_i(x) \langle p_i, A(x)(x - y) \rangle.$$ □

Invariant Subsets

We adapt to the "convex case" Theorem 1.6.2 on invariant subsets under a Lipschitzean map F.

Let $\Omega \subset X$ be a nonempty open subset and F be a set-valued map from Ω into X.

We recall the definition of invariant subsets (see Definition 4.6.1).

Definition 1. We shall say that a subset $K \subset \Omega$ is "*invariant*" under F if for all $x_0 \in K$, all the trajectories of the differential inclusion $x'(t) \in F(x(t))$, $x(0) = x_0$, remain in K. ▲

When K is a closed convex subset, we give a sufficient condition for invariance.

We recall that π_K denotes the projector of best approximation onto K.

Proposition 1. *Let us assume that a set-valued map F satisfies the "strong external tangential condition".*

(24) $$\forall x \in \Omega,\ \forall v \in F(x), \quad \langle x - \pi_K(x), v \rangle \leq 0.$$

Then K is invariant under F. ▲

Proof. Let us assume that the statement is false: There exists an absolutely continuous function $x(\cdot)$ satisfying $x'(t) \in F(x(t))$ a.e., $x(0) = x_0 \in K$ and, for some $t_1 \in]0, T[$, $x(t_1) \notin K$.

Let us consider the function $t \to d_K^2(x(t))$. Proposition 0.6.1 shows that $\nabla d_K^2(x) = 2(x - \pi_K(x))$. Since $x(\cdot)$ is absolutely continuous, we deduce that for almost all $t \in [0, T]$,

$$\frac{d}{dt} d_K^2(x(t)) = \langle \nabla d_K^2(x(t)), x'(t) \rangle = 2 \langle x(t) - \pi_K(x(t)), x'(t) \rangle.$$

If $x(t) \in K$, we obtain that $\frac{d}{dt} d_K^2(x(t)) = 0$. If $x(t) \notin K$, we deduce from the strong external tangential condition that $\frac{d}{dt} d_K^2(x(t)) \leq 0$ since $x'(t) \in F(x(t))$.

Hence

$$0 < d_K^2(x(t_1) - d_K^2(x(0)) = \int_0^{t_1} \frac{d}{dt} d_K^2(x(\tau))\, d\tau \leq 0.$$

We have obtained a contradiction. □

As a corollary, we obtain the following result on invariant subsets:

Theorem 7. *Let K be a closed convex subset of R^n and $F: K \to R^n$ be an upper semicontinuous set-valued map with compact convex images. If $F(K)$ is relatively compact and if the "strong tangential condition"*

$$\forall x \in K, \quad F(x) \subset T_K(x),$$

holds true, then, for any $x_0 \in K$, there exists a solution $x(\cdot)$ to the problem

(25) $$\begin{cases} \text{i) } \forall t \geq 0,\ x(t) \in K \ (\textit{and } x(0) = x_0), \\ \text{ii) } \textit{for a.e. } t > 0,\ x'(t) \in F(x(t)). \end{cases}$$

Moreover F can be extended to R^n so that K is invariant under F. ▲

Proof. We associate to F its extension $G: R^n \to R^n$ defined by $G(x) \doteq F(\pi_K(x))$, which is upper semicontinuous with compact convex images, such that $G(R^n) = F(K)$ is bounded. Hence there exist solutions to the differential inclusion $x'(t) \in G(x(t)) = F(\pi_K(x(t)); x(0) = x_0$. (See Theorem 2.1.4). But G satisfies the strong external tangential condition. Hence K is invariant under G thanks to the above proposition. □

3. Viability Implies the Existence of Periodic Trajectories

By the Approximation Theorem, upper semicontinuous maps with closed convex images are particular instances of σ-selectionable maps, the definition of which we recall

Definition 1. We say that a set-valued map F from X to Y is σ-selectionable if there exists a nonincreasing sequence of upper semicontinuous maps with compact values satisfying

$$(1) \qquad \begin{cases} \text{i) } \forall n \geq 0, \; F_n \text{ has a continuous selection,} \\ \text{ii) } \forall x \in X, \; F(x) = \bigcap_{n \geq 0} F_n(x). \end{cases}$$ ▲

We recall that the map that associates with any point the set of trajectories issued from this point is σ-selectionable, with values which are not generally convex (See Theorem 2.2.3).

We say that a map F from a subset $K \subset X$ to X is *strongly inward* if

$$(2) \qquad \forall x \in K, \quad F(x) \subset x + T_K(x).$$

Theorem 1. *Let K be a compact convex subset of a Fréchet space X and F be a σ-selectionable map from K to X. If F is strongly inward, then it has a fixed point $x_* \in K$.* ▲

Proof. a) Let d be a distance defining the topology of X, invariant by translation and defining convex balls. Let π_K be the projector onto K, associating with x its best approximation. This is a set-valued map with closed graph and convex values.

Let $f_n\colon K \to X$ be a continuous selection of F_n. Then $\pi_K \circ f_n$ maps K into K and, by Kakutani's Theorem, has a fixed point x_n. Let $v_n \doteq f_n(x_n)$ belong to $F_n(x_n)$. Therefore (v_n, x_n) belongs to graph (π_K) and (x_n, v_n) belongs to graph (F_n); which is contained in the graph of F_p for all $p \leq n$. Since the graph of F_0 is compact, F_0 being upper semicontinuous with compact values and K being compact, a subsequence (again denoted by) (x_n, v_n) converges to some (x_*, v_*), which belongs to graph (F_p) for all p. Hence $v_* \in \bigcap_{p > 0} F_p(x_*) \doteq F(x_*)$. We also infer that (v_*, x_*) belongs to graph (π_K). Then:

$$(3) \qquad x_* \quad \text{belongs to } \pi_K F(x_*).$$

b) Let v_* be a point of $F(x_*)$ such that $x_* = \pi_K(v_*)$. If v_* belongs to K, then $x_* = v_*$ and x_* is a fixed point of F. If we assume that v_* does not belong to K, we shall derive a contradiction. Let $r = \|v_* - x_*\| > 0$. The open ball $\mathring{B}(v_*, r)$ does not intersect K. The separation Theorem implies the existence of continuous linear form $p \in X_*$ such that

$$\sup_{y \in K} \langle p, y \rangle \leq \inf_{\|z - v_*\| \leq r} \langle p, z \rangle.$$

Since x_* belongs to K and to the closure of the ball $\mathring{B}(v_*, r)$, we deduce that $\langle p, x_*\rangle = \sup_{z\in K}\langle p, z\rangle$, i.e. that p belongs to the normal cone $N_K(x_*)$ to K at x_*. Since $v_* - x_*$ belongs to $F(x_*) - x_*$, which is contained in $T_K(x_*)$ because F is strongly inward, we have $\langle p, v_* - x_*\rangle \leq 0$. Hence $\langle p, v_*\rangle \leq \langle p, x_*\rangle \leq \inf_{\|z - v_*\| \leq r} \langle p, z\rangle < \langle p, v_*\rangle$, which is a contradiction. □

We deduce the following consequences.

Theorem 2. *Let K be a convex subset of a Fréchet space X and F be a σ-selectionable map from K to X satisfying*

$$\forall x \in K, \quad F(x) \subset T_K(x). \tag{4}$$

Then there exists an equilibrium $x_* \in K$ of F. ▲

Proof. We apply Theorem 1 to the map G defined by $G(x) \doteq F(x) - x$, which is strongly inward and thus, has a fixed point, an equilibrium of F.

Theorem 3. *Let K be a convex subset of a Fréchet space X and F be a σ-selectionable map from K to X, which is strongly outward in the sense that*

$$\forall x \in K, \quad F(x) \subset x - T_K(x). \tag{5}$$

Then

$$\begin{array}{l} \text{i) } F \text{ has a fixed point } x_* \in K, \\ \text{ii) For all } y \in K, \text{ there exists } x \in K \text{ such that } y \in F(x). \end{array} \tag{6}$$

Proof. a) The map G_1 defined by $G_1(x) \doteq 2x - F(x)$ is strongly inward: Then G_1 has a fixed point $x_* \in K$, which is a fixed point of F.

b) The map G_2 defined by $G_2(x) \doteq y + x - F(x)$ is strongly inward because $G_2(x) = y - x + G_1(x) \subset 2T_K(x) = T_K(x)$. Hence it has a fixed point x_*, which is a solution to the inclusion $y \in F(x_*)$. □

Application. Periodic Solutions of Differential Inclusions

We begin by a particular case.

Proposition 1. *Let K be a compact convex subset of R^n and F be an upper semicontinuous map from $R_+ \times K$ to the compact convex subsets of R^n satisfying*

$$\forall (t, x) \in [0, T] \times K, \quad F(t, x) \subset T_K(x). \tag{7}$$

Then there exist $x_* \in K$ and a trajectory such that

$$\begin{cases} \text{i) } x'(t) \in F(t, x(t)), \\ \text{ii) } x(0) = x(T) = x_*. \end{cases} \tag{8}$$

▲

Proof. Let r_T be the continuous map from $\mathscr{C}(0,\infty,R^n)$ to R^n defined by $r_T x \doteq x(T)$. Theorem 2.7 implies that every trajectory $x(\cdot)$ of $\mathscr{T}_\infty(x)$ is viable when $x\in K$. Then $r_T\circ\mathscr{T}_\infty$ maps K into K and thus, is strongly inward. Theorem 2.2.2 states that the map $r_T\circ\mathscr{T}_\infty$ is σ-selectionable and thus, by Theorem 1, we deduce that it has a fixed point x_*. This means that there exists a solution to the differential inclusion $x'(t)\in F(t,x(t))$ satisfying $x(0)=x(T)=x_*$. □

Now, we shall prove the main Theorem on periodic solutions to differential inclusions.

Theorem 4. *Let K be a compact convex subset of R^n and F be an upper semicontinuous map from $R_+\times K$ to the compact convex subsets of R^n satisfying*

$$(9)\qquad \begin{cases} \text{i)}\ \forall (t,x)\in R_+\times K,\ F(t,x)\cap T_K(x)\neq\emptyset, \\ \text{ii)}\ \exists T>0 \text{ such that, } \forall (t,x)\in R_+\times K,\ F(t,x)=F(t+T,x). \end{cases}$$

Then there exists a periodic viable trajectory of the differential inclusion $x'(t)\in F(t,x(t))$ of period T. ▲

Proof. We shall "approximate" the differential inclusion $x'(t)\in F(t,x(t))$ by differential equations satisfying the assumption (7) of Proposition 1.

1-st Step. We can suppose with no loss of generality that K has a nonempty interior in R^n.

Indeed, using an appropriate translation, we can first suppose that 0 belongs to K without changing the hypotheses. Then, we can replace R^n by the vector space $M(K)$ spanned by K (see § 5.1). Then

$$(10)\qquad \begin{cases} \text{i)}\ K \text{ has a nonempty interior in } M(K), \\ \text{ii)}\ \forall x\in K,\ T_K(x)\subset M(K). \end{cases}$$

We finally replace F by the set-valued map $(t,x)\to F(x)\cap M(K)$.

We can easily prove that all the hypotheses of Theorem 4 are satisfied.

2-nd Step. By Theorem 1.6.1, we can "approximate" the set-valued map F by set-valued maps F_n defined by

$$(11)\qquad \forall (t,x)\in R_+\times K,\quad F_n(t,x)=\sum_{\text{finite}}\psi_i^n(t,x)\,C_i^n$$

where ψ_i^n are locally Lipschitzean functions and C_i^n are compact convex subsets contained in $\operatorname{Im} F$. They satisfy

$$(12)\qquad \begin{cases} \forall n\geq 0, & F(t,x)\subset\ldots\subset F_n(t,x)\subset F_0(t,x), \\ \forall \varepsilon>0, & \exists N(\varepsilon,t,x)\,|\,\forall n\geq N(\varepsilon,t,x),\quad F_n(t,x)\subset F(t,x)+\varepsilon B. \end{cases}$$

Furthermore, assumption (9) i) implies that

$$(13)\qquad \forall n\geq 0,\ \forall (t,x)\in R_+\times K,\quad F_n(t,x)\cap T_K(x)\neq\emptyset.$$

We define now the set-valued maps G_n from $R_+ \times K$ to R^n by

(14) $$\forall (t,x) \in R_+ \times K, \quad G_n(t,x) = F_n(t,x) + \frac{1}{n} B.$$

It is obvious that these set-valued maps G_n are continuous with closed convex values. Moreover, since $\operatorname{Int}(K)$ is nonempty, then $v + \frac{1}{n} B$ belongs to $\operatorname{Int} T_K(x)$ for all $v \in T_K(x)$. Therefore (12) implies that

(15) $$\forall n > 0, \ \forall (t,x) \in R_+ \times K, \quad G_n(t,x) \cap \operatorname{Int} T_K(x) \neq \emptyset.$$

Since G_n is lower semicontinuous and since the map $x \to \operatorname{Int} T_K(x)$ has an open graph (see Proposition 5.1.4), then corollary 1 of Michael's selection Theorem (see Theorem 1.11.1) implies the existence of a *continuous* selection g_n:

(16) $$\forall n \geq 0, \ \forall (t,x) \in R_n \times K, \quad g_n(t,x) \in G_n(t,x) \cap \operatorname{Int} T_K(x).$$

The functions g_n satisfy the assumptions of the above Proposition 1. Consequently, there exist $x_{n*} \in K$ and solutions $x_n(\cdot)$ of the differential equation

(17) $$x_n'(t) = g_n(t, x_n(t))$$

satisfying

(18) $$x_n(0) = x_{n*} = x_n(T).$$

It is now a matter of routine to check that the solutions $x_n(\cdot)$ of (17) converge uniformly to a solution $x(\cdot)$ of the differential inclusion

(19) $$x'(t) \in F(t, x(t))$$

satisfying

(20) $$x(0) = x_* = x(T)$$

as in the proof of Theorem 2.2.3.

When F is periodic (in the sense that (9) ii) holds, we deduce the existence of a periodic solution. □

4. Regulation of Controled Systems Through Viability

Let us translate the Viability Theorem in the language of Control Theory. The dynamics of the system are described by a map

(1) $$f: K \times U \to X$$

where U is the "control set". The state of the system evolves according to the differential equation

(2) $$\begin{cases} \text{i)} \ x'(t) = f(x(t), u(t)), \\ \text{ii)} \ x(0) = x_0 \in K. \end{cases}$$

The regulation problem can be expressed in the following way:

a) Does there exist a function $t \to u(t)$ *(open loop control)* such that the differential equation (2) has viable trajectories?

b) Does there exist a continuous single-valued function $\underset{\sim}{u}$ from K to U (*closed loop control* or *feedback control*) such that the differential equation

(3) $$\begin{cases} \text{i)} \ x'(t) = f(x(t), \underset{\sim}{u}(x(t))), \\ \text{ii)} \ x(0) = x_0 \in K, \end{cases}$$

has viable trajectories?

c) Does there exist an *equilibrium* $(\bar{x}, \bar{u}) \in K \times U$, a solution to the non-linear equation

(4) $$f(\bar{x}, \bar{u}) = 0.$$

We introduce the *feedback map* C defined by

(5) $$\forall x \in K, \quad C(x) \doteq \{u \in U \mid f(x, u) \in T_K(x)\}.$$

We shall assume that:

(6) $$\begin{cases} \text{i)} \ U \text{ is compact,} \\ \text{ii)} \ f\colon K \times U \to Y \text{ is continuous,} \end{cases}$$

so that the set-valued map F defined by

(7) $$F(x) \doteq \{f(x, u)\}_{u \in U}$$

is continuous with compact values.

We summarize in the following statement the consequences of the Viability Theorem.

Theorem 1. *Let K be a locally compact subset of a Hilbert space X, U be a compact set and $f\colon K \times U \to X$ be a continuous map.*

We assume that the feedback map C is strict:

(8) $$\forall x \in K, \quad C(x) \doteq \{u \in U \mid f(x, u) \in T_K(x)\} \neq \emptyset.$$

We posit the assumption

(9) $$\forall x \in K, \quad f(x, U) \doteq \{f(x, u)\}_{u \in U} \text{ is convex.}$$

Then

(10) $$\begin{cases} \forall x_0 \in K, \ \exists T > 0, \text{ there exist a measurable function} \\ u(\cdot) \text{ and a viable trajectory of the differential equation (2)} \end{cases}$$

which are related by

(11) $$\text{for almost all } t \in [0, T], \quad u(t) \in C(x(t)).$$

If we assume moreover that K is convex and compact, then we can take $T = \infty$ *and infer the existence of an equilibrium* $(\bar{x}, \bar{u}) \in K \times U$:

(12) $$\exists (\bar{x}, \bar{u}) \in K \times U \quad \text{such that } f(\bar{x}, \bar{u}) = 0. \qquad \blacktriangle$$

Proof. a) We apply Theorem 4.2.1 to the differential inclusion

(13) $$\begin{cases} \text{i)} & x'(t) \in F(x(t)), \\ \text{ii)} & x(0) = x_0, \end{cases}$$

where $F(x) \doteq f(x, U)$. There exists a viable trajectory of the differential inclusion (13) i). Hence, for almost all $t \in [0, T]$, there exists $u(t) \in U$ such that $x'(t) = f(x(t), u(t))$ belongs to $F(x(t))$. Since $x(\cdot)$ is locally Lipschitzean, for almost all t the derivative $x'(t) = \lim_{h \to 0+} \frac{x(t+h) - x(t)}{h}$ exists and thus belongs to the contigent cone $T_K(x(t))$. Therefore,

$$\text{for almost all } t \in [0, T], \quad x'(t) \in F(x(t)) \cap T_K(x(t)).$$

This means that there exists, for almost all $t \in [0, T]$, $x'(t) \in f(x(t), U) \cap T_K(x(t))$. By Corollary 1.14.1, there exists a measurable function $t \to u(t) \in U$ such that $x'(t) = f(x(t), u(t))$ for almost all t. Hence $u(t)$ belongs to $C(x(t))$.

When K is convex and compact, Theorem 2.1 implies that there exists $\bar{x} \in K$ such that 0 belongs to $F(\bar{x}) = f(\bar{x}, U)$. Hence there exists $\bar{u} \in U$ such that $f(\bar{x}, \bar{u}) = 0$. □

When K is convex, we can use the *dual version* of the tangential condition (8).

We consider the set-valued map $N_K(\cdot)$: $x \to N_K(x)$ associating to each $x \in K$ the normal cone to K at x. We introduce the function d defined on the graph of $N_K(\cdot)$ by

(14) $$\forall (x, p) \in \text{graph } N_K(\cdot), \quad d(x, p) \doteq \inf_{v \in U} \langle p, f(x, v) \rangle.$$

Assumptions (6) imply that the function d is continuous and that the set-valued map D defined on the graph of $N_K(\cdot)$ by

(15) $$\forall (x, p) \in \text{graph } N_K(\cdot), \quad D(x, p) \doteq \{u \in U \mid \langle p, f(x, u) \rangle = d(x, p)\}$$

is strict and upper semicontinuous.

The dual tangential condition

(16) $$\forall (x, p) \in \text{graph } N_K(\cdot), \quad \sigma(F(x), -p) \geq 0$$

is equivalent to the tangential condition (8) because the set-valued map $F(x) \doteq f(x, U)$ has compact convex values. Since $\sigma(F(x), -p) = -d(x, p)$, we obtain the following statement.

Theorem 2. *Let K be a closed convex subset of a Hilbert space X, U be a compact set and f: $K \times U \to X$ be a continuous map such that $f(x,U)$ is convex for all $x \in K$. We assume that*

$$\forall (x,p) \in \operatorname{graph} N_K(\cdot), \quad d(x,p) \leq 0. \tag{17}$$

Then conclusions (10) *and* (11) *of Theorem* 1 *hold true. Moreover, if K is compact, there exists an equilibrium $(\bar{x}, \bar{u})$.* ▲

We mention explicitely the following corollary, which we shall use in applications to economics.

Corollary 1. *Let K be a closed convex subset of a Hilbert space X, U be a compact set and f: $K \times U \to X$ be a continuous map such that $f(x,U)$ is convex for all $x \in K$. We assume that there exists a map $\underset{\sim}{w}$ from the graph of N_K to the set of controls U satisfying*

$$\forall (x,p) \in \operatorname{graph} N_K(\cdot), \quad \langle p, f(x, \underset{\sim}{w}(x,p))\rangle \leq 0. \tag{18}$$

Then conclusions (11) *and* (12) *of Theorem* 1 *hold true. Moreover, if K is compact, there exists an equilibrium.* ▲

Let us consider the case when K is defined by linear constraints in the sense that

$$K \doteq \{x \in L \mid Ax \in M\}. \tag{19}$$

Theorem 1.2 tells us that, when $L \subset X$ and $M \subset Y$ are closed convex subsets, A is a continuous linear operator from X to Y and

$$0 \in \operatorname{Int}(A(L) - M), \tag{20}$$

then $T_K(x) = T_L(x) \cap A^{-1} T_M(Ax)$. Therefore, the feedback map can be written

$$C(x) = \{u \in U \mid f(x,u) \in T_L(x) \quad \text{and} \quad Af(x,u) \in T_M(Ax)\}. \tag{21}$$

Let us single out the following consequence for controled systems.

Corollary 2. *Let X and Y be Hilbert spaces, A belong to $\mathscr{L}(X,Y)$, $L \subset X$ and $M \subset Y$ be closed convex subsets satisfying*

$$0 \in \operatorname{Int}(A(L) - M). \tag{20}$$

Let U be a compact set and f: $K \times U \to X$ be a continuous map such that $f(x,U)$ is convex for all $x \in K$. We assume that

$$\forall x \in L, \ \forall u \in U, \quad f(x,u) \in T_L(x) \tag{22}$$

and that there exists a map $\underset{\sim}{w}$ from the graph of $N_M(\cdot)$ to the set of controls $u \in U$ satisfying

$$\forall x \in A^{-1}(M), \ \forall q \in N_M(Ax), \quad \langle q, Af(x, \underset{\sim}{w}(x,q))\rangle \leq 0. \tag{23}$$

Then

(24) $\forall x_0 \in K$, $\exists T > 0$, *a measurable function* $u\colon [0, T] \to U$ *and a trajectory of the controled differential equation* $x'(t) = f(x(t), u(t))$, $x(0) = x_0$ *which is viable in the sense that* $\forall t \in [0, T]$, $x(t) \in L$ *and* $A x(t) \in M$.

By denoting by

$$C(x) \doteq \{u \in U \mid A f(x, u) \in T_M(A x)\}, \tag{25}$$

the feedback map, then

$$\text{for almost all } t \in [0, T], \quad u(t) \text{ belongs to } C(x(t)). \tag{26}$$

Furthermore, if K is compact and convex, there exists an equilibrium $(\bar{x}, \bar{u}) \in K \times U$ *and we can take* $T \doteq \infty$ *in* (23). ▲

We give now examples of maps f satisfying the assumption that $f(x, U)$ is convex for every $x \in K$. This is obviously the case when

$$\begin{cases} \text{i)} & U \text{ is convex,} \\ \text{ii)} & \forall x \in K,\ u \to f(x, u) \text{ is affine.} \end{cases} \tag{27}$$

For this example, we can obtain the existence of a continuous feedback control $\underset{\sim}{u}$ yielding viable trajectories.

Theorem 3. *Let K and U be convex compact subsets of finite dimensional vector spaces, f be a continuous map from* $K \times U$ *to X which is affine with respect to* $u \in U$

We assume that there exists $\gamma > 0$ *such that*

$$\forall x \in K,\ \forall y \in \gamma B, \quad \exists u \in U \mid f(x, u) + y \in T_K(x). \tag{28}$$

Then conclusions (10), (11) *and* (12) *of Theorem* 1 *hold true and there exists a continuous feedback control* $\underset{\sim}{u}\colon K \to U$ *yielding viable trajectories of the differential equation* (3). ▲

Proof. By Theorem 1.1, the set-valued map $x \to T_K(x)$ is lower semicontinuous with convex values. Hence the assumptions of Theorem 1.2.3 are satisfied and thus, the feedback map $x \to C(x)$ is lower semicontinuous with closed convex values.

Michael's Selection Theorem 1.14.1 implies the existence of a continuous selection $\underset{\sim}{u}$ of the feedback map C. Hence for all $x \in K$, $f(x, \underset{\sim}{u}(x))$ belongs to $T_K(x)$, so that the Nagumo Theorem implies the existence of a viable trajectory of the differential equation (3). □

Application to Game Theory

We denote by $X=R^p$ the "environment" over which act n players labelled $i=1,\dots,n$.

The actual environment is a subset $K\subset X$, usually defined by such constraints as scarcity of resources. The action of player i is described by a map h_i from K to X associating with each state x of the environment the rate of change $h_i(x)$ that player i forces on the environment.

An important example is the case when $h_i(x)=\nabla U_i(x)$ is the gradient at x of a utility function U_i; in this case, action of player i amounts to the marginal increase of utility.

Let A be a coalition of players. We suppose that the action of coalition A is the sum of actions of players $i\in A$. We recall that a coalition A is characterized by its characteristic function $c^A\in\{0,1\}^n$, associating with any player i its rate of participation c_i^A, defined by

$$c_i^A=\begin{cases}1 & \text{if } i\in A\\ 0 & \text{if } i\notin A.\end{cases} \tag{29}$$

Actually, we introduce a "continuum" of coalitions, by taking the convex hull $[0,1]^n$ of the subset $\{0,1\}^n$ of coalitions: we interpret $c\in[0,1]^n$ as a "fuzzy coalition", associating with any player i its rate of participation $c_i\in[0,1]$.

We suppose that the action of a fuzzy coalition $c\in[0,1]^n$ on the environment is the sum of the actions of players i multiplied by their rates of participation $\left(\text{i.e., } \sum_{i=1}^{n} c_i h_i(x)\right)$.

Let $g\colon K\to X$ define the endogeneous evolution law of the environment (in the absence of players).

Therefore, we define the evolution law of states of environments by

$$\begin{cases}\text{i) } x'(t)=g(x(t))+\sum_{i=1}^{n} c_i(t)\,h_i(x(t)),\\ \text{ii) } x(0)=x_0.\end{cases} \tag{30}$$

We introduce the feedback map C associating to every state $x\in K$ the subset of fuzzy coalitions

$$C(x)\doteq\left\{c\in[0,1]^n \,\middle|\, g(x)+\sum_{i=1}^{n} c_i h_i(x)\in T_K(x)\right\}. \tag{31}$$

Theorems 1 and 3 imply the following corollary.

Corollary 3. *Let K be a nonempty compact convex subset and let us suppose that the maps g and f_i are continuous. If $C(x)\neq\emptyset$ for all $x\in K$, then*

a) *there exists an equilibrium* $(x^*, c^*) \in K \times [0,1]^n$:

(32) $$g(x^*) + \sum_{i=1}^{n} c_i^* h_i(x^*) = 0.$$

b) *For every initial state* $x_0 \in K$, *there exist fuzzy coalitions* $c(t)$ *for which the differential systems has a viable trajectory* $x(\cdot)$; *the fuzzy coalitions are related to the state by the feedback relation:*

(33) $$c(t) \in C(x(t)).$$

c) *If we assume that there exists* $\gamma > 0$ *such that*

(34) $$\forall x \in K,\ \forall y \in \gamma B,\ \exists c \in [0,1]^n \mid g(x) + \sum_{i=1}^{h} c_i h_i(x) + y \in T_K(x),$$

then there exists a feedback fuzzy coalition $\underset{\sim}{c}$: $K \to [0,1]^n$ *yielding viable trajectories.* ▲

This simple model may provide an explanation to the formation of coalitions (actually, of fuzzy coalitions) in a dynamical model, formation which is due to the scarcity constraints on the environment.

One can avoid the use of fuzzy coalitions by replacing a finite set of players by a "continuum" Ω of players $\omega \in \Omega$. By continuum of players, we mean a set Ω supplied with a σ-algebra $\mathscr{A}$ and a *nonatomic measure* μ. We regard the σ-algebra $\mathscr{A}$ as the set of coalitions A that players $\omega \in \Omega$ are allowed to form. As before, we assume that player ω "acts" on the environment X by associating to each state $x \in K$ a rate of change $h(x, \omega)$. If $g(x)$ still denotes the endogeneous dynamics of the environment, the evolution law is given by

(35) $$\begin{cases} \text{i) } x'(t) = g(x(t)) + \int_{A(t)} h(x(t), \omega)\, d\mu(\omega), \\ \text{ii) } x(0) = x_0. \end{cases}$$

If $K \subset X$ is the viability set, the question arises whether there exist coalitions $A(t) \in \mathscr{A}$ such that the differential equation (34) yields viable trajectories.

We introduce the feedback map C associating to every state $x \in K$ the subset of coalitions

(36) $$C(x) \doteq \{A \in \mathscr{A} \mid g(x) + \int_A h(x, \omega)\, d\mu(\omega) \in T_K(x)\}.$$

Corollary 4. *Let* K *be a closed subset of a finite dimensional space* X, g *be a continuous map from* K *to* X *and* h *be a map from* $K \times \Omega$ *to* X *satisfying*

(37) $$\begin{cases} \text{i) } \forall x \in K,\ H(x)\colon \omega \to h(x, \omega) \text{ belongs to } L^\infty(\Omega, X), \\ \text{ii) } H \text{ is continuous from } K \text{ to } L^\infty(\Omega, X). \end{cases}$$

We assume that the images $C(x)$ of the feedback map C defined by (35) *are nonempty.*

Then

(38) $\forall x_0 \in K,\ \exists T>0,\ \exists A\colon [0,T] \to \mathscr{A}$ *and a viable trajectory of the differential equation* (34)

and

(39) $$\text{for almost all } t \in [0,T], \quad A(t) \in C(x(t)).$$

If we assume moreover that K is convex and compact, we can take $T=+\infty$ in (38) *and obtain the existence of an equilibrium:*

(40) *there exist $\bar{x} \in K$ and $\bar{A} \in \mathscr{A}$ such that* $g(\bar{x}) + \int_{\bar{A}} h(\bar{x}, \omega)\, d\mu(\omega) = 0.$ ▲

Proof. We apply Theorem 1 to the case when the set of controls U is equal to the σ-algebra $\mathscr{A}$ and where the dynamics are described by

(41) $$f(x, A) \doteq g(x) + \int_A h(x, \omega)\, d\mu(\omega).$$

Since μ is a nonatomic measure, the Lyapunov convexity Theorem (See, for instance, Aubin [1979a], appendix C, p. 580) implies that

(42) $$\forall x \in K, \quad f(x, \mathscr{A}) \text{ is convex compact.}$$

Assumption (37) implies obviously that the map $x \to f(x, \mathscr{A})$ is upper semicontinuous. Then we can apply Theorem 1 and conclude. □

5. Walras Equilibria and Dynamical Price Decentralization

We apply the Viability Theorems for giving a possible explanation to the role of price systems in decentralizing the behavior of different consumers, in the sense that the knowledge of the price system allows each consumer to make his choice without knowing the global state of the economy and, in particular, without knowing (necessarily) the choices of his fellow consumers.

There is no doubt that Adam Smith is at the origin of what we now call *decentralization*, i.e. the ability for a complex system moved by different actions in pursuit of different objectives to achieve an allocation of scarce resources. He introduced this mysterious and quite paradoxical property in a poetic way: Let us quote the famous citation of the Wealth of Nations, published in 1776, two centuries ago.

"*Every individual endeavours to employ his capital so that its produce may be of greatest value. He generally neither intends to promote the public interest, nor knows how much he is promoting it. He intends only his own security, only his own gain. And he is in this led by an invisible hand to promote an end which was no part of his intention. By pursuing his own interest, he frequently thus promotes that of society more effectually that when he really intends to promote it*".

However, Adam Amith did not provide a careful statement of what the *invisible hand* manipulates nor, a fortiori, a rigorous argument for its existence.

We had to wait a century for Léon Walras to recognize that price systems are the elements on which the invisible hand acts and that actions of the different agents are guided by those price systems, providing enough information to all the agents for garanteeing the consistency of their actions with the scarcity of available commodities. He presented in 1874 the general equilibrium concept in "Elements d'économie politique pure", as a solution to a system of nonlinear equations. Equality between the number of equations and the number of unknowns led him and his followers quite optimistically to assume that a solution does exist. But it required one more century for Arrow, Debreu, Gale, Nikaido and others to provide rigorous statements and proofs of the existence of an equilibrium.

Before giving precise definitions of commodity bundles, price systems and equilibria, we shall try to convey briefly the main idea of what a Walras equilibrium is. We begin, for simplicity, by the problem of sharing between n consumers a commodity bundle w, the *supply*. So the problem is to find n commodity bundles x_i such that $\sum_{i=1}^{n} x_i \leq w$; *A* solution to this problem is called an *allocation of* w. The question adressed is to devise mechanisms which yield an allocation of the supply bundle w, *whatever is the available commodity* w.

The solution proposed by Walras and his followers consists in letting price systems play a crucial role by summarizing enough information on the economic system for the use of the n consumers. Namely, a consumer is defined as an *automaton* associating to every price p and every income r (in monetary units) its *demand* $\delta_i(p,r)$, which is the commodity bundle that he buys when the price system is p and its income is r. In other words, it is assumed that demand functions describe the behavior of consumers. We observe that it is possible to devise other mathematical descriptions – models – of the same behavior, and we shall do so. We also have to mention that neoclassical economists assume that demand functions derive from the maximization of a "utility function" in order to comply to the first sentence of Adam Smith's quotation recalled above. But this is by no means necessary.

Anyway, *assume for the time that consumers are just demand functions* $\delta_i(\cdot,\cdot)$, *independent of the supply bundle* w (which cannot be known by the consumers).

We need another assumption to define the mechanism:

We assume that an *income allocation of the gross income* w *is given*. This means the following. If p is the price system, the gross income is the value

$\langle p, w\rangle$ of the supply w. We then assume that gross income $r(p) \doteq \langle p, w\rangle$ is allocated among consumers in incomes $r_i(p)$:

$$\sum_{i=1}^{n} r_i(p) = r(p). \tag{1}$$

We insist on the fact that the *model does not provide this allocation of income*, but assumes *that it is given.*

One example of such an income allocation is supplied by "exchange economies", where the supply w is the sum of n supply bundles w_i brought to the market by the n consumers. Hence in this case, we take $r(p) \doteq \langle p, w\rangle$ and $r_i(p) \doteq \langle p, w_i\rangle$, the income derived by agent i from its supply bundle w_i.

In summary, the mechanism we are about to describe depends upon:

1) the description of each consumer i by its demand function $\delta_i(\cdot, \cdot)$,

2) an allocation $r(p) = \sum_{i=1}^{n} r_i(p)$ of the gross income.

Hence, when p is the price on the market, the income of consumer i is $r_i(p)$ and its demand is $\delta_i(p, r_i(p))$. The mechanism works if and only if demand balances supply, i.e. if and only if

$$\sum_{i=1}^{n} \delta_i(p, r_i(p)) \le w. \tag{2}$$

A solution $\bar{p}$ to this problem is called a "Walras equilibrium". So, we see that if Adam Smith's invisible hand does provide a Walras equilibrium $\bar{p}$, then consumers i are led to demand commodities $\delta_i(p, r_i(p))$ that permits to share w according to the desire of everybody.

The task is now to solve problem (2).

This requires to choose among all the sufficient conditions which can be devised the ones which have an economic interpretation.

It is remarkable that a "financial constraint" on the behavior of the consumers, known as the "Walras law", provides such a sufficient condition. The Walras law *forbids consumers to spend more than their incomes*, i.e., that

$$\forall i = 1, \ldots, \ldots, n, \quad \langle p, \delta_i(p, r)\rangle \le r. \tag{3}$$

Actually, this law can be less rigorous: it is sufficient to assume that

$$\sum_{i=1}^{n} \langle p, \delta_i(p, r_i)\rangle \le \sum_{i=1}^{n} r_i. \tag{4}$$

This latter law, the collective Walras law, allows financial transactions among consumers. We insist on the fact that the Walras laws (3) or (4) do not *involve the supply bundle w.*

We shall prove that Walras law (3) implies the existence of a Walras equilibrium, i.e., allows Adam Smith's invisible hand to provide a price system summarizing information on the state of this economy, i.e., the supply vector w and the behavior of all consumers described by the demand functions δ_i and their share $r_i(\cdot)$ of the gross income.

Actually, we shall study a more general model, in which the supply bundle is not given, but has to be chosen in a subset M of *available commodity bundles supplied to the market:* Thus, the income derived from this subset M is

$$r(p) \doteq \sup_{w \in M} \langle p, w \rangle \tag{5}$$

i.e., the maximum value yielded by the available commodities. When M is reduced to one supply vector w, we fall back to the case we have considered above.

We *keep the same mechanism*, described by

1) the n demand functions $\delta_i(\cdot, \cdot)$,

2) an income allocation $r(p) = \sum_{i=1}^{n} r_i(p)$ (which depends upon M via formula (5)).

The problem is to find a price $\bar{p}$, a Walras equilibrium, clearing the market in the sense that

$$\sum_{i=1}^{n} \delta_i(\bar{p}, r_i(\bar{p})) \in M, \tag{6}$$

i.e., that the sum of the demands lies among the set of available supplies. This illustrates better Adam Smith's quotation, because by choosing a Walras equilibrium $\bar{p}$, the invisible hand promotes an end, $\sum_{i=1}^{n} \delta_i(\bar{p}, r_i(\bar{p}))$, which was no part of the intentions of the consumers.

The workability of this mechanism is guaranteed by either Walras laws (3) or (4), which, we stress again, do not depend on the set M of available supplies.

After the pioneering work of Wald, von Neumann, Kakutani, etc., started in the 1930's, the first proof of the existence of a Walras equilibrium is due to Arrow and Debreu in 1954. Further works on this problem were due to McKenzie, Nikaido, Uzawa, and many others.

We shall propose a statement and a very simple proof based on the Ky Fan inequality.

Walras defined not only the concept of equilibrium, but proposed a process known as "Walras's tâtonnement" (Tâtonnement means "tentative process", "trial and error", etc. – literally, clumbersily walking in obscurity by touch –). Indeed, a Walras equilibrium is an equilibrium for the "excess demand map" E defined by

$$\forall p, \; E(p) = \sum_{i=1}^{n} \delta_i(p, r_i(p)) - M. \tag{7}$$

So, the idea was to associate the dynamical system

$$p'(t) \in E(p(t)), \quad p(0) = p_0 \tag{8}$$

and to study under which conditions one can prove that $p(t)$ converges to a Walras equilibrium $\bar{p}$, solution to the inclusion $0 \in E(\bar{p})$.

We observe that if $p(t)$ is a price supplied by the Walras tâtonnement process (8) and if $p(t)$ is not a Walras equilibrium, it *cannot be implemented* because, the associated total demand $\sum_{i=1}^{n} \delta_i(p(t), r_i(p(t)))$ is *not necessarily available.*

Hence this model *forbids consumers to transact as long as the price $p(t)$ is not an equilibrium.* It is as if there was a super-auctioneer calling prices and receiving transaction offers from the consumers. If the offers do not match, he calls another set of price according to the rule (8), but *does not allow transactions* to take place as long as the offers are not consistent.

Tâtonnement does not provide a model on how prices are actually evolving. The fundamental nature of the Walras world is *static*, while we live in a dynamical environment where no equilibria have been observed.

We are going to propose a dynamical model that keeps the essential ideas underlying Adam Smith and Léon Walras's proposals. For this, we slightly modify the definition *of a consumer and regard a price* system not as the state of a dynamical system whose evolution law is known, as in the differential inclusion (8), but *as a control which* evolves as a function of the consumptions according to a feedback law.

To take into account the dynamical nature of the behavior of a consumer i, we describe it as an *automaton* d_i which associates to each price system p and his own consumption x_i its rate of change $d_i(x_i, p)$. Therefore, when the price $p(t)$ evolves, the consumption $x_i(t)$ of consumer i evolves according the differential equation

$$x_i'(t) = d_i(x_i(t), p(t)), \quad x_i(0) = x_i^0. \tag{9}$$

So, a viability problem arises: *does there exist a price function $p(t)$ such that the sum $\sum_{i=1}^{n} x_i(t)$ of the consumptions remains available?* In other words, do the trajectories $x_i(\cdot)$ of the n coupled differential equations satisfy the viability condition

$$\forall t \geq 0, \quad \sum_{i=1}^{n} x_i(t) \in M? \tag{10}$$

We observe that this mechanism shares with the Walras model the *decentralization property:* The price systems $p(t)$ summarize enough information on the economic system to allow each consumer to change his own consumption independently of the other consumers and by ignoring the set M of available supplies.

We still have a concept of equilibrium. It is a sequence $(\bar{x}_1, \ldots, \bar{x}_n, \bar{p})$ of n consumptions $\bar{x}_i$ and of a price system $\bar{p}$ such that

$$(11) \qquad \begin{cases} \text{i)} \ \forall i=1, \ldots, n, \ d_i(\bar{x}_i, \bar{p})=0, \\ \text{ii)} \ \sum_{i=1}^{n} \bar{x}_i \in M. \end{cases}$$

But contrary to the Walras tâtonnement, the associated dynamical system (9) yields implementable (i.e., viable) trajectories.

In order to keep all the good features of the Walras model, it remains to check that there are sufficient conditions which have an economic interpretation. This is still the case, since we shall prove that equilibria of (11) and viable trajectories of (9) do exist if the instantaneous demand functions d_i satisfy the "*instantaneous Walras law*"

$$(12) \qquad \forall p, \ \forall x_i, \ \langle p, d_i(x_i, p)\rangle \leq 0.$$

This is a financial rule that requires that at each instant, the value of the rate of change of each consumer is not positive, i.e., that each consumer *does not spend more than he earns in an instantaneous exchange of goods.*

As in the Walras model, the instantaneous Walras law does not involve the subset M of available supplies.

Furthermore, we shall observe that the price system evolves as a feedback control: it depends upon time, but through the state of the system in the sense that there exists a set-valued map C from the set of allocations to the set of prices such that

$$(13) \qquad \text{for almost all } t \geq 0, \quad p(t) \in C(x_1(t), \ldots, x_n(t)).$$

In this framework, we see Adam Smith's invisible hand (which should be called more to the point, invisible brain) setting prices as functions of allocations for the purpose of promoting consumers to respect scarcity constraints. It is in this sense that we may regard such a dynamical system as a *regulation mechanism.* So the feedback condition (13) involves the price system, *but does not influence its variation.* This is the second important difference with dynamical systems such as (8), which require, so to speak, that Adam Smith's invisible hand acts in an active way, as in mechanics, to set the prices for adjusting demand to supply. In the model we suggest, only consumers are supposed to have the ability to act dynamically according to differential equation (9) and the prices "follow" the consumption according to the relation (13).

Description of an Economy

In order to represent an economy, we begin by introducing l types of commodities, which are endowed with an "unit", so that we can speak of x units of a commodity. A commodity may involve not only its physical properties, but also the place where it is available, the date when it is available and, in the case of uncertainty about the future, the elementary event that will be realized (for instance, 100 hundred grams of bread which will be available in New-York in 32 days when there is a strike of the subway). *Services* also may be included, as long as units are perfectly well defined.

So, a *commodity bundle* is a vector $x \in R^l$, which describes the quantity x_h of each commodity $h=1,\dots,l$.

The description of an exchange economy begins with

(14) the subset $M \subset R^l$ of *available commodities*.

It continues with the specification of n *consumers* $i=1,\dots,n$. A first definition of a consumer i starts with

(15) *the consumption set* $L_i \subset R^l$

which is interpreted as the set of commodity bundles he "needs". If $x \in L_i$, then x_h is i's demand of commodity h when $x_h > 0$ and $|x_h|$ is the i's supply of commodity h when $x_h < 0$.

The question now arises: Can the consumers share an available commodity?

We define an allocation $\vec{x} \in (R^l)^n$ as n commodity bundles $x_i \in L_i$ such that their sum $\sum_{i=1}^{n} x_i$ is available. We denote by

(16) $$K \doteq \left\{ \vec{x} \in \prod_{i=1}^{n} L_i \,\middle|\, \sum_{i=1}^{n} x_i \in M \right\}$$

the set of allocations.

The next question that arises is whether one can devise mechanisms that provide allocations. Indeed, when the number of consumers is large, finding an allocation is difficult because it requires the knowledge of a large amount of informations. Can we find a way of summarizing enough information to allow each consumer to choose his consumption in a decentralised manner?

This is possible by using price systems. A price system is a linear functional p that associates with any commodity $x \in R^l$ its value $\langle p, x \rangle \in R$. We denote by

$$S^l \doteq \left\{ p \in R^l_+ \,\middle|\, \sum_{i=1}^{l} p_i = 1 \right\}$$

the price simplex.

We now regard a consumer i as an automaton that can associate with any price system p and any income r a consumption $\delta_i(x, r) \in L_i$. In other words, a consumer i characterized by its *demand function* δ_i: $S^l \times R \to L_i$.

We next regard the support function

(17) $$\sigma_M(p) \doteq \sup_{y \in M} \langle p, y \rangle$$

as the *gross income*, which is the maximum value of the available commodity bundles for the price system p.

We assume that this gross income is allocated among the n consumers: there exist n functions r_i: $S^l \to R$ such that

$$\sum_{i=1}^{l} r_i(p) = \sigma_M(p). \tag{18}$$

Does this mechanism work; namely, does there exist a price system $\bar{p} \in S^l$ such that the sum of the demands is available?, i.e., such that

$$0 \in \sum_{i=1}^{n} \delta_i(\bar{p}, r_i(\bar{p})) - M. \tag{19}$$

Definition 1. A *price system* $\bar{p} \in S^l$ such that (19) holds true is said to be *Walras equilibrium*, or that "it clears the market". The map E defined on S^l by

$$E(p) \doteq \sum_{i=1}^{n} \delta_i(p, r_i(p)) - M \tag{20}$$

is called the *excess demand map*. ▲

Observe that this mechanism is *decentralized:* the choice of i-th consumer depends only upon the price system p (through its income function r_i) and does not require the knowledge of the choices of the other consumers.

We shall make the following assumption on the demand functions, known as the *collective Walras law*

$$\forall p \in S^l, \quad \left\langle p, \sum_{i=1}^{n} \delta_i(p, r_i) \right\rangle \leq \sum_{i=1}^{n} r_i. \tag{21}$$

It states that the value of the sum of the demands cannot exceed the sum of the incomes.

A stronger rule, which is decentralized, is the (individual) Walras law:

$$\forall p \in S^l, \quad \langle p, \delta_i(p, r) \rangle \leq r. \tag{22}$$

The Walras law, together with mild technical assumptions, implies the existence of a Walras equilibrium.

Theorem 1. *Let us assume that*

(23) $\quad M = M_0 - R_+^l$ *is closed and convex, where* M_0 *is compact*

that

(24) $\quad \forall i = 1, \ldots, n,\ L_i$ *is closed, convex and bounded below*

that

(25) *the demand functions δ_i are continuous and satisfy the collective Walras law*

and that

(26) *the income functions r_i are continuous.*

Then there exists a Walras equilibrium $\bar{p} \in S^l$. ▲

Proof. It follows from the Ky Fan inequality (see Theorem 2.2) with $K \doteq S^l$ and φ defined on $S^l \times S^l$ by

$$\varphi(p, q) \doteq \sum_{i=1}^{n} \langle q, \delta_i(p, r_i(p)) \rangle - \sigma_M(q).$$

The function φ is obviously lower semicontinuous with respect to p and concave with respect to q. The collective Walras law implies that

$$\varphi(q, q) \leq \sum_{i=1}^{n} r_i(q) - \sigma_M(q) \leq 0$$

since the gross income is allocated among the consumers. Then Ky Fan's Theorem 2.2 implies the existence of $\bar{p} \in S^l$ such that

$$\forall q \in S^l, \quad \left\langle q, \sum_{i=1}^{n} \delta_i(\bar{p}, r_i(\bar{p})) \right\rangle \leq \sigma_M(q).$$

We deduce that

$$\forall q \in R^l, \quad \left\langle q, \sum_{i=1}^{n} \delta_i(\bar{p}, r_i(\bar{p})) \right\rangle \leq \sigma(M - R^l_+, q).$$

Since $M - R^l_+ = M$ is closed and convex, we infer that $\sum_{i=1}^{n} \delta_i(\bar{p}, r_i(\bar{p}))$ does belong to M, i.e., that $\bar{p}$ is a Walras equilibrium. ■

Decentralization by a Price Regulation Mechanism

Now, we describe the behaviour of each agent by its instantaneous demand function d_i: $L_i \times S^l \to R^l$ which sets the variation in consumer's i demand when the price is p and its consumption is x.

We posit the following assumptions on the instantaneous demand functions d_i:

$$\text{(27)} \quad \begin{cases} \text{i) } \forall i = 1, \ldots, n, \text{ the function } d_i\colon L_i \times S^l \to R^l \text{ is continuous,} \\ \text{ii) } \forall x \in L_i, \ \forall p \in S, \ d_i(x, p) \in T_{L_i}(x), \end{cases}$$

and

(28) $$\forall x \in L_i, \quad p \to d_i(x, p) \text{ is affine.}$$

Above all, we shall assume that the *instantaneous collective Walras law*:

$$\forall x \in \prod_{i=1}^{n} L_i, \quad \forall p \in S^l, \quad \left\langle p, \sum_{i=1}^{n} d_i(x_i, p) \right\rangle \leq 0 \tag{29}$$

holds true. We observe that this law does not depend upon the choice of the set M of available commodities. It is weaker than the *instantaneous Walras laws*

$$\forall x \in L_i, \quad \forall p \in S^l, \quad \langle p, d_i(x, p)\rangle \leq 0 \quad \text{for all } i=1, \ldots, n. \tag{30}$$

We also assume that

$$0 \in \operatorname{Int}\left(\sum_{i=1}^{n} L_i - M\right). \tag{31}$$

We shall say that the following set-valued map C from the set K of allocations to the set S^l of prices defined by

$$\forall x \in K, \quad C(x) \doteq \left\{p \in S^l \,\middle|\, \sum_{i=1}^{n} d(x_i, p) \in T_M\left(\sum_{i=1}^{n} x_i\right)\right\} \tag{32}$$

is the *set-valued feedback control.*

Theorem 2. *We posit assumptions* (23), (24) *and* (31) *on the economy, and assumptions* (27) *and* (28) *on the instantaneous demand functions* d_i.

We assume that the instantaneous collective Walras law (29) *holds true.*

a) *There exists an equilibrium* $(\bar{x}, \bar{p}) \in K \times S^l$ *satisfying:*

$$\forall i=1, \ldots, n, \quad d_i(\bar{x}_i, \bar{p}) = 0. \tag{33}$$

b) *For any initial allocation* $x^0 \in K$, *there exist* n *absolutely continuous functions* $x_i(\cdot)$: $[0, \infty[\to R^l$ *and a measurable function* $p(\cdot)$: $[0, \infty[\to S^l$ *solutions to the differential system*

$$\begin{aligned} &\text{for almost all } t \geq 0, \quad x_i'(t) = d_i(x_i(t)), p(t)) \\ &\text{and } x_i(0) = x_i^0 \quad (i=1, \ldots, n), \end{aligned} \tag{34}$$

which satisfy the viability conditions

$$\begin{cases} \text{i) } \forall t \geq 0, \ \forall i=1, \ldots, n, \quad x_i(t) \in L_i, \\ \text{ii) } \forall t \geq 0, \quad \sum_{i=1}^{n} x_i(t) \in M, \end{cases} \tag{35}$$

and the budget constraint

$$\text{for almost all } t \geq 0, \quad \left\langle p(t), \sum_{i=1}^{n} x_i'(t) \right\rangle \leq 0. \tag{36}$$

c) *The price* $p(t)$ *plays the role of a feedback control:*

$$\text{for almost all } t \geq 0, \quad p(t) \in C(x_1(t), \ldots, x_n(t)). \tag{37}$$

▲

Proof. It follows from Corollary 4.2 with $X \doteq (R^l)^n$, $Y \doteq R^l$, $A\vec{x} = \sum_{i=1}^{n} x_i$, $\vec{L} \doteq \prod_{i=1}^{n} L_i$, $U \doteq S^l$ and $f(x, p) \doteq (d_i(x_i, p))_{i=1,\dots,p}$.

Assumption (23) on M implies that

$$\forall y \in M, \quad N_M(y) \subset R^l_+.$$

Then, assumption (4.22) of Corollary 4.2 is satisfied by taking

$$\underset{\sim}{w}(x, p) = p \Big/ \sum_{h=1}^{l} p_h \in S^l \doteq U$$

since the collective Walras law implies that

$$\langle p, A f(x, \underset{\sim}{w}(x, p)) \rangle = \left\langle p, \sum_{i=1}^{n} d_i \left(x_i, \frac{p}{\sum_h p_h} \right) \right\rangle \leq 0$$

for all $p \in N_M \left(\sum_{i=1}^{n} x_i \right) \subset R^l_+$.

Moreover, the set K of allocations is *compact* and *convex* (by assumption (24)). Hence there exists an equilibrium $(\bar{x}, \bar{p}) \in K \times S^l$ and, for any initial allocation $x^0 = (x_1^0, \dots, x_n^0) \in K$, a function $p(\cdot)$ from $[0, \infty[$ to S^l such that the differential equation (34) yields viable trajectories $x_i(\cdot)$. They satisfy

$$\left\langle p(t), \sum_{i=1}^{n} x_i'(t) \right\rangle = \left\langle p(t), \sum_{i=1}^{n} d_i(x_i(t), p(t)) \right\rangle \leq 0$$

thanks to the instantaneous collective law. □

Remarks. a) We can assume that the instantaneous demand functions d_i are set-valued and, in this way, allow incertainty in the model.

b) We can take in account the "satisficing" behaviour of consumers by introducing preorder relations on L_i defined by closed set-valued map P_i: $L_i \to L_i$ with convex values satisfying

(38) $$\forall x \in L_i, \ x \in P_i(x) \text{ and } \forall y \in P_i(x), \quad P_i(y) \subset P_i(x)$$

by replacing assumption (27) ii) by

(39) $$\forall x \in L_i, \ \forall p \in S^l, \quad d_i(x, p) \in T_{P_i(x)}(x).$$

Then we can replace conclusion (35) by

(40) $$\forall t \geq s \geq 0, \ \forall i = 1, \dots, n, \quad x_i(t) \in P_i(x_i(s)).$$ □

We exhibit now a class of instantaneous demand functions. We associate to any consumer i:

(41) three continuous maps f_i, g_i, h_i from L_i to R^l and a strictly positive continuous real-valued function θ_i.

We set

(42) $$d_i(x,p) = \theta_i(x)\,(\langle p, f_i(x)\rangle\, g_i(x) - \|f_i(x)\|\,\|g_i(x)\|\,p - h_i(x)).$$

We observe that assumptions (27) i) and (28) are satisfied.

The assumption

(43) $$\forall i = 1, \ldots, n,\ \forall x \in L_i, \quad h_i(x) \in R^l_+$$

implies the instantaneous Walras laws and the assumption

(44) $$\forall x \in \prod_{i=1}^{n} L_i, \quad \sum_{i=1}^{n} h_i(x_i) \in R^l_+$$

implies the instantaneous collective Walras law.

Indeed, the Cauchy-Schwarz inequality yields inequality

(45) $$\langle p, d_i(x,p)\rangle \leq -\theta_i(x)\,\langle p, h_i(x)\rangle.$$

The tangential condition (27) ii) is satisfied wherever we assume that:

(46) $$\forall x \in \partial L_i \text{ (boundary of } L_i), \quad g_i(x) = h_i(x) = 0.$$

We also observe that:

$$\langle f_i(x), d_i(x,p)\rangle = -\theta_i(x)\,\langle f_i(x), h_i(x)\rangle + \theta_i(x)\,(\langle p, f_i(x)\rangle\,(\langle f_i(x), g_i(x)\rangle - \|f_i(x)\|\,\|g_i(x)\|)).$$

Therefore, the Cauchy-Schwarz inequality implies that

(47) $$\text{if } \forall x \in L_i, \quad h_i(x) \in R^l_+, \quad f_i(x) \in -R^l_+$$

then

(48) $$\langle f_i(x), d_i(x,p)\rangle \geq 0.$$

An important case is when

(49) $$f_i(x) = \nabla u_i(x) \quad \text{belongs to } -R^l_+ \text{ for all } x \in L_i$$

where u_i is a continuously differentiable utility function defined on a neighborhood of L_i.

Indeed, assumption (47) and (49) imply that the trajectories of the differential system (34) satisfy

$$\frac{d}{dt} u_i(x_i(t)) = \langle f_i(x(t)), d_i(x_i(t)), p(t))\rangle \geq 0$$

and consequently,

(50) the functions $t \to u_i(x_i(t))$ are not decreasing.

If we take

(51) $$f_i(x) = g_i(x) = \nabla u_i(x) \in -R^l_+, \quad h_i(x) = 0,$$

we obtain

$$\frac{d}{dt}u_i(x_i(t))=\langle f_i(x_i(t)), d_i(x_i(t), p(t))\rangle=0$$

and consequently,

$$\forall t\geq 0, \quad u_i(x_i(t))=u_i(x_i^0). \tag{52}$$

Properties of the Set-Valued Feedback Control

Two other questions arise: what is the regularity of the set-valued feedback control C? Do single valued continuous feedback controls exist?

We answer these questions by mentioning a stronger instantaneous collective Walras law and n stronger tangential conditions

$$\begin{cases} \text{i) } \exists \gamma>0 \text{ such that, } \forall p\in S^l,\ \forall x\in K,\ \sum_{i=1}^{n}\langle p, d_i(x_i, p)\rangle\leq -\gamma, \\ \text{ii) } \forall i=1,\dots,n,\ \forall x\in L_i,\ \forall p\in S^l,\ \forall y\in\gamma B,\ c_i(x,p)+y\in T_{L_i}(x) \end{cases} \tag{53}$$

implies that the set-valued feedback control C is lower semicontinuous (see Theorem 1.2.3) and, consequently, that there exists a continuous selection π from K to S^l such that the differential equations

$$x_i'(t)=d_i(x_i(t), \pi(x_1(t), \dots, x_i(t), \dots, x_n(t))) \tag{54}$$

has *viable trajectories*. □

Other Viability Conditions

We already mentioned that the collective Walras law provides only one sufficient condition for the dynamical system (34) to have viable trajectories. One can investigate other sufficient conditions.

It may be relevant to consider the case when

$$\begin{aligned} &R^l=R^{l_1}\times R^{l_2}, \quad M=M^1\times M^2 \text{ with } M^1\subset R^{l_1} \text{ and } M^2\subset R^{l_2}, \\ &L_i=L_i^1\times L_i^2 \text{ with } L_i^1\subset R^{l_1} \text{ and } L_i^2\subset R^{l_2}. \end{aligned} \tag{55}$$

This means that the commodities are classified in two categories (we can say for instance that R^{l_1} is a space of *physical commodities* and that R^{l_2} is a space of *fiduciary commodities*, in which the set M^2 may be "controled").

We set $x=(x^1, x^2)\in R^{l_1}\times R^{l_2}$ or $(R^{l_1})^n\times(R^{l_2})^n$, $p=(p^1,p^2)$, $d_i(x,p) =(d_i^1(x^1,x^2,p^1,p^2),\ d_i^2(x^1,x^2,p^1,p^2))$, etc. ... The set of feasible allocations is the product of K^1 and K^2 where

$$K^j \doteq \left\{x\in\prod_{i=1}^{n} L_i^j \,\middle|\, \sum_{i=1}^{n} x_i^j\in M^j\right\}. \tag{56}$$

The set-valued feedback control can be written

$$(57)\quad \begin{aligned} C(x^1,x^2) \doteq \Big\{(p^1,p^2) \,\Big|\, &\sum_{i=1}^n d_i^1(x_i^1,x_i^2,p^1,p^2) \in T_{M^1}\Big(\sum_{i=1}^n x_i^1\Big) \\ \text{and } &\sum_{i=1}^n d_i^2(x_i^1,x_i^2,p^1,p^2) \in T_{M^2}\Big(\sum_{i=1}^n x_i^2\Big)\Big\}. \end{aligned}$$

Now, we can devise a law weaker than the instantaneous collective Walras law, which requires the two kinds of commodities to play a different role, namely:

$$(58)\quad \begin{aligned} &\forall x_i^1 \in L_i,\ \forall x_i^2 \in L_i^2,\ \forall (p^1,p^2) \in S^l \\ &\langle p^1, d_i^1(x_i^1,x_i^2,p^1,p^2)\rangle \le \langle p^2, x_i^2\rangle \end{aligned}$$

and

$$(59)\quad \begin{aligned} &\forall x^1 \in K^1,\ \forall x^2 \in K^2 \text{ satisfying } \sum_{i=1}^n x_i^2 \in \partial M^2, \\ &\forall (p^1,p^2) \in S^l,\quad \Big\langle p^2, \sum_{i=1}^n d_i^2(x_i^1,x_i^2,p^1,p^2)\Big\rangle \le -\sum_{i=1}^n \langle p^2, x_i^2\rangle. \end{aligned}$$

Theorem 3. *We assume that the commodity space is split according to* (55) *and that assumptions* (23), (24) *and* (31) *apply to* M^1, M^2, L_i^1 *and* L_i^2. *We posit assumptions* (27) *and* (28) *on the instantaneous demand functions. We replace the instantaneous collective Walras law* (29) *by the laws* (58) *and* (59). *Then, for every allocation* $x=(x^1,x^2) \in K^1 \times K^2$, *there exist trajectories of the differential system*

$$(60)\quad \begin{cases} \dfrac{d}{dt} x_i^j = d_i(x_i^1(t), x_i^2(t), p^1(t), p^2(t)) \quad (i=1,\dots,n;\ j=1,2) \\ x_i^j(o) = x_i^{jo} \end{cases}$$

satisfying the viability constraint

$$(61)\quad \begin{cases} \text{i) } \forall t \ge 0, \quad x_i^j(t) \in L_i \quad (i=1,\dots,n;\ j=1,2), \\ \text{ii) } \forall t \ge 0, \quad \sum_{i=1}^n x_i^j(t) \in M^j \quad (j=1,2), \end{cases}$$

and the budget constraint

$$(62)\quad \begin{cases} \text{i) for almost all } t \ge 0,\ \Big\langle p^1(t), \dfrac{d}{dt} x_i^1(t)\Big\rangle \le \langle p^2(t), x_i^2(t)\rangle, \\ \text{ii) for almost all } t \ge 0 \text{ such that } \sum_{i=1}^n x_i^2(t) \in \partial M, \\ \quad \text{then } \Big\langle p^2(t), \dfrac{d}{dt} \sum_{i=1}^n x_i^2(t)\Big\rangle \le -\Big\langle p^2(t), \sum_{i=1}^n x_i^2(t)\Big\rangle. \end{cases}$$

▲

In other words, the *new laws* (58) and (59) *involve different "budgetary rules" which allow some "transfer of value" from the fiduciary commodities to the physical commodities while viability constraints on the physical goods are still satisfied.*

Regulation Mechanism Involving Coalitions

Formula (32) yielding the set-valued feedback control shows that the larger is the set K of allocations, the larger are the subsets $C(x_1, \dots, x_n)$ of "feedback prices" and, consequently, the *more flexible* is the economy.

The instantaneous demand functions being given, is it possible to "enlarge" the set of allocations? For instance, we can allow several "subeconomies" to form, involving smaller coalitions of consumers and preventing the other consumers to gain access to the sharing of available commodities.

Namely, we introduce

(63) a family $\mathscr{A}$ of feasible coalitions $A \subset \{1, \dots, n\}$ containing the set $N \doteq \{1, \dots, n\}$ of all consumers.

We associate with any A a subset $M(A) \subset R^l$ of available commodities for the coalition A which satisfies

$$(64) \quad \begin{cases} \text{i) } M(A) = M_0(A) - R^l_+ \text{ is closed and convex, where } M_0(A) \text{ is compact,} \\ \text{ii) } 0 \in \operatorname{Int}\left(\sum\limits_{i \in A} L_i - M(A)\right). \end{cases}$$

We still assume that

$$(65) \qquad \forall i = 1, \dots, n, \quad L_i \text{ is closed, convex and bounded below.}$$

We denote by

$$(66) \qquad \mathscr{A}(x) = \{A \in \mathscr{A} \mid \sum_{i \in A} x_i \in M(A)\}.$$

For each A, we denote by

$$(67) \qquad K(A) \doteq \left\{ x \in \prod_{i=1}^{n} L_i \,\middle|\, \sum_{i \in A} x_i \in M(A) \right\}$$

the set of feasible allocations for the coalition A and by

$$(68) \qquad \hat{K} \doteq \bigcup_{A \in \mathscr{A}} K(A)$$

the enlarged set of allocations.

We assume that the instantaneous demand functions satisfy

$$(69) \quad \begin{cases} \text{i) } d_i \text{ is continuous from } L_i \times S^l \text{ to } R_+, \\ \text{ii) } \forall x \in L_i, \quad p \to d_i(x_i, p) \text{ is affine,} \\ \text{iii) } \forall x \in L_i, \ \forall p \in S^l, \quad d_i(x, p) \in T_{L_i}(x). \end{cases}$$

We now relax the instantaneous Walras collective law and replace it by the:

Instantaneous coalitionnal Walras law:

(70) $$\forall x \in \hat{K},\ \exists A \in \mathscr{A}(x) \text{ such that } \forall p \in S^l, \quad \langle p, \sum_{i \in A} d_i(x_i, p)\rangle \leq 0.$$

We set for all $x \in K$,

(71) $$\hat{C}(x) \doteq \{(p, A) \in S^l \times \mathscr{A}(x) \mid \sum_{i \in A} d_i(x_i, p) \in T_{M(A)}(\sum_{i \in A} x_i)\}.$$

Theorem 4. *We posit the assumptions* (64), (65), (69) *and the instantaneous coalitionnal Walras law* (70). *Therefore for any initial allocation* $x^0 \in \hat{K}$ *there exist* n *absolutely continuous functions* $x_i(\cdot)$: $[0, \infty[\to R$, *a function* $p(\cdot)$: $[0, \infty[\to S^l$ *and a function* $A(\cdot)$: $[0, \infty[\to \mathscr{A}$ *such that*

(72) $$\begin{cases} \text{i) for almost all } t \geq 0,\ \forall i \in N, & x_i'(t) = d_i(x_i(t), p(t)), \\ \text{ii) for almost all } t \geq 0,\ \forall i \in N, & x_i(t) \in L_i, \\ \text{iii) for almost all } t \geq 0, & \sum_{i \in A(t)} x_i(t) \in M(A(t)). \end{cases}$$

Furthermore, for almost all $t > 0$, *we have*

(73) $$(p(t), A(t)) \in \hat{C}(x_1(t), \ldots, x_n(t))$$

where $\hat{C}$ *is the set-valued feedback control map from* $\hat{K}$ *to* $S^l \times \mathscr{A}$ *defined by* (71).

Proof. We apply Theorem 4.1 where $\hat{K} \doteq \bigcup_{A \in \mathscr{A}} K(A)$.

Since the contingent cone $T_{\hat{K}}(x)$ is the union of the tangent cones $T_{K(A)}(x)$ when A belongs to $\mathscr{A}(x)$ and since $T_{K(A)}(x)$ is the set of $v \in (R^l)^n$ such that $\sum_{i \in A} v_i$ belongs to $T_{M(A)}(\sum_{i \in A} x_i)$, then the viability conditions can be written

$$\forall x \in \hat{K}, \quad \hat{C}(x) \neq \emptyset.$$

But the instantaneous coalitionnal Walras law implies that when $A \in \mathscr{A}(x)$, there exists $p \in S^l$ such that when $(d_i(x_i, p))_{i=1,\ldots,n}$ belongs to $T_{K(A)}(x)$. Hence our Theorem 3 follows from the viability Theorem. □

Models Involving Durable Commodities

We can devise models where the evolutionary laws (34) takes in account not only the present state of the consumption of each consumer, but the whole history of its consumption, including consequences of past decisions (investment, pollution, etc. ...).

We introduce the following notations:

a) $\mathscr{C} \doteq \mathscr{C}(-\infty, 0; R^l)$ is the space of continuous functions from $]-\infty, 0]$ to R^l supplied with the topology of uniform convergence on compact sets,

b) $T(t) \in \mathscr{L}(\mathscr{C}(-\infty, +\infty; R^l), \mathscr{C})$ is the translation operator defined by

$$T(t)\, x(s) \doteq x(t+s) \qquad \forall s \leq 0,$$

c) $\mathscr{L}_i \doteq \{\psi \in \mathscr{C} \mid \psi(0) \in L_i\}$,

d) $\mathscr{K} \doteq \{(\psi_1, \ldots, \psi_n) \in \mathscr{C}^n \mid \psi(0) \in K\}$.

The *infinitesimal demand functions with memory* are functions $d_i \colon \mathscr{L}_i \times S^l \to R^l$. We assume that:

$$\text{(74)} \qquad \begin{cases} \text{i)} \ \forall i = 1, \ldots, n, \quad d_i \text{ is continuous and bounded,} \\ \text{ii)} \ \forall \psi \in \mathscr{L}_i, \ \forall p \in S^l, \quad d_i(\psi, p) \in T_{L_i}(\psi(0)), \end{cases}$$

and

$$\text{(75)} \qquad \forall \psi \in \mathscr{L}_i, \quad p \to d_i(\psi, p) \text{ is affine.}$$

The set-valued feedback control map C is the map from $\mathscr{K} \subset \mathscr{C}^n$ to S defined by

$$\text{(76)} \qquad C(\psi) \doteq \left\{ p \in S \,\middle|\, \sum_{i=1}^n d_i(\psi_i, p) \in T_M\left(\sum_{i=1}^n \psi_i(0)\right) \right\}.$$

Theorem 5. *We posit assumptions* (23), (24) *and* (31) *on the economy and assumptions* (74) *and* (75) *on the instantaneous demand functions with memory. Furthermore, we assume that the instantaneous collective Walras law with memory*

$$\text{(77)} \qquad \forall \psi \in \prod_{i=1}^n \mathscr{L}_i, \ \forall p \in S, \quad \left\langle p, \sum_{i=1}^n d_i(\psi_i, p) \right\rangle \leq 0$$

holds true.

Then, for all $\psi^0 \in \mathscr{K}$, there exist n absolutely continuous functions $x_i(\cdot)$: $R \to R^l$ and a function $p(\cdot)$: $R_+ \to S^l$ such that

$$\text{(78)} \qquad \begin{cases} \text{i) for almost all } t \geq 0, \ \forall i = 1, \ldots, n, \quad x_i'(t) = d_i(T(t)\, x_i, p(t)), \\ \text{ii)} \ T(0)\, x_i = \psi_i^0, \end{cases}$$

satisfying the viability constraints

$$\text{(79)} \qquad \begin{cases} \text{i)} \ t \geq 0, \ \forall i = 1, \ldots, n, \quad x_i(t) \in L_i, \\ \text{ii)} \ t \geq 0, \quad \sum_{i=1}^n x_i(t) \in M, \end{cases}$$

and the budget constraints

$$\text{(80)} \qquad \text{for almost all } t \geq 0, \quad \left\langle p(t), \sum_{i=1}^n x_i'(t) \right\rangle \leq 0.$$

Furthermore, the price $p(t)$ plays the role of a feedback control:

(81) $$\text{for almost all } t \geq 0, \quad p(t) \in C(T(t)x_1, \ldots, T(t)x_n).$$ ▲

Proof. We deduce it from Corollary 4.7.1. We write the system (78) in the form of the differential inclusion with memory

(82) $$\begin{cases} \text{i) for almost all } t \geq 0, & x'(t) \in D(T(t)x), \\ \text{ii) } T(0)x = \psi^0, \end{cases}$$

where

(83) $$\begin{cases} \text{i) } d(\psi, p) \doteq (d_1(\psi_1, p), \ldots, d_n(\psi_n, p)), \\ \text{ii) } D(\psi) \doteq \{d(\psi, p)\}_{p \in S}. \end{cases}$$

By Corollary 4.7.1, we have to check that

(84) $$\forall \psi \in \mathscr{K}, \quad D(\psi) \cap T_K(\psi(0)) \neq \emptyset.$$

This is equivalent to say, thanks to (74) ii)

(85) $$\forall \psi \in \mathscr{K}, \quad C(\psi_1, \ldots, \psi_n) \neq \emptyset.$$

The latter condition follows from the instantaneous Walras law (77) by a separation argument. □

This theorem can be used in a variety of situations. We shall briefly sketch an application to durable goods.

Let T be the number of future periods τ ($\tau = 1, \ldots, T$). A durable commodity is represented by an element $\vec{x} := (x^0, \ldots, x^\tau, \ldots, x^T)$ of $(R^l)^{(T+1)}$, where x^τ denotes the state of the durable commodity $\vec{x}$ at the τ-th period after its acquisition.

An element $\vec{p} \doteq (p^0, \ldots, p^\tau, \ldots, p^T)$ of $(R^l)^{(T)*}$ denotes a system of *discounted prices.* We denote by $\vec{S}$ the set of normalized non negative systems of *discounted prices.*

A trajectory $\vec{x}(\cdot)$ describes, at each instant t, the durable commodities $\vec{x}(t) \doteq (x^0(t), \ldots, x^\tau(t), \ldots, x^T(t))$ where $x^\tau(t)$ is the state of the commodity at time $t + \tau$.

We describe the economy by

(86) $$\begin{cases} \text{i) a subset } \vec{M} \subset (R^l)^{(T+1)} \text{ of available durable commodities,} \\ \text{ii) } n \text{ consumptions subsets } \vec{L}_i = \prod_{\tau=0}^{T} L_i^\tau \subset (R^l)^{(T+1)}, \end{cases}$$

that satisfy

(87) $$\vec{M} = \vec{M}_0 - R_+^{l(T+1)} \quad \text{where } \vec{M}_0 \text{ is compact and convex,}$$

(88) $$\forall i = 1, \ldots, n, \ \forall \tau = 0, \ldots, T, \quad L_i^\tau \text{ is closed convex and bounded below}$$

and

(89) $$0 \in \operatorname{Int}\left(\sum_{i=1}^{n} \vec{L}_i - \vec{M}\right).$$

For simplicity, we shall consider only examples of instantaneous demand functions of the type

(90) $$d_i^r(\vec{\psi}, \vec{p}) \doteq \left\langle p^r, \sum_{\tau=r}^{T} \psi^\tau(r-\tau) \right\rangle g_i^r(\psi^r(0)) - \left\| \sum_{\tau=r}^{T} \psi^\tau(r-\tau) \right\| \|g_i^r(\psi^r(0))\| \, p^r.$$

We observe that $\sum_{\tau=r}^{T} \psi_\tau(r-\tau)$ is the state at time r of the durable commodities acquired during the past and that $\psi^r(0)$ is the state at time r of the commodity acquired at time 0.

We assume that

(91) the functions $g_i^r\colon L_i^r \to R^l$ are continuous and vanish on the boundary of L_i^r.

Theorem 6. *We posit assumptions* (87), (88), (89) *and* (91). *Then, for all* $\psi^0 \in \mathscr{K}$, *there exist n absolutely continuous functions* $x_i(\cdot)\colon R \to R^{l(T+1)}$ *and a function* $\vec{p}(\cdot)\colon R_+ \to \vec{S}$ *such that, for almost all* $t \geq 0$, *for all* $i = 1, \ldots, n$, *and for all* $r = 0, \ldots, T$, *we have*

(92) $$\begin{cases} \dfrac{d}{dt} x_i^r(t) = \left\langle p^r(t), \sum_{\tau=r}^{n} x_i^\tau(t+r-\tau) \right\rangle g_i^r(x_i^r(t)) \\ \qquad - \left\| \sum_{\tau=r}^{T} x_i^\tau(t+r-\tau) \right\| \|g_i^r(x_i^r(t))\| \, p^r(t)) \end{cases}$$

and

(93) $$T(0)\, x_i^r = \psi_i^r.$$

These trajectories satisfy the viability constraint: for all $t \geq 0$

(94) $$\begin{cases} \text{i) } \forall i = 1, \ldots, n, \ \forall r = 0, \ldots, T, \quad x_i^r(t) \in L_i^r, \\ \text{ii) } \sum_{i=1}^{n} \vec{x}_i(t) \in \vec{M}, \end{cases}$$

and the budget constraints: for almost all $t \geq 0$,

(95) $$\begin{gathered} \text{for all } i = 1, \ldots, n, \quad \text{for all } r = 0, \ldots, T, \\ \left\langle p^r(t), \frac{d}{dt} x_i^r(t) \right\rangle \leq 0. \end{gathered}$$ ▲

This system shows that when time t evolves, consumers inherit from past consumptions and also predetermine the future by acquiring durable commodities.

6. Differential Variational Inequalities

We have seen that a necessary and sufficient condition for a trajectory of the differential inclusion

(1) $$\begin{cases} \text{i)} & x'(t) \in F(x(t)), \\ \text{ii)} & x(0) = x_0 \in K \end{cases}$$

to remain in a *closed convex* subset K (to be "viable") is that F satisfies the tangential condition

(2) $$\forall x \in K, \quad F(x) \cap T_K(x) \neq \emptyset$$

where $T_K(x)$ is the tangent cone to K at x_0.

What can be done when this assumption is no longer satisfied and we still want a dynamic system to provide viable trajectories and be "as close as possible" to the original dynamic system? The natural answer to that is to replace $F(x)$ by its projection onto the tangent cone $T_K(x)$: we introduce the set-valued map G defined by

(3) $$\forall x \in K, \quad G(x) \doteq \pi_{T_K(x)} F(x)$$

(where π_T is the orthogonal projector onto the closed convex cone T). We note that $G(x) = F(x)$ whenever $x \in \operatorname{Int} K$; $F(x)$ is only modified on the boundary of K. So, we replace the original differential inclusion by the "projected differential inclusion"

(4) $$\begin{cases} \text{i)} & x' \in \pi_{T_K(x)}(F(x)) \quad (\doteq G(x)), \\ \text{ii)} & x(0) = x_0. \end{cases}$$

The only trouble is that when F is upper hemi-continuous with closed convex values, G does not inherit these properties; hence, we cannot apply the theorem of existence of viable trajectories.

This is only an apparent drawback, thanks to the properties of the orthogonal projectors on convex cone. We shall begin by checking that viable solutions to the "projected differential inclusion" are viable solutions to

(5) $$\begin{cases} \text{i)} & x'(t) \in F(x) - N_K(x) \quad (N_K(x) \doteq T_K(x)^- \text{ is the normal cone to } K \text{ at } x), \\ \text{ii)} & x(0) = x_0 \in K. \end{cases}$$

Differential inclusions of this type are called "differential variational inequalities".

Differential variational inequalities appear naturally in many problems. We mention only that the point of view we presented above was introduced by Henry for constructing planning procedures in mathematical economics, where the use of viable trajectories is essential.

We shall also note that the *slow solutions* to the projected differential inclusions and to the variational differential inequalities coincide.

We shall give an application that is fundamental in economics: we consider m convex twice continuously differentiable functions on a neighborhood of K and we set

(6) $$F(x) \doteq -\operatorname{co}\left(\bigcup_{j=1}^{m} \nabla V_j(x)\right).$$

We conclude that there are viable slow solutions of the associated differential variational inequalities; they satisfy

(7) $$\forall j=1,\ldots,m, \quad t \to V_j(x(t)) \text{ decreases when } t \to \infty$$

and the cluster points x_* of $x(t)$ are Pareto minima. In other words, this result provides a *dynamical system that has viable monotone trajectories, the cluster points* x_* *of which are weak Pareto minima* (i.e., there is no $x \in K$ such that $\forall j=1,\ldots,m$, $V_j(x) < V_j(x_*)$). Recall that $x \to N_K(x) = \partial\psi_K(x)$ is a maximal monotone operator. Then, if $x \to N_K(x) - F(x)$ is a maximal monotone operator, we know that the differential variational inequalities have unique solutions, that are slow solutions (see Section 3.2).

Differential Variational Inequalities

Let $K \subset X$ be a *closed convex* subset and F be a set map from K into X.

Definition 1. A problem of the type: Find an absolutely continuous function $x(\cdot)$ from $[0, T]$ into X satisfying

(8) $$\begin{cases} \text{i) } \forall t \in [0, T], & x(t) \in K \\ \text{ii) for a.a. } t \in [0, T], & x'(t) \in F(x(t)) - N_K(x(t)) \end{cases}$$

is called a *differential variational inequality*. ▲

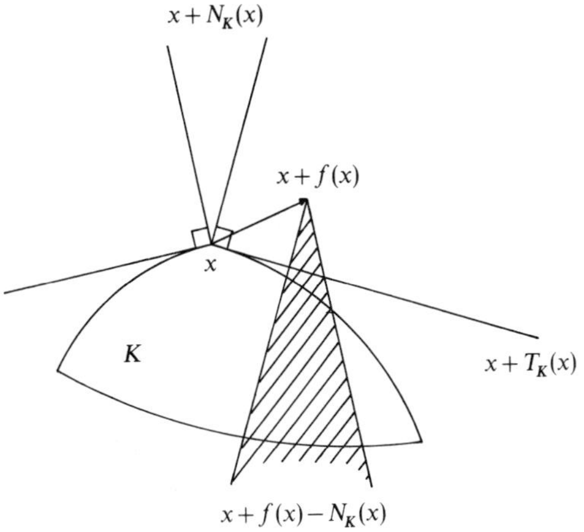

Fig. 26. Modification of a velocity in a variational inequality

Therefore, differential variational inequalities are special cases of differential inclusions. The terminology comes from the definition of the normal cone. (See Definition 1.2.)

Proposition 1. *An absolutely continuous function $x(\cdot)$ is a solution to the differential variational inequality* (8) *if and only if for almost all $t \in [0, T]$, there exists an element $f(x(t)) \in F(x(t))$ such that*

$$(9) \qquad \begin{cases} \text{i)} \ \forall t \in [0, T], & x(t) \in K, \\ \text{ii)} \ \forall \text{ a.a. } t \in [0, T], & \forall y \in K, \ \langle x'(t) - f(x(t)), x(t) - y \rangle \leq 0. \end{cases}$$ ▲

Naturally, since $F(x) \subset F(x) - N_K(x)$, any *viable* solution $x(\cdot)$ to the differential inclusion $x'(t) \in F(x(t))$ is a solution to the differential variational inequality. Also, if $\operatorname{Int} K \neq \emptyset$, any solution to the differential variational inequality satisfies $x'(t) \in F(x(t))$ when $x(t) \in \operatorname{Int}(K)$.

Remark. If K is a closed convex cone, inequalities (9) can be written

$$(10) \qquad \begin{cases} \text{i)} \ \forall t \in [0, T], & x(t) \in K, \\ \text{ii)} \ \forall \text{ a.a. } t \in [0, T], & x'(t) \in f(x(t)) - K^-, \\ \text{iii)} \ \forall \text{ a.a. } t \in [0, T], & \langle x'(t) - f(x(t)), x(t) \rangle = 0. \end{cases}$$

In particular, if $X = R^n$ and $K = R^n_+$, if x_i denotes the i-th component of x and $f_i(x)$ the i-th component of $f(x)$, then (10) takes the form: $\forall i = 1, \ldots, n$

$$(11) \qquad \begin{cases} \text{i)} \ \forall t \in [0, T], & x_i(t) \geq 0, \\ \text{ii)} \ \forall \text{ a.a. } t \in [0, T], & x_i'(t) \geq f_i(x(t)), \\ \text{iii)} \ \forall \text{ a.a. } t \in [0, T], & \forall i = 1, \ldots, n, \ (x_i'(t) - f_i(x(t))) \, x_i(t) = 0. \end{cases}$$

When K is a closed subspace, the inequalities become

$$(12) \qquad \begin{cases} \text{i)} \ \forall t \in [0, T], & x(t) \in K, \\ \text{ii)} \ \forall \text{ a.a. } t \in [0, T], & x'(t) \in F(x(t)) + K^\perp. \end{cases}$$

Proposition 2. *The solutions to the differential variational inequalities* (8) *are the solutions to the "projected differential inclusion"*

$$(13) \qquad \begin{cases} \text{i)} \ \forall t \in [0, T], & x(t) \in K, \\ \text{ii)} \ \text{for a.a. } t \in [0, T], & x'(t) \in \pi_{T_K(x(t))} F(x(t)) \end{cases}$$

and conversely. ▲

Proof. By Proposition 0.6.4, $\pi_{T_K(x)}(F(x)) \subset F(x) - N_K(x)$. So viable solutions to the projected differential inclusion are solutions to the differential variational inequality.

It remains to prove that any solution to the differential variational inequality (8) is a solution of the projected differential inclusion (13).

Since the trajectory $x(\cdot)$ remains in K, we know that for almost all $t \in [0, T]$, $x'(t) \in N_K(x(t))^\perp \subset T_K(x(t))$. Hence if $x'(t) = f(t) - p(t)$ where $f(t) \in F(x(t))$ and $p(t) \in N_K(x(t))^\perp$, we deduce that $\langle x'(t) - f(t), x'(t) \rangle = 0$. Hence $x'(t)$ is the projection of $f(t)$ onto $T_K(x(t))$. □

Remark. If F satisfies the strong tangential condition on K, i.e.

$$\forall x \in K, \quad F(x) \subset T_K(x),$$

the set of trajectories to the differential variational inequalities (9) coincides with the set of trajectories of the differential inclusion $x'(t) \in F(x(t))$ that remain in K and the solutions $x_* \in K$ to the variational inequalities $0 \in F(x_*) - N_K(x_*)$ are the stationary points of F in K. □

We denote by

$$m(F(x) - N_K(x)) \in F(x) - N_K(x)$$

the velocity of minimal norm in $F(x) - N_K(x)$. We begin by examining some properties of "slow solutions" to the differential variational inequalities (8), i.e., the solutions to the differential equation

$$(14) \qquad \begin{cases} \text{i)} \ \forall t \in [0, T], & x(t) \in K, \\ \text{ii)} \ \text{for a.a. } t \in [0, T], & x'(t) = m(F(x(t)) - N_K(x(t))). \end{cases}$$

Proposition 3. *The slow solutions of the differential variational inequalities* (8) *and the slow solutions of the projected differential inclusion* (13) *coincide.*

If $x(\cdot) \in \mathscr{C}(0, T; R^n)$ is such a slow solution, it satisfies the following inequality

$$(15) \qquad \text{for a.a. } t \in [0, T], \quad \sigma(-F(x(t)), x'(t)) + \|x'(t)\|^2 = 0$$

(where $\sigma(-F(x), p)$ denotes the support function of $-F(x)$). ▲

Proof. The first assertion follows from equality $m(F(x) - N_K(x)) = m(\pi_{T_K(x)}(F(x)))$ (see Proposition 0.6.4). Since $x'(t)$ is the projection of 0 onto $F(x) - N_K(x)$, we deduce that $\langle x'(t), -v + x'(t) \rangle \leq 0$ for all $v \in F(x) - N_K(x)$. By taking the supremum, we obtain inequality (15). □

Existence of Solutions to Differential Variational Inequalities

We state the theorem on existence of trajectories and stationary points of differential variational inequalities.

Theorem 1. *Let K be a closed convex subset of R^n and F be an upper semicontinuous set-valued map with non-empty compact convex values from K to R^n. Then*

a) *If K is compact, there exists a solution $\bar{x} \in K$ to the inclusion $0 \in F(\bar{x}) - N_K(\bar{x})$.*

b) *If $m(F(\cdot))$ is bounded, then for any $x_0 \in K$, there exists a solution to the differential variational inequalities*

$$(8) \qquad \begin{cases} \text{i)} \ x'(t) \in F(x(t)) - N_K(x(t)), & x(0) = x_0, \\ \text{ii)} \ \forall t \geq 0, & x(t) \in K. \end{cases}$$

▲

Proof. Let $c \doteq \sup_{x \in K} \|m(F(x)\|$, which is finite when K is compact, and B be the unit ball. We set:

$$H(x) \doteq F(x) - (c\,B \cap N_K(x)).$$

The subsets $H(x)$ are obviously convex compact; the set-valued map $x \to c\,B \cap N_K(x)$ is upper semicontinuous being the intersection of $N_K(\cdot)$, which has a closed graph, and of $x \to c\,B$, which is constant with compact images. Hence H is also upper semicontinuous. It satisfies the tangential condition: indeed, take

$$\pi_{T_K(x)}(m(F(x))) = m(F(x)) - \pi_{N_K(x)}\, m(F(x)).$$

It belongs to $T_K(x)$ and to $F(x) - N_K(x)$. It also belongs to $c\,B$ since $\|\pi_{T_K(x)}\, m(F(x))\| \leq \|m(F(x))\| \leq c$. Then $\pi_{T_K(x)}(m(F(x))) \in H(x) \cap T_K(x)$. Therefore, Theorem 3.1 implies the existence of an equilibrium $\bar{x} \in K$ of H. The existence of viable trajectories to the differential inclusion $x'(t) \in H(x(t))$ follows from the Viability Theorem. Since $H(x) \subset F(x) - N_K(x)$, the theorem ensues. □

Slow Solutions to Differential Variational Inequalities

Theorem 2. *Let F be a continuous set-valued map with non-empty compact convex values defined on a closed convex subset $K \subset R^n$. Assume that* $\sup\{\|m(F)\|,\ x \in K\} < +\infty$. *For any initial state $x_0 \in K$, there exists a solution* $x(\cdot) \in \mathscr{C}(0, \infty; R^n)$ *to*

$$(16) \qquad \begin{cases} \text{i)}\ \forall t \geq 0, & x(t) \in K, \\ \text{ii)}\ \text{for a.a. } t \geq 0, & x'(t) = m(F(x(t)) - N_K(x(t))) \end{cases}$$

starting at x_0. ▲

We recall that by Proposition 3, any solution to (16) is a solution to

$$(17) \qquad \begin{cases} \text{i)}\ \forall t \geq 0, & x(t) \in K, \\ \text{ii)}\ \text{for a.a. } t \geq 0, & x'(t) = m(\pi_{T_K(x(t))}(F(x(t)))). \end{cases}$$

Proof. If $p \in R^{n*}$, we denote by $p^{--} = p\,R_+$ the cone spanned by p and by $m(F(x) - (p)^{--})$ the element of $F(x) - (p)^{--}$ with minimal norm. Let $B \subset R^{n*}$ be the unit ball and S the unit sphere. We introduce the set-valued map H defined by

$$(18) \qquad H(x) \doteq \text{co}\,\{m(F(x) - (p)^{--}\}_{p \in N_K(x) \cap B}.$$

We note that

$$(19) \qquad H(x) = \text{co}\Big(m(F(x)) \cup \bigcup_{p \in N_K(x) \cap S} m(F(x) - (p)^{--})\Big).$$

We shall prove the following lemmas.

Lemma 1. *The set-valued map H is upper hemi-continuous with compact convex values;* ▲

and

Lemma 2

(20) $$\forall x \in K, \quad H(x) \cap T_K(x) = m(F(x) - N_K(x)).$$ ▲

Let us assume that these lemmas are proved. We can deduce that Theorem 2 holds true.

Proof of Theorem 2. Lemmas 1 and 2 imply that H is upper hemicontinuous with compact convex values and that it satisfies the tangential condition. By the viability theorem, there exists a viable trajectory of the differential inclusion $x'(t) \in H(x(t))$. Since $\forall t \geq 0$, $x(t) \in K$, we deduce that $x'(t) \in T_K(x(t))$. Hence $x'(t)$ belongs to $H(x(t)) \cap T_K(x(t))$ and, by Lemma 2 again, $x'(t) = m(F(x(t)) - N_K(x(t)))$. □

Proof of Lemma 1. Since F is continuous, $m(F(\cdot))$ is a continuous map. (See Theorem 1.7.1.) Let us prove that $(x, p) \in K \times S \to m(F(x) - (p)^{--})$ is continuous. Since $\|m(F(x) - (p)^{--})\| \leq \|m(F(x))\|$, we can write

$$\|m(F(x) - (p)^{--})\| = \min_{(v, a) \in F(x) \times [0, m(F(x))]} \|v - a p\|.$$

Now, since $x \to F(x) \times [0, m(F(x))]$ is continuous with compact values and since $(v, a) \to \|v - a p\|$ is continuous, the single value map $(x, p) \to m(F(x) - (p)^{--})$ is continuous on $K \times S$.

We infer that H is upper hemicontinuous: indeed, we can write its support function in the form

$$\sigma(H(x), q) = \max(\langle q, m(F(x))\rangle, \sup_{p \in N_K(x) \cap S} \langle q, m(F(x) - (p)^{--})\rangle).$$

Since $x \to N_K(x) \cap S$ is upper semicontinuous, the function $x \to \sigma(H(x), q)$ is upper semicontinuous.

Finally, S being compact, the subsets $\{m(F(x) - (p)^{--})\}_{p \in N_K(x) \cap S}$ are compact. Therefore, $H(x)$ being the convex hull of a compact set in a finite dimensional space, is also compact. (See Aubin [1979a], Proposition 2.7.7, p. 77.)

Proof of Lemma 2. a) Proposition 0.6.4 shows that $g(x) \doteq m(F(x) - N_K(x))$ can be written $g(x) = f_0 - p_0$, where $f_0 \in F(x)$, $p_0 \in N_K(x)$ satisfy

(21) $$\begin{cases} \text{i)} & g(x) = \pi_{T_K(x)}(f_0) \in T_K(x), \\ \text{ii)} & \langle g(x), p_0 \rangle = 0. \end{cases}$$

b) If $p_0=0$, then $H(x)=\{g(x)\}$ and thus, the statement is true. Indeed, for any $p\in N_K(x)$,

$$g(x)=f_0\in F(x)-(p)^{--} \quad \text{and} \quad \|g(x)\|\leq \min_{f\in F(x)} \min_{\alpha>0} \|f-\alpha p\|.$$

In other words, for any $p\in N_K(x)$, $g(x)=m(F(x)-(p)^{--})$.

In the rest of the proof, we assume $p_0\neq 0$ and we set

$$q_0=p_0/\|p_0\|.$$

c) We prove that

$$\forall x\in K, \quad g(x)\in H(x)\cap T_K(x). \tag{22}$$

By (21) i), it is sufficient to prove that $g(x)\in H(x)$. Indeed,

$$g(x)=f_0-\|p_0\|\, q_0\in F(x)-(q_0)^{--} \quad \text{and} \quad \|g(x)\|\leq \min_{f\in F(x)} \min_{\alpha>0} \|f-\alpha q_0\|.$$

Hence $g(x)=m(F(x)-(q_0)^{--})\in H(x)$.

d) To prove that $g(x)$ is the only element of the intersection, we need to prove first that

(23) if $p\in N_K(x)$ and if $m(F(x)-(p)^{--})\neq g(x)$, then $\langle m(F(x)-(p)^{--}), q_0\rangle>0$.

For that purpose, assume that $\langle m(F(x)-(p)^{--}), q_0\rangle\leq 0$, i.e., that $m(F(x)-(p)^{--})\in(q_0)^{-}$. By Proposition 0.6.4 with $T=(q_0)^{-}$, we deduce that

$$m(F(x)-(p)^{--})=m(F(x)-(q_0,p)^{--})$$

[for $(p_0)^{--}+(p)^{--}$ is the closed convex cone $(q_0,p)^{--}$ spanned by p_0 and p]. But

$$f_0-p_0\in F(x)-(q_0,p)^{--} \quad \text{and} \quad \|f_0-p_0\|\leq \min_{f\in F(x)} \min_{q\in(q_0,p)^{--}} \|f-q\|$$

since both p_0 and p belong to $N_K(x)$. Hence $g(x)=m(F(x)-(q_0,p)^{--})$, that is a contradiction.

e) Let us take $v\in H(x)\cap T_K(x)$ and show that $v=g(x)$. Since $p_0\in N_K(x)$, then $\langle v,p_0\rangle\leq 0$. We can write $v=\sum_{i=1}^{n}\alpha_i m(F(x)-(p_i)^{--})$ where $\alpha_i>0$ ($i=1,\ldots,n$) and $\sum_{i=1}^{n}\alpha_i=1$. Let us assume that the subset I of $i=1,\ldots,n$ satisfying $m(F(x)-(p_i)^{--})\neq g(x)$ is not empty. Then we can write by (21) i) and (23), that

$$\begin{aligned}\langle v,p_0\rangle &= \sum_{i\notin I}\alpha_i\langle m(F(x)-(p_i)^{--}),p_0\rangle\\ &\quad+\sum_{i\in I}\alpha_i\langle m(F(x)-(p_i)^{--}),p_0\rangle\\ &=\sum_{i\in I}\alpha_i\langle m(F(x)-(p_i)^{--}),p_0\rangle>0\end{aligned}$$

which is a contradiction. Hence $I=\emptyset$ and thus

$$v=\sum_{i=1}^{n}\alpha_i m(F(x)-(p_i)^{--})=\sum_{i=1}^{n}\alpha_i g(x)=g(x). \qquad \square$$

Corollary 1. *If $K\subset R^n$ is convex and compact and if f is a continuous map from K into X, then for any $x_0\in K$, there exists a slow solution to the differential variational inequality*

$$(8)\qquad \begin{cases} \text{i) } \forall t\geq 0, & x(t)\in K, \quad x(0)=x_0\in K,\\ \text{ii) for a.a. } t\geq 0, & x'(t)\in f(x(t))-N_K(x(t)),\end{cases}$$

which is a solution to the differential equation

$$(24)\qquad \begin{cases} \text{i) } \forall t\geq 0, & x(t)\in K, \quad x(0)=x_0\in K,\\ \text{ii) for a.a. } t\geq 0, & x'(t)=\pi_{T_K(x(t))}f(x(t)).\end{cases} \qquad \blacktriangle$$

Application. Construction of Viable Monotone Trajectories Converging to Pareto Minima

Let $K\subset\Omega\subset R^n$, where Ω is an open subset. Let us consider m functions U_j: $\Omega\to R$ satisfying

$$(25)\qquad \forall j=1,\ldots,m, \quad U_j \text{ is convex and twice continuously differentiable.}$$

Theorem 3. *Let $K\subset R^n$ be closed and convex. We posit assumption* (25). *There exist slow solutions of the differential inclusion*

$$(26)\qquad \begin{cases} \text{i) } \forall t\geq 0, & x(t)\in K; \quad x(0)=x_0 \text{ given in } K,\\ \text{ii) for a.a. } t\geq 0, & x'(t)\in\pi_{T_K(x(t))}\left(-\operatorname{co}\left(\bigcup_{j=1}^{m}\nabla U_j(x(t))\right)\right)\end{cases}$$

that satisfy, for any $s\geq t$,

$$(27)\qquad \max_{j=1,\ldots,m}(U_j(x(s))-U_j(x(t))+\int_t^s\|x'(\tau)\|^2\,d\tau\leq 0.$$

Furthermore, when $t\to\infty$, the cluster points x_* of $x(t)$ are weak Pareto minima that satisfy

$$(28)\qquad \forall j=1,\ldots,m, \quad U_j(x_*)=\lim_{t\to\infty}U_j(x(t)). \qquad \blacktriangle$$

Proof. We consider the set-valued map F defined by

$$(29)\qquad \forall x\in K, \quad F(x)=-\operatorname{co}\left(\bigcup_{j=1}^{m}\nabla U_j(x)\right)$$

which is clearly continuous with compact images.

Let $g(x) \doteq m(F(x) - N_K(x))$ be the slow velocity and H be the set-valued map defined by (18). By Lemmas 1 and 2, H is upper hemicontinuous with compact convex values and satisfy $H(x) \cap T_K(x) = g(x)$. Since $g(x)$ is the projection from 0 onto $F(x) - N_K(x)$, $g(x)$ does satisfy $\sigma(-F(x), g(x)) + \|g(x)\|^2 \leq 0$. i.e.

$$\max_{j=1,\dots,m} \langle \nabla U_j(x), g(x) \rangle + \|g(x)\|^2 \leq 0 \tag{30}$$

because $\sigma\left(\overline{\text{co}} \bigcup_{j=1}^{n} \nabla U_j(x), g(x)\right) = \sup_{i=1,\dots,n} \langle \nabla U_j(x), g(x) \rangle$.

So assumptions of Theorem 2 are satisfied and therefore, there exist trajectories of $x'(t) = g(x(t))$ satisfying (27). Therefore, cluster points x_* of $x(t)$ when $t \to \infty$ do exist and satisfy (28).

Since $-x'(t) = \sum_{j=1}^{m} \lambda_j \nabla U_j(x(t)) + n(t)$ where $n(t) \in N_K(x(t))$, since the functions U_j are convex and since $x_* \in K$, we deduce that

$$\langle -x'(t), x(t) - x_* \rangle \geq \sum_{j=1}^{m} \lambda_j (U_j(x(t)) - U_j(x_*)) + \langle n(t), x(t) - x_* \rangle \geq 0.$$

Therefore

$$\frac{d}{dt} \frac{1}{2} \|x(t) - x_*\|^2 = \langle x'(t), x(t) - x_* \rangle \leq 0$$

and thus, x_* belongs to the subset N of elements y such that $\lim_{t \to \infty} \|x(t) - y\|^2$ exists.

By Proposition 0.4.3, we deduce that $x(t)$ converges to the asymptotic center x_∞ of $x(t)$. By Theorem 6.5.2, the limit x_∞ is an equilibrium, i.e., satisfies:

$$0 \in \text{co}\left(\bigcup_{j=1}^{m} \partial U_j(x_\infty)\right) + N_K(x_\infty).$$

In other words,

$$0 \in \sum_{j=1}^{m} \lambda_j \partial U_j(x_\infty) + p = \partial\left(\sum_{j=1}^{m} \lambda_j U_j\right)(x_\infty) + p$$

where $\lambda_j \geq 0$, $\sum_{j=1}^{m} \lambda_j = 1$ and $p \in N_K(x_\infty)$.

Since the functions U_j are convex, this implies that

$$\sum_{j=1}^{m} \lambda_j U_j(x_\infty) = \min_{x \in K} \sum_{j=1}^{m} \lambda_j U_j(x).$$

Therefore, x_∞ is a weak Pareto minimum. □

Remark. Let us denote by

$$g(x) \doteq m\left(-\text{co}\left(\bigcup_{j=1}^{m} \nabla U_j(x) - N_K(x)\right)\right) = m\left(\pi_{T_K(x)}\left(-\text{co}\left(\bigcup_{j=1}^{m} \nabla U_j(x)\right)\right)\right) \tag{31}$$

the unique element of minimal norm. Slow solutions of differential inclusion (26) are the solutions to the differential equation

$$x'(t) = g(x(t)). \tag{32}$$

We shall characterize the right-hand side $g(x)$ defined by (31) as a solution of a two-person game.

Let

$$\phi(\lambda, v) \doteq \sum_{i=1}^{m} \lambda_i \langle \nabla U_i(x), v \rangle \tag{33}$$

be defined on $S^m \times M_K(x)$, where

$$\begin{cases} \text{i)} \ S^m \doteq \left\{ \lambda \in R_+^m \,\middle|\, \sum_{i=1}^{m} \lambda_i = 1 \right\}, \\ \text{ii)} \ M_K(x) \doteq \{ v \in T_K(x) \mid \|v\| \le 1 \}. \end{cases} \tag{34}$$

Since those two subsets are convex compact, then there exists a minimax $(\bar{\lambda}, \bar{v})$ of $\phi(\lambda, v)$, i.e. a solution to the inequalities; $\forall \lambda \in S^m$, $\forall v \in M_K(x)$,

$$\sum_{i=1}^{m} \lambda_i \langle \nabla U_i(x), \bar{v} \rangle \le \sum_{i=1}^{n} \bar{\lambda}_i \langle \nabla U_i(x), \bar{v} \rangle \le \sum_{i=1}^{n} \bar{\lambda}_i \langle \nabla U_i(x), v \rangle. \tag{35}$$

Then, we shall prove that

$$g(x) = \pi_{T_K(x)} \left(- \sum_{i=1}^{m} \bar{\lambda}_i \nabla U_j(x) \right) \quad \text{and} \quad \bar{v} = g(x)/\|g(x)\|. \tag{36}$$

Let us set for the time $\bar{w} = \pi_{T_K(x)} \left(- \sum_{i=1}^{m} \bar{\lambda}_i \nabla U_i(x) \right)$. This is equivalent to

$$\begin{cases} \text{i)} \ \left\langle \bar{w} + \sum_{i=1}^{m} \bar{\lambda}_i \nabla U_i(x), \bar{w} \right\rangle = 0, \\ \text{ii)} \ \left\langle \bar{w} + \sum_{i=1}^{m} \bar{\lambda}_i \nabla U_i(x), \ v \right\rangle \ge 0 \quad \forall v \in T_K(x). \end{cases} \tag{37}$$

Therefore, by (37) ii), (35) and (37) i), we obtain

$$\begin{aligned} \langle \bar{w}, \bar{v} \rangle &\ge - \sum_{i=1}^{m} \langle \bar{\lambda}_i \nabla U_i(x), \bar{v} \rangle \\ &\ge - \sum_{i=1}^{m} \left\langle \bar{\lambda}_i \nabla U_i(x), \frac{\bar{w}}{\|\bar{w}\|} \right\rangle \qquad \text{because } \frac{\bar{w}}{\|\bar{w}\|} \in M_K(x) \\ &= \frac{\|\bar{w}\|^2}{\|\bar{w}\|} = \|\bar{w}\|. \end{aligned}$$

Consequently,

$$\left\| \bar{v} - \frac{\bar{w}}{\|\bar{w}\|} \right\|^2 \le \|\bar{v}\|^2 + \left\| \frac{\bar{w}}{\|\bar{w}\|} \right\|^2 - 2 \left\langle \bar{v}, \frac{\bar{w}}{\|\bar{w}\|} \right\rangle \le 2 - 2 = 0$$

and thus, $\bar{v}=\bar{w}/\|\bar{w}\|$. Now, (37) i) implies that for all $\lambda\in S^m$,

$$\begin{aligned}\|\bar{w}\|^2 &= -\sum_{i=1}^{m}\bar{\lambda}_i\langle \nabla U_i(x),\bar{w}\rangle\\ &\le -\sum_{i=1}^{m}\lambda_i\langle \nabla U_i(x),\bar{w}\rangle \quad \text{(by (35))}\\ &\le\left\langle \pi_{T_K(x)}\left(-\sum_{i=1}^{n}\lambda_i \nabla U_i(x)\right),\bar{w}\right\rangle \quad (\text{because } \bar{w}\in T_K(x))\\ &\le\left\|\pi_{T_K(x)}\left(-\sum_{i=1}^{n}\lambda_i \nabla U_i(x)\right)\right\|\,\|\bar{w}\|.\end{aligned}$$

This proves that $\bar{w}=m\left(\pi_{T_K(x)}\left(\mathrm{co}\bigcup_{j=1}^{m}\nabla U_j(x)\right)\right)$. Therefore, $\bar{w}=g(x)$ and $\bar{v}=g(x)/\|g(x)\|$. □

Inequalities (35) and equations (36) imply that

(38) $$\frac{g(x)}{\|g(x)\|} \text{ minimizes } v\to \max_{i=1,\dots,m}\langle \nabla U_i(x),v\rangle \text{ over } M_K(x).$$

7. Rate Equations and Inclusions

Rate equations or inclusions are those differential equations or inclusions which yield viable trajectories in the simplex

(1) $$S^n \doteq \left\{x\in R^n_+ \,\middle|\, \sum_{i=1}^{n}x_i=1\right\}.$$

The Logistic Equation

The most celebrated example is the Verhust-Pearl *logistic* equation which yield viable trajectories for $n=2$:

(2) $$\begin{cases}\text{i) } x_1'(t)=x_1(t)\,(a_1-a_1x_1(t)-a_2x_2(t)),\\ \text{ii) } x_2'(t)=x_2(t)\,(a_2-a_1x_1(t)-a_2x_2(t)).\end{cases}$$

the solutions of which are given by

(3) $$x_1(t)=\frac{e^{a_1t+b_1}}{e^{a_1t+b_1}+e^{a_2t+b_2}},\qquad x_2(t)=\frac{e^{a_2t+b_2}}{e^{a_1t+b_1}+e^{a_2t+b_2}}.$$

Observe that, by setting $x_2=1-x_1$, equation (2) i) can be written

(4) $$x_1'(t)=(a_1-a_2)\,x_1(t)\,(1-x_1(t))$$

and its solution can be written

(5) $$x_1(t)=\frac{1}{1+e^{-(a_1-a_2)t}\,e^{-(b_1-b_2)}}.$$

This is the *logistic function*. Its graph is the S-shaped logistic curve.

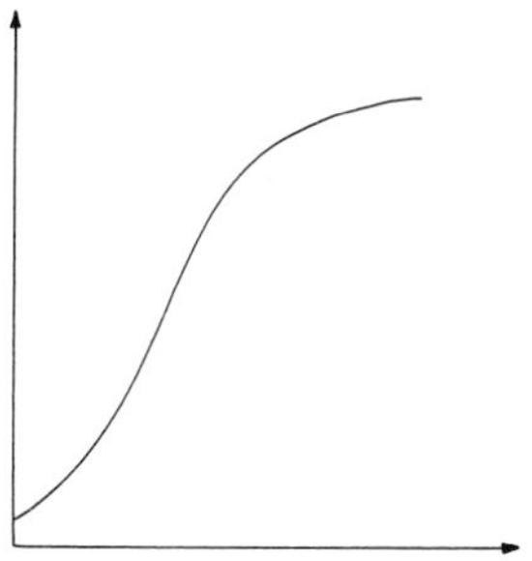

Fig. 27. Graph of a logistic curve

Because of the basic importance of the logistic curve in population dynamics, it is worth mentioning lines of argument that lead to it.

In the simple birth and death process it is assumed that an organism will reproduce with a constant growth rate

(6) $$x_1'(t)/x_1(t)=a_1.$$

Obviously, this can only be true only for a population so small that there is no interference among its members and that there is no shortage of resources. If there is another species and if the environment is restricted then there exists a stage which is reached when the demands made by the existing population preclude further growth. This amounts to saying that the growth rate, is negative for large x_1: the larger the population, the greater its inhibitory effect on further growth. The simplest rate is affine:

(7) $$x_1'(t)/x_1(t)=r(1-x_1(t)).$$

This is equation (4) with $r=a_1-a_2$. We can regard differential equation (6) as the growth equation describing the growth of the population if the resources were unlimited and the individuals did not affect one another. Here r is the *intrinsic* rate of natural increase. Now, assume that the actual growth rate is the product of this intrinsic rate times the proportion of the maximum attainable population size (normalized to 1) that is still unrealized (which is equal to $1-x_1$).

The historical argument advanced by Lotka in 1925 involves the concept of equilibria of the dynamical system. He assumed that the growth rate is a function $f(x_1)$ of the size x_1, which may be expanded as a power series in x_1 so that

$$x_1' = a_0 + a_1 x_1 + a_2 x_1^2 + \dots.$$

Obviously, the growth rate is zero at $x_1 = 0$, since at least one individual must be present for the population to grow at all. Thus $a_0 = 0$.

If we were to set $x_i' = a_1 x_1$, we should have $x_1' = 0$ only at $x_1 = 0$ and $x_1' > 0$ for all other population sizes. Lotka required the existence of another equilibrium at the point where x_1 reaches its saturation level. The simplest formula is then

$$x_1' = a_1 x_1 + a_2 x_1^2$$

which is again the logistic equation.

The n-Dimensional Logistic Equation

Consider the system of differential equations

(8) $$x_i'(t)/x_i(t) = a_i - \sum_{j=1}^{n} a_j x_j(t) \quad (i = 1, \dots, n).$$

One can check easily that the functions

(9) $$x_i(t) = \frac{\exp(a_i t + b_i)}{\sum_{j=1}^{n} \exp(a_j t + b_j)}$$

are solutions to this dynamical system. It is easy to check that the vector $x(t) = (x_1(t), \dots, x_n(t))$ ranges over the simplex S^n. Again, one can regard this system of differential equations as a perturbation of the individual linear systems

(10) $$x_i'(t)/x_i(t) = a_i$$

(whose solutions are $\exp(a_i t + b_i)$) which forces the solutions $x_i(t)$ to satisfy the constraint $\sum_{i=1}^{n} x_i(t) = 1$.

Rate Equations and Inclusions

We consider now general "rate equations" of the form

(11) $$x_i'(t) = x_i(t)\, G_i(x_1(t), \dots, x_n(t)) \qquad i = 1, \dots, n$$

where G_i are continuous maps from S^n to R_+^n.

Theorem 1. *Assume that the maps* G_i *are continuous and satisfy the condition*

$$\forall x\in S^n, \quad \sum_{i=1}^{n} x_i G_i(x)=0 \tag{12}$$

as well as

$$\exists \alpha>0 \text{ such that } \forall x\in\partial S^n, \quad \sum_{i=1}^{n} G_i(x)>\alpha. \tag{13}$$

Then, for any $x^0\in S^n$, *the rate equation* (11) *has viable trajectories in the sense that*

$$\forall t\geq 0, \quad x(t)\in S^n, \tag{14}$$

S^n *is invariant and there exists a nontrivial equilibrium* $\bar{x}\in S^n$ *in the sense that*

$$\forall i=1,\dots,n, \quad \bar{x}_i>0 \text{ and } G_i(\bar{x})=0. \tag{15}$$ ▲

We shall prove actually the set-valued version of the above Theorem. For that purpose, we introduce the following definition.

Definition 1. If x is a function from $[0, \infty[$ to R^n, we set

$$x\cdot G(x) \doteq \{(x_i v_i)_{i=1,\dots,n}\}_{v\in G(x)} \tag{16}$$

A differential inclusion

$$x'(t)\in x(t)\cdot G(x(t)) \tag{17}$$

is called a *rate inclusion.* ▲

Theorem 2. *Assume that the set-valued map* G *from* S^n *to* R^n *is upper semicontinuous with compact convex values and that it satisfies*

$$\forall x\in S^n, \ \forall v\in G(x), \quad \sum_{i=1}^{n} x_i v_i=0. \tag{18}$$

Assume also that

$$\exists \alpha>0 \,|\, \forall x\in\partial S^n, \ \forall v\in G(x), \quad \sum_{i=1}^{n} v_i\geq\alpha. \tag{19}$$

Then for all $x^0\in S^n$, *the rate inclusion* (17) *has a viable solution and there exists a nontrivial equilibrium* $\bar{x}\in S^n$ *in the sense that*

$$\forall i=1,\dots,n, \quad \bar{x}_i>0 \text{ and } 0\in G(\bar{x}). \tag{20}$$ ▲

Proof. We set

$$F(x)\doteq\{(x_1 v_1,\dots,x_n v_n)\}_{v\in G(x)}. \tag{21}$$

The rate inclusion (17) is the differential inclusion

$$x'(t)\in F(x(t)). \tag{22}$$

The set-valued map is obviously upper semicontinuous with compact convex values.

Since $T_{S^n}(x)=\left\{v\in R^n \,\middle|\, \sum_{i=1}^{n} v_i=0 \text{ and } v_i\geq 0 \text{ when } x_i=0\right\}$ by Proposition 1.7, we deduce that

$$\text{(23)} \qquad \forall x\in S^n, \quad F(x)\subset T_{S^n}(x).$$

Hence assumptions of Theorem 2.7 are satisfied and, consequently, the differential inclusion has viable trajectories, S^n is invariant under F and there exists an equilibrium $\bar{x}\in S^n$, solution to the inclusion $0\in F(\bar{x})$: there exists $\bar{v}\in G(\bar{x})$ such that $\bar{x}_i\bar{v}_i=0$ for all $i=1,\dots,n$.

We shall prove that the boundary condition (19) implies the existence of a nontrivial equilibrium. Since S^n is compact and G is upper semicontinuous with compact images, $G(S^n)$ is compact and therefore, there exist $a\in R^n_-$ and $b\in R^n_+$ such that

$$\text{(24)} \qquad \forall x\in S^n, \quad G(x)\subset(a+R^n_+)\cap(b-R^n_+).$$

We set $c_i\doteq a_i-\sum_{j=1}^{n} b_j\leq 0$, $c\doteq(c_1,\dots,c_n)$, and we fix $\varepsilon>0$ such that

$$\varepsilon\leq\min\left(\alpha,-n\sum_{i=1}^{n} c_i\right).$$

Let us put:

$$\text{(25)} \qquad H(x)\doteq\begin{cases}\left\{\left(v_i-\frac{1}{n}\sum_{j=1}^{n} v_j\right)_{i=1,\dots,n}\right\}_{v\in G(x)} & \text{if } x\notin\partial S^n,\\ \left\{y\in c+R^n_+ \,\middle|\, \sum_{i=1}^{n} y_i=0 \text{ and } \langle y,x\rangle\leq-\varepsilon/n\right\} & \text{if } x\in\partial S^n.\end{cases}$$

The images $H(x)$ of the set-valued map are obviously convex and compact and remain in a compact subset. Then it is sufficient to prove that its graph is closed to claim that it is upper semicontinuous. For that purpose, let us consider a sequence of elements (x_p, y_p) of the graph of H converging to (x, y).

When x does not belong to ∂S^n, then y belongs to $H(x)$ because the graph of G is closed.

When x belongs to ∂S^n and when an infinity of x_p''s belong to ∂S^n, we check again that y belongs to $H(x)$.

It remains the case when x belongs to ∂S^n and all (but a finite) x_p''s do not belong to ∂S^n. Then there exists $v_p\in G(x_p)$ such that $y_{p_i}=v_{p_i}-\frac{1}{n}\sum_{j=1}^{n} v_{p_j}$. We observe that $y_p\geq c$ and that $\sum_{j=1}^{n} y_{p_j}=0$. Since G is upper semicontinuous and $\varepsilon<\alpha$, then $\varepsilon\leq\sum_{i=1}^{n} v_{p_i}$ for p large enough thanks to assumption (19), and property (18) implies that

$$(26) \qquad \langle y_p, x_p \rangle = \sum_{i=1}^{n} x_i v_{p_i} - \frac{1}{n} \sum_{j=1}^{n} v_{p_j} = -\frac{1}{n} \sum_{j=1}^{n} v_{p_j} \leq -\varepsilon/n.$$

Therefore, by taking the limit when p goes to ∞, we deduce that $y \geq c$, $\sum_{j=1}^{n} y_j = 0$ and $\langle y, x \rangle \leq 0$. In other words, y belongs to $H(x)$.

We check now that $H(x)$ intersects the tangent cone to S^n at x. It is obvious that for all $x \in S^n$, for all $y \in H(x)$, we have $\sum_{i=1}^{n} y_i = 0$. Let us set $I(x) \doteq \{i \mid x_i = 0\}$, which is nonempty when x belongs to the boundary ∂S^n. We define y by

$$(27) \qquad y_i \doteq \varepsilon \frac{n - |I(x)|}{n|I(x)|} \quad \text{when } i \in I(x), \qquad y_i \doteq -\frac{\varepsilon}{n} \quad \text{when } i \notin I(x)$$

(where $|I(x)|$ is the number of elements of $I(x)$). Therefore

$$(28) \qquad \sum_{i=1}^{n} y_i = \varepsilon \frac{n - |I(x)|}{n} - \varepsilon \frac{n - |I(x)|}{n} = 0$$

and

$$(29) \qquad \langle y, x \rangle = \sum_{i \notin I(x)} x_i \left(-\frac{\varepsilon}{n} \right) = -\frac{\varepsilon}{n}.$$

Hence y belongs to $H(x)$ and to $T_{S_n}(x)$ since $y_i \geq 0$ for all $i \in I(x)$.

Theorem 2.1 implies the existence of $\bar{x} \in S^n$ such that $0 \in H(\bar{x})$. Such a $\bar{x}$ cannot belong to ∂S^n; otherwise, we would have $\langle 0, \bar{x} \rangle \leq -\varepsilon/n$, which is impossible. Therefore, $\bar{x}_i > 0$ for all i and there exists $\bar{v} \in G(\bar{x})$ such that $\bar{v}_i - \frac{1}{n} \sum_{j=1}^{n} \bar{v}_j = 0$ for all $i = 1, \ldots, n$. By multiplying by $\bar{x}_i$ and adding these equations, we deduce from condition (18) that $\frac{1}{n} \sum_{j=1}^{n} \bar{v}_j = 0$. Consequently, $\bar{v}_i = 0$ for all $i = 1, \ldots, n$ and $\bar{x}$ is a nontrivial equilibrium.

Construction of Rate Equations

We can associate with a system of n independent growth equations

$$(30) \qquad x_i'(t) = x_i(t)\, H_i(x_i)$$

the rate equation

$$(31) \qquad x_i'(t) = x_i(t) \left(H_i(x_i) - \sum_{j=1}^{n} x_j H_j(x_j) \right)$$

which satisfies the condition (12) because

$$(32) \qquad \forall x \in S^n, \quad \sum_{i=1}^{n} x_i \left(H_i(x) - \sum_{j=1}^{n} x_j H_j(x_j) \right) = 0.$$

Corollary 1. *Let us assume that the n functions H_i from R_+ to R are continuous and satisfy*

$$\min_{i=1,\dots,n} H_i(0) > (n-1) \max_{i=1,\dots,n} \sup_{x\geq 0} \left(x-\frac{1}{n}\right) H_j(x). \tag{33}$$

Then the rate equation (31) *has viable trajectories, S^n is invariant and there exists a nontrivial equilibrium $\bar{x}\in S^n$:*

$$\forall i=1,\dots,n, \quad \bar{x}_i>0 \quad \text{and} \quad H_i(\bar{x}) = \sum_{j=1}^{n} \bar{x}_j H_j(\bar{x}_j). \tag{34}$$ ▲

Remark. We observe that

$$\sum_{i=1}^{n} H_i(x_i(t))\, x_i'(t) \geq 0 \tag{35}$$

because

$$\left(\sum_{i=1}^{n} x_i H_i(x_i)\right)^2 \leq \left(\sum_{i=1}^{n} x_i\right) \sum_{i=1}^{n} x_i H_i(x_i)^2 = \sum_{i=1}^{n} x_i H_i(x_i)^2.$$

Chapter 6. Liapunov Functions

Introduction

We shall investigate whether differential inclusions

(1) $$x'(t)\in F(x(t)), \quad x(0)=x_0$$

do have trajectories satisfying the property

(2) $$\forall t>s, \quad V(x(t))-V(x(s))+\int_s^t W(x(\tau),x'(\tau))dx\le 0$$

where

(3) $$\begin{cases}\text{i) } V \text{ is a function from } K \doteq \operatorname{Dom} F \text{ to } R_+\\ \text{ii) } W \text{ is a function from Graph } (F) \text{ to } R_+\end{cases}$$

Trajectories $x(\cdot)$ of differential inclusion (1) satisfying (2) will be called "monotone trajectories" (with respect to V and W).

We shall answer the following questions:

1. What are the *necessary and sufficient* conditions linking F, V and W for the differential inclusion (1) to have monotone trajectories with respect to V and W?

2. Does these necessary and sufficient conditions imply the existence of pairs $(x_*, v_*)\in\operatorname{graph}(F)$ satisfying $W(x_*, v_*)=0$? Observe that if the values $W(x, v)$ are strictly positive whenever v is different from 0, then such an x_* is an *equilibrium*.

3. Are the cluster points x_* and v_* of the functions $t\to x(t)$ and $t\to x'(t)$, when $t\to\infty$, solutions to the equation $W(x_*, v_*)=0$?

4. The set-valued map F and the function W from graph(F) to R_+ being given, can we construct a function V such that these necessary and sufficient conditions are satisfied?

For answering these questions positively, we have to introduce the concept of *upper contingent derivative* of a proper function V from a Hilbert space X to $R\cup\{+\infty\}$.

Why not use simply the concept of contingent derivative of the map V from Dom(V) to R? This is because we are using the order relation of R for finding monotone trajectories. We take in account this order by associating to a proper function V: $X\to R\cup\{+\infty\}$ the set-valued map V_+: $X\to R$ defined by

$$V_+(x) \doteq \begin{cases} V(x) + R_+ & \text{when } x \in \text{Dom}(V) \\ \emptyset & \text{when } x \notin \text{Dom}(V) \end{cases} \tag{4}$$

whose *graph* is the *epigraph of* V.

Instead of using the contingent derivative of V, we consider the contingent derivative of the set-valued map V_+ at points $(x, V(x))$. We observe that the values $DV_+(x, V(x))(u)$ are either empty, or half-lines, or the whole line R. We set

$$D_+ V(x)(u) \doteq \inf\{v \mid v \in DV_+(x, V(x))(u)\} \tag{5}$$

and we obtain the formula

$$D_+ V(x)(u_0) = \liminf_{\substack{h \to 0_+ \\ u \to u_0}} \frac{V(x+hu) - V(x)}{h}. \tag{6}$$

The function $u \to D_+ V(x)(u)$ is called the *upper contingent derivative* of V at x.

We remark the following facts:

a) When V is Gâteaux-differentiable, $D_+ V(x)$ coïncides with the gradient $\nabla V(x)$:

$$D_+ V(x)(u) = \langle \nabla V(x), u \rangle \quad \text{for all } u \in X. \tag{7}$$

b) When V is convex, the upper contingent derivative is related to the derivative from the right by the formula

$$D_+ V(x)(u_0) = \liminf_{u \to u_0} DV(x)(u). \tag{8}$$

They coincide when the latter is lower semicontinuous.

c) When V is locally Lipschitz, the upper contingent derivative coïncides with a Dini derivative:

$$D_+ V(x)(u_0) = \liminf_{h \to 0_+} \frac{V(x+hu_0) - V(x)}{h}. \tag{9}$$

d) When V is Gâteaux-differentiable on a neighborhood of a subset K, then

$$D_+(V|_K)(x)(u) = \begin{cases} \langle \nabla V(x), u \rangle & \text{when } u \in T_K(x) \\ +\infty & \text{when } u \notin T_K(x). \end{cases} \tag{10}$$

This means that the upper contingent derivative of the restriction of a function to a subset K is the restriction of its gradient to the contingent cone.

The main justification for the introduction of the upper contingent derivatives is the following characterization:

Assume that F is an upper semicontinuous map from a locally compact subset K of R^n to the convex compact subsets of R^n, V is a continuous function from K to R_+ and W is a lower semicontinuous function from $\text{Graph}(F)$ to R_+, convex with respect to the second argument. A necessary and sufficient condition for the differential inclusion (1) to have monotone trajectories with respect to V and W is that:

(11) $$\forall x \in K, \quad \exists v \in F(x) \text{ such that } D_+ V(x)(v) + W(x, v) \leq 0.$$

We shall say that a function V from K to R_+ satisfying the above condition is a *Liapunov function for F with respect to W*. Indeed, we recognize that when K is open, V is differentiable and F is single-valued, this condition is nothing other than the usual property

(12) $$\forall x \in K, \quad \langle \nabla V(x), F(x) \rangle + W(x, F(x)) \leq 0$$

used in Liapunov's method for studying the stability of solutions to differential equations. We also point out that conditions (11) implies the existence of a pair $(x_*, v_*) \in \text{graph}(F)$ satisfying $W(x_*, v_*) = 0$.

The next problem we investigate is the construction of Liapunov functions. Let $\mathscr{T}_\infty(x)$ denote the set of trajectories of the differential inclusion (1) starting at x.

We define the function V_F by

(13) $$V_F(x) \doteq \inf_{x(\cdot) \in \mathscr{T}_\infty(x)} \int_0^\infty W(x(\tau), x'(\tau)) d\tau.$$

We shall observe that V_F is smaller than or equal to any Liapunov function V for F with respect to W and that the monotone trajectories of (1) with respect to V_F and W are the trajectories that minimize on $\mathscr{T}_\infty(x)$ the functional

$$x(\cdot) \to \int_0^\infty W(x(\tau), x'(\tau)) d\tau.$$

We then state a result whose origin can be traced back to Carathéodory, Jacobi and Hamilton: If for all initial state x there exists a trajectory $x(\cdot) \in \mathscr{T}_\infty(x)$ that minimizes the above functional, then V_F is a Liapunov function for F with respect to W.

We proceed by investigating properties of monotone trajectories $x(\cdot)$ of (1) with respect to V and W. We observe that

(14) $$\begin{aligned} &\text{a) } t \to V(x(t)) \text{ is non increasing,} \\ &\text{b) } \int_0^\infty W(x(\tau), x'(\tau)) d\tau < +\infty. \end{aligned}$$

We show also that the cluster points x_* and v_* of the functions $x(\cdot)$ and $x'(\cdot)$ when $t \to \infty$ solve the equation

(15) $$(x_*, v_*) \in \text{Graph}(F) \quad \text{and} \quad W(x_*, v_*) = 0.$$

But we have to be careful, because $x'(\cdot)$ is not defined everywhere. So, we have to make precise the notion of "almost cluster point" of a measurable fonction.

We single out important instances:

a) $W(x, v) \doteq \|v\|$.

Condition (14) states that the *length* $\int_0^\infty \|x'(\tau)\| d\tau$ of the trajectory is finite and that $x(t)$ *has a limit when* $t \to \infty$, *which is an equilibrium of* F.

b) $W(x, v) \doteq \phi(V(x))$ where ϕ: $[0, \infty[\to R$ is a bounded continuous function. Let w be a solution to the differential equation:

$$w'(t) + \phi(w(t)) = 0, \qquad w(0) = V(x_0). \tag{16}$$

Then monotone trajectories do enjoy the estimate

$$V(x(t)) \leq w(t) \qquad \text{for all } t \geq 0. \tag{17}$$

c) $W(x, v) \doteq 0$. We shall investigate the stability and asymptotic stability of the monotone trajectories.

d) $V(x) \doteq U(x, x_*)$ where U is a kind of distance function and x_* is an equilibrium of F. Available estimates on the behavior of $V(x(t)) = U(x(t), x_*)$ yield some information on the convergence of the trajectory to an equilibrium.

1. Upper Contingent Derivative of a Real-Valued Function

We associate with the function V: $X \to R \cup \{+\infty\}$ the set-valued map V_+ defined by $V_+(x) \doteq V(x) + R_+$ when $V(x) < +\infty$ and $V_+(x) \doteq \emptyset$ when $V(x) = +\infty$. Its domain is the domain of V and its graph is the epigraph of V.

The latter is closed when V is lower semicontinuous and convex when V is convex.

We consider the contingent derivative $DV_+(x, V(x))$ of V_+ at the pair $(x, V(x))$. We observe that for all $u \in X$, $DV_+(x, V(x))(u)$ is either R, or a half line $[v_0, \infty[$, or empty. We set

$$D_+ V(x)(u) \doteq \inf\{v \mid v \in DV_+(x, V(x))(u)\}. \tag{1}$$

It is equal to $-\infty$ if $DV_+(x, V(x)) = R$, to v_0 if $DV_+(x, V(x))(u) = [v_0, \infty[$ and to $+\infty$ if $DV_+(x, V(x))(u) = \emptyset$.

Definition 1. We shall say that $D_+ V(x)(u)$ is the "upper contingent derivative" of V at x in the direction u. ▲

Proposition 1. *Let* x_0 *belong to the domain of* V. *Then*

$$D_+ V(x_0)(u_0) = \liminf_{\substack{h \to 0+ \\ u \to u_0}} \frac{V(x_0 + hu) - V(x_0)}{h}. \tag{2}$$ ▲

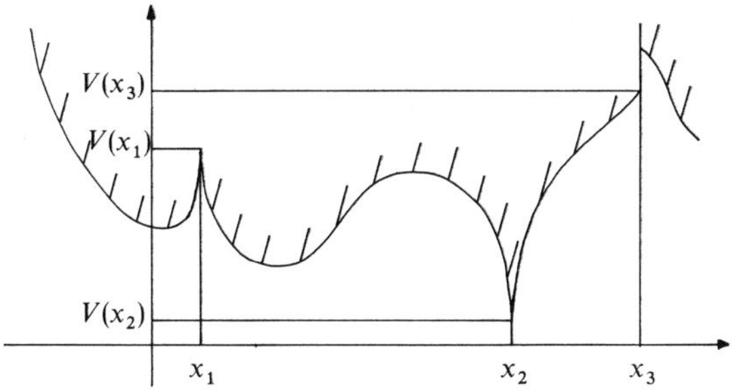

Fig. 28. Example of upper contingent derivative

$$D_+V(x_1)(u) = -\infty \qquad \forall u \in \mathbb{R}$$

$$D_+V(x_2)(u) = \begin{cases} +\infty & \forall u \neq 0 \\ 0 & \text{if } u = 0 \end{cases}$$

$$D_+V(x_3)(u) = \begin{cases} +\infty & \forall u \geq 0 \\ u \lim\limits_{h\to 0_+} \dfrac{V(x_3) - V(x_3 - h)}{h} & \forall u < 0 \end{cases}$$

Proof. Indeed, let $v_0 \in DV_+(x_0, V(x_0))(u_0)$; then $\forall \varepsilon_1 > 0$, $\varepsilon_2 > 0$, $\forall \alpha > 0$, there exist $u \in u_0 + \varepsilon_2 B$ and $h < \alpha$ such that

$$v_0 \in \frac{V_+(x_0 + hu) - V(x_0)}{h} + \varepsilon_1 B.$$

This implies that

$$v_0 \geq \frac{V(x_0 + hu) - V(x_0)}{h} - \varepsilon_1 \geq \inf_{h \leq \alpha} \inf_{\|u - u_0\| < \varepsilon_2} \frac{V(x_0 + hu) - V(x_0)}{h} - \varepsilon_1.$$

Therefore $v_0 \geq \liminf\limits_{\substack{h \to 0_+ \\ u \to u_0}} \dfrac{V(x_0 + hu) - V(x_0)}{h} - \varepsilon_1$. Let us set for the time

$$a \doteq \liminf_{\substack{h \to 0_+ \\ u \to u_0}} \frac{V(x_0 + hu) - V(x_0)}{h}.$$

So, we have proved that $a \leq D_+V(x_0)(u_0)$. On the other hand, we know that for any $M > a$,

$$\sup_{\substack{\alpha > 0 \\ \delta > 0}} \inf_{h \leq \alpha} \inf_{\|u_0 - u\| \leq \delta} \frac{V(x_0 + hu) - V(x_0)}{h} < M.$$

This shows that for all $\delta > 0$, there exist $h \leq \alpha$ and $u \in u_0 + \delta B$ such that:

$$\frac{V(x_0 + hu) - V(x_0)}{h} \leq M.$$

Hence $M \in \dfrac{V_+(x_0 + hu) - V(x_0)}{h}$. This proves that $a \in DV_+(x_0, V(x_0))(u_0)$ and consequently, that $a = D_+V(x_0)(u_0)$. □

Proposition 2. *If the function V is locally Lipschitzean, or if V is defined on a right open interval containing $x_0 \in R$,*

$$D_+ V(x_0)(v_0) = \liminf_{h \to 0_+} \frac{V(x_0 + hv_0) - V(x_0)}{h} \tag{3}$$

coincides with one of the Dini derivatives.

If the function V is convex, then

$$D_+ V(x_0)(v_0) = \liminf_{v \to v_0} DV(x_0)(v) \tag{4}$$

where we set

$$Dv(x_0)(v) \doteq \inf_{h \to 0_+} \frac{V(x_0 + hv) - V(x_0)}{h}. \tag{5}$$ ▲

Since the epigraph of the function $v \to D_+ V(x_0)(v)$ is a closed cone, then

(6) the function $v \to D_+ V(x)(v)$ is positively homogeneous and lower semicontinuous.

Definition 2. The indicator ψ_K: $X \to]-\infty, +\infty]$ of a subset $K \subset X$ is the function defined by $\psi_K(x) \doteq 0$ when $x \in K$ and $\psi_K(x) \doteq +\infty$ when $x \notin K$.

Proposition 3. *The upper contingent derivative of the indicator ψ_K of $K \subset X$ is the indicator of the contingent cone $T_K(x)$:*

$$D_+ \psi_K(x, 0)(\cdot) = \psi_{T_K(x)}(\cdot). \tag{7}$$ ▲

When V is a function from X to $R \cup \{+\infty\}$, we set

$$V|_K(x) \doteq \begin{cases} V(x) & \text{if } x \in K \\ +\infty & \text{if } x \notin K. \end{cases} \tag{8}$$

Proposition 4. *Let V and W be two functions from X to $R \cup \{+\infty\}$ and x_0 belong to* $\operatorname{Dom} V \cap \operatorname{Int} \operatorname{Dom} W$. *If W is continuously differentiable at x_0, then*

$$D_+(V + W)(x_0)(v_0) = D_+ V(x_0)(v_0) + \nabla W(x_0) \cdot v_0. \tag{9}$$

Let K be a subset of X and x_0 belong to $K \cap \operatorname{Int} \operatorname{Dom} W$. *If W is continuously differentiable at x_0, then*

$$D_+(W|_K)(x_0)(v_0) = \begin{cases} \langle \nabla W(x_0), v_0 \rangle & \text{if } v_0 \in T_K(x_0) \\ +\infty & \text{if } v_0 \notin T_K(x_0). \end{cases} \tag{10}$$ ▲

Variational Principle

We begin by stating the following well known fact.

Proposition 5. *Let V be a function from X to $R \cup \{+\infty\}$. If $x_0 \in \operatorname{Dom} V$ minimizes V on X, then*

$$\forall u_0 \in X, \quad 0 \le D_+ V(x_0)(u_0). \tag{11}$$

The converse is true when V is convex. ▲

We shall use the following version of Ekeland's Theorem couched in terms of the contingent derivative.

Theorem 1. *Let $V: X \to R \cup \{+\infty\}$ be a proper lower semicontinuous function, x_0 belong to the domain of V and $\varepsilon > 0$ be fixed. Then there exists $\bar{x} \in \operatorname{Dom} V$ satisfying*

$$(12) \quad \begin{cases} \text{i)} \ \forall u_0 \in X, \ 0 < D_+ V(\bar{x})(u_0) + \varepsilon \|u_0\|, \\ \text{ii)} \ V(\bar{x}) + \varepsilon \|\bar{x} - x_0\| \leq V(x_0). \end{cases}$$

▲

Proof. By Ekeland's Theorem, there exists $\bar{x} \in \operatorname{Dom} V$ such that

$$(13) \quad \begin{cases} \text{i)} \ \forall x \in \operatorname{Dom} V, \quad V(\bar{x}) \leq V(x) + \varepsilon \|\bar{x} - x\|, \\ \text{ii)} \ V(\bar{x}) + \varepsilon \|\bar{x} - x_0\| \leq V(x_0). \end{cases}$$

Let u_0 belong to the domain of $D_+ V(\bar{x})$. Then, for all $\eta > 0$, $\delta > 0$ and $\alpha > 0$, there exist $h \in]0, \alpha[$ and $u \in \delta B$ such that

$$\frac{V(\bar{x} + hu_0 + hu) - V(\bar{x})}{h} \leq D_+ V(\bar{x})(u_0) + \eta.$$

Hence, (13) i) with $x = \bar{x} + hu_0 + hu$ implies that

$$0 \leq D_+ V(\bar{x})(u_0) + \varepsilon \|u_0\| + \eta + \varepsilon\delta.$$

We obtain inequality (12) i) by letting η and δ converge to 0. □

Remark. We can define as well

$$V_-(x) = V(x) - R_+ \quad \text{and} \quad D_- V(x)(u) \doteq \sup \{v \mid v \in D V_-(x, V(x))(u)\}.$$

We say that $D_- V(x)(u)$ is the *lower contingent derivative* of V at x in the direction u.

We can compute in the same way the lower contingent derivative of V: we obtain

$$(14) \quad D_- V(x_0)(u_0) = \limsup_{\substack{h \to 0+ \\ u \to u_0}} \frac{V(x_0 + hu) - V(x_0)}{h}.$$

Therefore, we always have

$$(15) \quad D_+ V(x_0)(u_0) \leq D_- V(x_0)(u_0).$$

We shall say that the interval-valued map:

$$u_0 \to [D_+ V(x_0)(u_0), D_- V(x_0)(u_0)]$$

is the *contingent gap* map.

Let us mention also that

$$(16) \quad D_+ V(x_0)(-u_0) = \limsup_{\substack{h \to 0+ \\ u \to u_0}} \frac{V(x_0) - V(x_0 - hu)}{h}.$$

□

We estimate now the upper contingent derivative of the sum of two functions.

Proposition 6. *Let V and W be two functions from K to R and $U \doteq V+W$. Let $L \subset K$ be a subset of K. Then*

$$(17)\quad \begin{cases} \text{i)}\ D_+ U(x_0)(u_0) \le D_+ V(x_0)(u_0) + D_- W(x_0)(u_0), \\ \text{ii)}\ \forall u_0 \in T_L(x_0), \quad D_+ V(x_0)(u_0) \le D_+ V|_L(x_0)(u_0) \le D_- V(x_0)(u_0). \end{cases} \qquad \blacktriangle$$

Proof. Inequality (17) i) follows from the fact that

$$\liminf_{\substack{h\to 0_+ \\ v\to v_0}} (f(h,v)+g(h,v)) \le \liminf_{\substack{h\to 0_+ \\ v\to v_0}} f(h,v) + \limsup_{\substack{h\to 0_+ \\ v\to v_0}} g(h,v)$$

where we set $f(h,v) = \dfrac{V(x_0+hv)-V(x_0)}{h}$ and $g(h,v) = \dfrac{W(x_0+hv)-W(x_0)}{h}$. We deduce inequality (17) ii) by taking for function W the indicator $\psi_L(\cdot)$ of L. Inequality $D_+ V(x_0)(v_0) \le D_+ V|_L(x_0)(v_0)$ is straightforward.

Chain Rule Formulas

We begin by the nonconvex case.

Propositions 7. *Let A be a differentiable map from an open subset Ω of X to Y and V be a proper function from Y to $R \cup \{+\infty\}$.* We assume that

$$(18)\qquad 0 \in A(\Omega) - \operatorname{Dom} V \quad (\text{or } A(\Omega) \cap \operatorname{Dom} V \neq \emptyset).$$

Then, for all $x_0 \in \Omega \cap A^{-1}(\operatorname{Dom} V)$, we have

$$(19)\qquad \forall u_0 \in X, \quad D_+ V(Ax_0)(\nabla A(x_0)u_0) \le D_+ (V\circ A)(x_0)(u_0).$$

Furthermore, if V is Lipschitzean around $A(x_0)$ then

$$(20)\qquad \forall u_0 \in X, \quad D_+ V(Ax_0)(\nabla A(x_0)u_0) = D_+ (V\circ A)(x_0)(u_0).$$

Proof. The first inequality is straightforward. The opposite inequality follows from the fact that

$$\begin{aligned} \frac{V(A(x_0+hu)) - V(Ax_0)}{h} &\le \frac{V(A(x_0)+hv)) - V(A(x_0))}{h} \\ &\quad + \|\nabla A(x_0)u + O(h) - v\| \\ &\le \frac{V(A(x_0)+hv)) - V(A(x_0))}{h} + \eta \end{aligned}$$

whenever $\|u-u_0\| \le \eta/3\|\nabla A(x_0)\|$, $\|v - \nabla A(x_0)u_0\| \le \eta/3$ and $O(h) \le \eta/3$ if $\nabla A(x_0) \neq 0$ or $\|v\| \le \eta/2$ and $O(h) \le \eta/2$ when $\nabla A(x_0) = 0$. We deduce that $D_+(V\circ A)(x_0)(u_0) \le D_+ V(Ax_0)(Au_0)$. $\square$

When the functions are convex, we obtain the following important result.

Theorem 2. *Let A be a continuous linear operator from a Banach space X to a Banach space Y and $U: X \to R \cup \{+\infty\}$, $V: Y \to R \cup \{+\infty\}$ be proper lower semicontinuous convex functions. We assume that*

$$0 \in \operatorname{Int}(A \operatorname{Dom} U - \operatorname{Dom} V) \tag{21}$$

Then, for all $x_0 \in \operatorname{Dom} U \cap A^{-1} \operatorname{Dom} V$, we have

$$\forall u_0 \in X, \quad D_+(U + V \circ A)(x_0)(v_0) = D_+ U(x_0)(u_0) + D_+ V(Ax_0)(Au_0). \tag{22}$$

(See Aubin-Ekeland [1984].) ▲

Let us make explicit several consequences:

a) If $0 \in \operatorname{Int}(\operatorname{Im} A - \operatorname{Dom} V)$, then

$$D_+(V \circ A)(x_0) = D_+ V(Ax_0) \circ A. \tag{23}$$

b) If $0 \in \operatorname{Int}(\operatorname{Dom} U - \operatorname{Dom} V)$, then

$$D_+(U + V)(x_0) = D_+ U(x_0) + D_+ V(x_0). \tag{24}$$

c) If K is closed and convex and $0 \in \operatorname{Int}(K - \operatorname{Dom} V)$, then

$$D_+ V|_K(x_0)(v_0) = \begin{cases} D_+ V(x_0)(v_0) & \text{if } v_0 \in T_K(x_0) \\ +\infty & \text{if } v_0 \notin T_K(x_0). \end{cases} \tag{25}$$ □

We can "integrate" inequalities involving contingent derivatives.

Proposition 8. *Let V be a continuous function from $[0, T]$ to R and W be an upper semicontinuous function from $[0, T]$ to R which is bounded above. We assume that*

$$\forall t \in [0, T[, \quad D_+ V(t) + W(t) \le 0. \tag{26}$$

Then, for all $0 < a < b < T$, we obtain the inequality

$$V(b) - V(a) + \int_a^b W(\tau) d\tau \le 0. \tag{27}$$ ▲

Proof. Assume that for some $\varepsilon > 0$, $V(b) - V(a) + \int_a^b W(\tau) d\tau > \varepsilon(b-a)$. Define g as $g(t) \doteq V(t) - V(a) + \int_a^t W(\tau) d\tau - \varepsilon(t-a)$. Let us compute $D_+ g(t)$. By proposition 2

$$D_+ g(t) = \liminf_{s \to t_+} \left\{ \frac{V(s) - V(t)}{s-t} + \frac{1}{s-t} \int_s^t W(\tau) d\tau - \varepsilon \right\}.$$

Fix $\eta > 0$. Let $\delta > 0$ be such that $(\tau - t) < \delta$ implies $W(t) + \eta \ge W(\tau)$. Hence

$$\begin{aligned} D_+ g(t) &\le \liminf_{s \to t_+} \left\{ \frac{V(s) - V(t)}{s-t} + W(t) + \eta - \varepsilon \right\} \\ &= D_+ V(t) + W(t) - \varepsilon + \eta. \end{aligned}$$

Being η arbitrary,

(28) $$D_+ g(t) \leq D_+ V(t) + W(t) - \varepsilon < 0.$$

In particular the above holds at a. Hence, since $g(a)=0$, for some t^0 near a, $t^0 > a$, we have $g(t^0) < 0$. Set $t^1 \doteq \sup\{t < b \mid g(t) = 0\}$. Being g continuous and $g(b) > 0$, we infer that $a < t^0 < t^1 < b$. Since at the right of t^1, $g(t) > 0$,

$$D_+ g(t^1) \geq 0$$

a contradiction to (28). □

In particular, we obtain the following useful consequence.

Proposition 9. *Let V be a continuous function from $[0, T]$ to R satisfying*

(29) $$\forall t \in]0, T[, \quad D_+ V(t) \leq 0.$$

Then the function V is non-increasing. ▲

2. Liapunov Functions and Existence of Equilibria

We shall use the existence of Liapunov functions which we define below for proving the existence of equilibria. In the next sections, we shall prove that the existence of these Liapunov functions imply also that solutions to the associated differential inclusion do converge in some sense to equilibria of F when $t \to \infty$. Finally, we shall construct Liapunov functions.

We begin with a particular case, which is an extension to the case of set-valued maps of the Caristi fixed point theorem.

Theorem 1. *Let K be a closed subset of the Hilbert space X, $F: K \to X$ be a setvalued map and $V: K \to [0, \infty[$ be a lower semicontinuous function satisfying*

(1) $$\forall x \in K, \quad \exists v \in F(x) \quad \textit{such that } D_+ V(x)(v) + \|v\| \leq 0.$$

Then there exists a stationary point $\bar{x} \in K$ of F. ▲

Proof. We apply the differential version of Ekeland's Theorem (Theorem 1.1).

Take $\varepsilon < 1$ and $\bar{x} \in K$ satisfying,

$$\forall u \in X, \quad 0 \leq D_+ V(\bar{x})(u) + \varepsilon \|u\|.$$

By assumption, we can take $u \in F(\bar{x})$ satisfying $D_+ V(\bar{x})(u) \leq -\|u\|$. Hence $(1-\varepsilon)\|u\| \leq 0$, i.e., $u = 0 \in F(\bar{x})$. □

Remark. This theorem can be regarded as the "equilibrium" version of the "Caristi's fixed point theorem":

Let K be a closed subset, $g: K \to K$ be a single-valued map and V be a lower semicontinuous map from K to R_+. If

$$\forall x \in K, \quad V(g(x)) - V(x) + \|g(x) - x\| \le 0,$$

then g has a fixed point. □

So, we have proved that if there exists a lower semicontinuous function V such that $D_+V(x)(v) + \|v\| \le 0$ for all $x \in K$, there exists an equilibrium. The question arises whether existence of stationary points implies the existence of a function satisfying the above condition.

Proposition 1. *Let K be a closed subset and F be a set-valued map from K to X satisfying*

(2) $$-F \text{ is monotone}$$

We assume that the set $F^{-1}(0)$ of equilibria of F has a non-empty interior. Let us associate with any $x_0 \in \operatorname{Int} F^{-1}(0)$ the function V defined by $V(x) = \frac{1}{2\rho}\|x - x_0\|^2$ where $\rho = d(x_0, \complement F^{-1}(0)) > 0$. Then

(3) $$\forall x \in K, \quad \forall v \in F(x), \quad DV(x)(v) + \|v\| \le 0.$$

and, consequently, for any $x \in x_0 + \rho \operatorname{Int} B$, $F(x) = \{0\}$. ▲

Proof. Let us take $x \in K$, $\tilde{\rho} < \rho$, $v \in F(x)$. Since $x_0 + \tilde{\rho} B \subset F^{-1}(0)$, then $x_0 - \frac{\tilde{\rho} v}{\|v\|}$ is an equilibrium of F. The monotonicity of $-F$ implies that

(4) $$\langle v, x - x_0 \rangle + \tilde{\rho}\|v\| = \left\langle v - 0, x - x_0 + \frac{\tilde{\rho} v}{\|v\|} \right\rangle \le 0.$$

By letting $\tilde{\rho}$ converge to ρ, we obtain (3). Take u in $\operatorname{Int} B$. Let $x = x_0 + \rho u \in x_0 + \rho \operatorname{Int} B$, and $v \in F(x)$. We infer that $\rho\langle v, u \rangle + \rho\|v\| \le 0$. Hence $\|v\| \le \|v\|\,\|u\| < \|v\|$, which is impossible when $v \neq 0$. □

We may generalize Theorem 1 by introducing the concept of Liapunov function V with respect to a set-valued map F and a given function W defined on graph (F).

Definition 1. Let F be a set-valued map from X to X and W be a non-negative function defined on graph (F). We shall say that the function V defined on $K \doteq \operatorname{Dom} F$ is a Liapunov function for F with respect to W if it satisfies the following "Liapunov property"

(5) $$\forall x \in K, \quad \exists v \in F(x) \quad \text{such that } D_+V(x)(v) + W(x, v) \le 0.$$

When $W \equiv 0$, we say simply that V is a Liapunov function for F. ▲

Theorem 2. *Let F be a set-valued map from a closed subset $K \subset X$ to X, W be a nonnegative function from graph (F) to R and V be a lower semicontinuous function from K to R_+. Assume that V is a Liapunov function for F with respect to W. Then,*

$$(6) \qquad \forall \varepsilon > 0, \quad \exists x_\varepsilon \in K \ \text{ and } \ v_\varepsilon \in F(x_\varepsilon) \quad \text{satisfying } W(x_\varepsilon, v_\varepsilon) \leq \varepsilon \|v_\varepsilon\|.$$

If we assume moreover that V is inf-compact *(i.e., for all $x \in R$ the subsets $\{x \in K \mid V(x) \leq \lambda\}$ are relatively compact), then*

$$(7) \qquad \exists x_* \in K, \quad \exists v_* \in F(x_*) \quad \text{such that } W(x_*, v_*) = 0. \qquad \blacktriangle$$

Proof. a) We apply Theorem 1.1: $\forall \varepsilon > 0$, $\exists x_\varepsilon \in K$ such that, $\forall v \in X$, $0 \leq D_+ V(x_\varepsilon)(v) + \varepsilon \|v\|^2$. Since V is a Liapunov function with respect to W, there exists $v_\varepsilon \in F(x_\varepsilon)$ such that $D_+ V(x_\varepsilon)(v_\varepsilon) \leq -W(x_\varepsilon, v_\varepsilon)$. Hence property (6) holds true.

b) When there exists $x_* \in K$ that achieves the minimum of V, then, $\forall v \in X$, $0 \leq D_+ V(x_*)(v)$ by the variational principle. Since V is a Liapunov function, we choose $v_* \in F(x_*)$ such that $D_+ V(x_*)(v_*) \leq -W(x_*, v_*)$. □

So, Theorem 1 is the particular case when $W(x, v) \doteq \|v\|$. By taking $W(x, v) \doteq \|v\|^\alpha$, $\alpha > 1$, we obtain the existence of approximate equilibria.

Corollary 1. *Let F be a set-valued map from a closed subset $K \subset X$ to X and V be a lower semicontinuous Liapunov function for F with respect to $\|\cdot\|^\alpha$, $\alpha > 1$. Then,*

$$(8) \qquad \forall \varepsilon > 0, \quad \exists x_\varepsilon \in K \quad \text{such that } F(x_\varepsilon) \cap \varepsilon B \neq \emptyset. \qquad \blacktriangle$$

Note also that when W satisfies the condition

$$(9) \qquad \forall x \in K, \quad \forall v \neq 0, \quad W(x, v) > 0$$

we obtain the existence of stationary points.

Corollary 2. *Let F be a set-valued map from a closed subset $K \subset X$ to X, W: graph$(F) \to R_+$ be a nonnegative function satisfying property (9) and V be a lower semicontinuous and inf-compact Liapunov function for F with respect to W. Then there exists an equilibrium x_* of F.* ▲

Remarks. We shall prove that in this case, under some supplementary continuity assumptions, solutions to the differential inclusion $x' \in F(x)$, $x(0) = x_0$, converge to a stationary point of F in some sense when $t \to \infty$. Note that the tangential condition

$$(10) \qquad \forall x \in K, \quad F(x) \cap T_K(x) \neq \emptyset$$

are involved in the Liapunov condition (5), since the domain of the upper contingent derivative $D_+ V(x)(\cdot)$ is contained in the contingent cone $T_K(x)$ to the domain K of V: Indeed, if V is a Liapunov function, there exists $v \in F(x)$ such that $D_+ V(x)(v) \le -W(x, v) \le 0$. Hence $v \in \operatorname{Dom} D_+ V(x)(\cdot) \subset T_K(x)$. Therefore the tangential condition (10) holds true.

Actually, if we take $V \doteq \psi_K$ the indicator of K, defined by $\psi_K(x) \doteq 0$ when $x \in K$ and by $\psi_K(x) \doteq \infty$ when $x \notin K$, the Liapunov condition can be written

$$\text{(11)} \qquad \forall x \in K, \quad \exists v \in F(x) \cap T_K(x) \quad \text{satisfying } W(x, v) \le 0.$$

We recall also that when K is convex and compact and when F is upper hemicontinuous with closed convex values, Theorem 5.2.1 states that the tangential condition (10) is sufficient for establishing the existence of an equilibrium $x_* \in K$ of F.

3. Monotone Trajectories of a Differential Inclusion

We take $X = R^n$. We consider a set-valued map F from X to X, V a function from $\operatorname{Dom} F$ to R_+ and W a function from graph (F) to R_+. We set $K \doteq \operatorname{Dom} F$.

Definition 1. We say that a trajectory $x(\cdot)$ of the differential inclusion

$$\text{(1)} \qquad x'(t) \in F(x(t)); \quad x(0) = x_0$$

is *monotone* (with respect to V and W) if

$$\text{(2)} \qquad \forall s \ge t \ge 0, \quad V(x(s)) - V(x(t)) + \int_t^s W(x(\tau), x'(\tau)) d\tau \le 0. \qquad \blacktriangle$$

Note that this condition implies that

$$\text{(3)} \qquad \forall t \ge 0, \quad x(t) \in K \quad \text{(i.e., } x(\cdot) \text{ is "viable")}$$

since V is defined on K.

Propositions 1. *Let* $x(\cdot)$ *be a monotone trajectory. Then*

$$\text{(4)} \qquad t \to V(x(t)) \quad \textit{decreases and converges to } \alpha \doteq \lim_{t \to \infty} V(x(t))$$

and

$$\text{(5)} \qquad \int_0^\infty W(x(\tau), x'(\tau)) d\tau \doteq \lim_{t \to \infty} \int_0^t W(x(\tau), x'(\tau)) d\tau < +\infty. \qquad \blacktriangle$$

Remark. Note that (4) implies that when V is continuous, all the cluster points x_* of the trajectory when $t \to \infty$ (if any) satisfy $\alpha = V(x_*)$.

Proof. The first statement is obvious.

Since

$$\int_t^s W(x(\tau), x'(\tau)) d\tau \le V(x(t)) - V(x(s)) \to \alpha - \alpha = 0$$

when $t, s \to \infty$, the Cauchy criterion implies that

$$\int_0^\infty W(x(\tau), x'(\tau))d\tau \doteq \lim_{t\to\infty} \int_0^t W(x(\tau), x'(\tau))d\tau < +\infty. \qquad \square$$

We shall see in Theorem 5.3 that this latter condition implies that $W(x(t), x'(t))$ converges to 0 in some sense when $t \to \infty$.

Necessary Condition for the Existence of Monotone Trajectories

We shall prove that the existence of monotone trajectories with respect to V and W of the differential inclusion $x' \in F(x)$, $x(0) = x_0$ for all initial values $x_0 \in K$ implies that V is a Liapunov function for F with respect to W.

Proposition 2. *Let F be an upper semicontinuous map from $K \subset R^n$ to R^n with compact convex values. Let V be a nonnegative function defined on* Dom(F) *and W be a lower semicontinuous nonnegative function on graph(F) which is convex with respect to the second variable. We assume that for all $x_0 \in K$, there exist $T > 0$ and a trajectory $x(\cdot)$ on $]0, T[$ of the differential inclusion $x' \in F(x)$, $x(0) = x_0$ satisfying*

(2) $$\forall s \geq t, \quad V(x(s)) - V(x(t)) + \int_t^s W(x(\tau), x'(\tau))d\tau \leq 0.$$

Then, V is a Liapunov function for F with respect to W:

(6) $$\forall x \in K, \quad \exists v \in F(x) \quad \text{such that } D_+ V(x)(v) + W(x, v) \leq 0. \qquad \blacktriangle$$

Proof. It is analogous to the proof of Proposition 4.2.1. Since F is upper semicontinuous, we can associate to $\varepsilon > 0$ an $\eta > 0$ such that, for all $\tau \leq \eta$, $F(x(\tau)) \subset F(x_0) + \varepsilon B$, which is compact and convex. Since

$$\frac{x(h) - x_0}{h} = \frac{1}{h}\int_0^h x'(\tau)d\tau$$

and since, for almost all τ, $x'(\tau)$ belongs to $F(x_0) + \varepsilon B$, the mean-value theorem implies that $\frac{x(h) - x_0}{h}$ belongs to this compact subset.

Hence there exists a subsequence $h_n \to 0$ such that $v_n \doteq \frac{x(h_n) - x_0}{h}$ converges to some v_0 in $F(x_0) + \varepsilon B$.

Since this inclusion holds true for all $\varepsilon > 0$, we deduce that $v_0 \in F(x_0)$.

We also observe that $x_0 + h_n v_n = x(h_n)$ belongs to K, the domain of V.

Hence property (2) implies that

$$\frac{V(x_0+h_n v_n)-V(x_0)}{h_n}+\frac{1}{h_n}\int_0^{h_n} W(x(\tau),x'(\tau))d\tau\le 0. \tag{7}$$

Let us assume for a while that the following result holds true:

Proposition 3. *Let W be a continuous function on graph(F) which is convex with respect to the second variable. Let $x\in\mathscr{C}(0,T;K)$ and $v\in L^\infty(0,T;M)$ be given, where M is compact and convex. Let*

$$x_0 \doteq \lim_{t\to\infty} x(t) \quad \text{and} \quad v_0\doteq \lim_{h_n\to 0}\frac{1}{h_n}\int_0^{h_n} v(t)dt. \tag{8}$$

Then,

$$W(x_0,v_0)\le \liminf_{h_n\to 0}\frac{1}{h_n}\int_0^{h_n} W(x(\tau),v(\tau))d\tau. \tag{9}$$ ▲

End of the proof of Proposition 2. By Proposition 3,

$$W(x_0,v_0)\le \liminf_{h_n\to 0}\int_0^{h_n} W(x(\tau),x'(\tau))d\tau.$$

Hence

$$\begin{aligned}D_+V(x_0)(v_0)+W(x_0,v_0)&\le \liminf_{\substack{h_n\to 0^+\\ v_n\to v_0}}\frac{V(x_0+h_n v_n)-V(x_0)}{h_n}+\\ &\quad+\liminf_{h_n\to 0^+}\frac{1}{h_n}\int_0^{h_n} W(x(\tau),x'(\tau))d\tau\\ &\le \liminf_{\substack{h_n\to 0^+\\ v_n\to v_0}}\left(\frac{V(x_0+h_n v_n)-V(x_0)}{h_n}+\frac{1}{h_n}\int_0^{h_n} W(x(\tau),x'(\tau))d\tau\right)\le 0.\end{aligned}$$ □

Proof of Proposition 3. By Proposition 1.2.3, $x\to\mathscr{E}p\,W_M(x,\cdot)$ is ε-upper semicontinuous. Then there exists h_ε such that, for all $t\in[0,h_\varepsilon]$, $\mathscr{E}p\,W_M(x(t),\cdot)\subset\mathscr{E}p\,W_M(x_0,\cdot)+\varepsilon(B\times B)$. Therefore, for all $\tau\in[0,h_\varepsilon]$, $(v(\tau),W(x(\tau),v(\tau)))\in\mathscr{E}p\,W_M(x(\tau),\cdot)\subset\mathscr{E}p\,W_M(x_0,\cdot)+\varepsilon(B\times B)$. Hence, by the mean-value theorem, we deduce that, for all $h_n\le h_\varepsilon$,

$$\left(\frac{1}{h_n}\int_0^{h_n} v(\tau)d\tau,\frac{1}{h_n}\int_0^{h_n} W(x(\tau),v(\tau))d\tau\right)\in\overline{\text{co}}(\mathscr{E}p\,W_M(x_0,\cdot)+\varepsilon(B\times B)).$$

Therefore, by taking the limit, we obtain:

$$\left(v_0,\liminf_{h_n\to 0}\frac{1}{h_n}\int_0^{h_n} W(x(\tau),v(\tau))d\tau\right)\in\overline{\text{co}}\,(\mathscr{E}p\,W_M(x_0,\tau)+\varepsilon(B\times B)).$$

Since this is true for all $\varepsilon\ge 0$ and since $\mathscr{E}p\,W_M(x_0,\cdot)$ is closed and convex, we get:

$$W(x_0,v_0)\le \liminf_{h_n\to 0}\frac{1}{h_n}\int_0^{h_n} W(x(\tau),v(\tau))d\tau.$$ □

Sufficient Conditions for the Existence of Monotone Trajectories

We shall prove that, conversely, if V is a Liapunov function for F with respect to W there exist trajectories of the differential inclusion $x' \in F(x)$ that are monotone with respect to V and W.

Theorem 1. *Let K be a locally compact subset of R^n and F be an upper semicontinuous map from K to the nonempty compact convex subsets of R^n. Let W be a function defined on graph (F) satisfying*

$$\text{(10)} \qquad \begin{aligned} &\text{i) } W \text{ is nonnegative and continuous,}\\ &\text{ii) } \forall x \in K,\ v \to W(x,v) \text{ is convex.}\end{aligned}$$

Let $V: K \to R_+$ be a Liapunov function with respect to F and W:

$$\text{(11)} \qquad \forall x \in K, \quad \exists v \in F(x) \quad \text{such that } D_+ V(x)(v) + W(x,v) \leq 0.$$

We also assume that

$$\text{(12)} \qquad V \text{ is continuous (for the topology induced on } K).$$

Then, for every $x_0 \in X$, there exist $T > 0$ and a monotone trajectory $x(\cdot)$ on $[0, T[$ of the differential inclusion $x' \in F(x)$ and $x(0) = x_0$.

If K is closed and $F(K)$ is bounded, then we can take $T = \infty$.

Proof. It is analogous to the proof of the Viability Theorem. (See Theorem 4.2.1.) Since K is locally compact, there exists $r > 0$ such that $K_0 \doteq K \cap (x_0 + rB)$ is compact. We set $T \doteq r/(\|F(K_0)\| + 1)$. If $F(K)$ is bounded, we take T arbitrary and we set $K_0 \doteq K \cap cl(x_0 + TF(K) + B)$, which is compact.

Let us take $y \in K$ and $v_y \in F(y)$ satisfying

$$\text{(13)} \qquad D_+ V(y)(v_y) + W(y, v_y) \leq 0.$$

This is possible thanks to assumption (11).

By the very definition of $D_+ V(y)(v_y)$, we know that there exist $h_y \leq 1/k$ and $u_y \in v_y + (1/k)B$ such that

$$\text{(14)} \qquad \frac{V(y + h_y u_y) - V(y)}{h_y} < D_+ V(y)(v_y) + 1/k.$$

Hence

$$\text{(15)} \qquad \frac{V(y + h_y u_y) - V(y)}{h_y} + W(y, v_y) < \frac{1}{k}.$$

We set

$$\text{(16)} \qquad N(y) \doteq \left\{ x \in K \ \ \frac{V(x + h_y u_y) - V(x)}{h_y} + W(y, v_y) < \frac{1}{k} \right\}.$$

The sets $N(y)$ are open because the function V is continuous and y belongs to $N(y)$. So we can find a ball of radius $\alpha_y \in \left]0, \frac{1}{k}\right[$ such that $(y + \alpha_y B) \cap K \subset N(y)$.

Since K_0 is compact, it can be covered by q such balls $y_j+\alpha_{y_j}B$. We set $\alpha_j \doteq \alpha_{y_j}$, $h_j \doteq h_{y_i}$, $u_j \doteq u_{y_j}$ and $v_j \doteq v_{y_j}$. Let us take now any $x \in K_0$. It belongs to a ball $y_j+\alpha_j B$ for some $j=1, \ldots, q$. Then there exists $u_j \in C \doteq F(K_0)+B$ such that

$$(17)\qquad \begin{aligned} &\text{i)}\ \frac{V(x+h_j u_j)-V(x)}{h_j}+W(y_j, v_j)<\frac{1}{k},\\ &\text{ii)}\ \|x-y_j\| \leq \alpha_j \leq \frac{1}{k}, \qquad \|u_j-v_j\| \leq \frac{1}{k}. \end{aligned}$$

Let $h_0(k) \doteq \min_{j=1,\ldots,q} h_j>0$. So, cancelling the index j, we have proved that for all $x \in K_0$, there exist $h \in \left[h_0(k), \frac{1}{k}\right]$ and $u \in C$ satisfying the two following properties

$$(18)\qquad \begin{cases} \text{i)}\ \exists y \in K \text{ and } v \in F(y) \text{ such that } \|x-y\| \leq \frac{1}{k},\ \|u-v\| \leq \frac{1}{k},\\ \text{ii)}\ \frac{V(x+hu)-V(x)}{h}+W(y, v) \leq \frac{1}{k}. \end{cases}$$

Therefore, we can construct inductively a sequence of elements $x_p \in K_0$, $v_p \in F(y_p)$, $h_p \in \left[h_0(k), \frac{1}{k}\right]$ and $u_p \in C$ satisfying

$$(19)\qquad \begin{cases} \text{i)}\ x_{p+1}=x_p+h_p u_p \in K_0,\\ \text{ii)}\ (x_p, u_p) \in (y_p, v_p)+\frac{1}{k}(B \times B) \subset \operatorname{graph}(F)+\frac{1}{k}(B \times B),\\ \text{iii)}\ \frac{V(x_{p+1})-V(x_p)}{h_p}+W(y_p, v_p) \leq \frac{1}{k}. \end{cases}$$

We are sure that there exists an integer m such that

$$h_0+h_1+\ldots+h_m \leq T<h_1+\ldots+h_m+h_{m+1}$$

for $h_p \in \left[h_0(k), \frac{1}{k}\right]$ for all p.

Let us set $\tau_k^q=h_0+\ldots+h_{q-1}$. We interpolate this sequence by the piecewise linear function $x_k(\cdot)$ defined on each interval $]\tau_k^{p-1}, \tau_k^p[$ by $x_k(t) \doteq x_{p-1}+(t-\tau_k^{p-1})u_{p-1}$. We denote by $y_k(\cdot)$ and $v_k(\cdot)$ the step functions defined on this interval by $y_k(t)=y_p$ and $v_k(t)=v_p$.

When t is fixed in $]\tau_k^{p-1}, \tau_k^p[$, we have $|t-\tau_k^p| \leq 1/k$ and there exists $(y, v) \in \operatorname{graph}(F)$ such that $\|x_k'(t)-v\|=\|u_{p-1}-v\| \leq 1/k$ and

$$\begin{aligned} \|x_k(t)-y\| &\leq \|x_k(t)-y_{p-1}\|+\|x_{p-1}-y\| \leq |t-\tau_k^p|\|u_{p-1}\|+\|x_{p-1}-y\| \\ &\leq 1/k(\|F(x_0)\|+2). \end{aligned}$$

We have proved that $\forall t>0$,

$$(20)\qquad (x_k(t), x_k'(t)) \in (y_k(t), v_k(t))+\varepsilon(k)(B \times B) \subset \operatorname{graph}(F)+\varepsilon(k)(B \times B)$$

where $\varepsilon(k)\to 0$ when $k\to\infty$. We also know that $\|x_k'(t)\|\leq\|F(K_0)\|+1$ and $x_k(t)\in\overline{\mathrm{co}}(K_0)$, which is compact. Hence the assumptions of Theorem 0.3.4 and the *Convergence Theorem* are satisfied: A subsequence of $x_k(\cdot)$ converges uniformly over compact intervals to a solution $x(\cdot)$ of the differential inclusion $x'\in F(x)$. Moreover, the sequence of derivatives $x_k'(\cdot)$ converges to $x'(\cdot)$ in $L^\infty(0,T;R^n)$ supplied with the weak topology $\sigma(L^\infty,L^1)$.

On each point τ_k^p of the grid, the following inequality holds:

$$V(x_k(\tau_k^{p+1}))-V(x_k(\tau_k^p))+h^p W(x_k(\tau_k^p),v_k(\tau_k^p))\leq h^p/k. \tag{21}$$

By summing these inequalities from $p=q$ to $p=r-1$, we obtain,

$$V(x_k(\tau_k^r))-V(x_k(\tau_k^q))+\sum_{p=q}^{r-1} h^p W(x_k(\tau_k^p),v_k(\tau_k^p))\leq\frac{(\tau_k^r-\tau_k^q)}{k}.$$

We remark that $v_k(\tau)=x_k'(\tau_k^p)$ on the interval $[\tau_k^p,\tau_k^{p+1}[$. So, we can write the above inequality in the form:

$$V(y_k(\tau_k^r))-V(y_k(\tau_k^q))+\int_{\tau_k^q}^{\tau_k^r} W(y_k(\tau),v_k(\tau))d\tau\leq\frac{(\tau_k^r-\tau_k^q)}{k}. \tag{22}$$

We recall that $x_k(\cdot)$ converges to $x(\cdot)$ uniformly on compact intervals; so does $y_k(\cdot)$. We also know that $x_k'(\cdot)$ converges weakly to $x'(\cdot)$ in $L^\infty(0,T;X)$; so does $v_k(\cdot)$.

We assume for a while that the following Proposition holds true:

Proposition 4. *Assume that the function W: graph$(F)\to R_+$ is nonnegative, continuous and convex with respect to v. Then, for all compact convex subset M, the functional $\underset{\sim}{W}$ defined by*

$$\underset{\sim}{W}(x,v)\doteq\int_0^\infty W(x(\tau),v(\tau))d\tau\in[0,\infty] \tag{23}$$

is lower semicontinuous on $\mathscr{C}(0,\infty;K)\times L^\infty(0,\infty;M)$ when $\mathscr{C}(0,\infty;X)$ is supplied with the topology of uniform convergence on compact intervals and $L^\infty(0,\infty;M)$ with the topology induced by the weak topology $\sigma(L^\infty,L^1)$ on $L^\infty(0,\infty;X)$.* □

End of the proof of Theorem 4. By Proposition 4 we deduce that for all $s\geq t$,

$$\int_t^s W(x(\tau),x'(\tau))d\tau\leq\liminf_{k\to\infty}\int_t^s W(y_k(\tau),v_k(\tau))d\tau. \tag{24}$$

Since we can approximate s by τ_k^r and t by τ_k^q with $\tau_k^r>\tau_k^q$ for k large enough, we deduce from the continuity of V that $V(x_k(\tau_k^r))$ converges to $V(x(s))$ and $V(x_k(\tau_k^q))$ converges to $V(x(t))$. Also, since W is continuous, it is bounded on $K_0\times F(K_0)$ and thus,

$$\int_{\tau_k^q}^{\tau_k^r} W(x(\tau),x'(\tau))d\tau \quad \text{converges to} \int_t^s W(x(\tau),x'(\tau))d\tau.$$

Hence, we can take the limit when $k \to \infty$ in inequalities (22): we find that $x(\cdot)$ satisfies the monotonicity condition

$$\forall s \geq t, \quad V(x(s)) - V(x(t)) + \int_t^s W(x(\tau), x'(\tau)) d\tau \leq 0.$$

When $F(K)$ is bounded, we have chosen T independent of x_0. So we can extend the trajectory $x(\cdot)$ on $[0, T]$ to a trajectory on $[0, 2T]$, $[0, 3T]$, etc. Hence, there exists a trajectory defined on $[0, \infty[$. □

Proof of Proposition 4. Let $x_k(\cdot)$ converge to $x(\cdot)$ uniformly on compact intervals and $v(\cdot)$ converge weakly* to $v(\cdot)$ in $L^\infty(0, \infty; X)$. Hence

$$\text{(25)} \qquad \begin{cases} \text{i)} \ \forall t \in [0, \infty[, \ \{x_k(t)\} \text{ converges to } x(t), \\ \text{ii)} \ \forall T > 0, \ \{v_k(\cdot)\} \text{ converges weakly to } v(\cdot) \text{ in } L^1(0, T; X). \end{cases}$$

Actually, it is sufficient to suppose the latter properties (25) to hold true. If $\liminf_{k \to \infty} \underset{\sim}{W}(x_k, v_k) = +\infty$, the theorem is true; if not, let

$$\text{(26)} \qquad c \doteq \liminf_{k \to \infty} \underset{\sim}{W}(x_k, v_k).$$

There exist subsequences (again denoted x_k and v_k) of x_k and v_k such that

$$\text{(27)} \qquad \forall k \in N, \quad \underset{\sim}{W}(x_k, v_k) \leq c + \frac{1}{k}.$$

By Mazur's theorem, there exists a sequence of elements

$$\text{(28)} \qquad w_h(\cdot) = \sum_{k=h}^{\infty} a_h^k v_k(\cdot). \quad a_h^k \geq 0, \quad \sum_{k=h}^{\infty} a_h^k = 1,$$

(where $a_h^k = 0$ but for a finite number of indexes k) that converges strongly to $v(\cdot)$ in $L^1(0, T; X)$:

$$\forall h \in N, \quad \|w_h - v\|_{L^1(0,T;X)} \leq 1/h.$$

Hence, a subsequence (again denoted) w_h converges almost everywhere to $v(\cdot)$. Let $t > 0$ be any point where

$$\text{(29)} \qquad v(t) = \lim_{h \to \infty} w_h(t).$$

Let W_M denote the restriction of $v \to W(x, v)$ to M.

By Proposition 1.2.3 there exists η such that $\mathscr{E}p\, W_M(x, \cdot) \subset \mathscr{E}p\, W_M(x(t), \cdot) + \varepsilon(B \times B)$ when $\|x - x(t)\| \leq \eta$. Let k_0 be such that $\|x_k(t) - x(t)\| \leq \eta$ whenever $k \geq k_0$. Since $\mathscr{E}p\, W_M(x_k(t), \cdot)$ is convex, we deduce that

$$\left(w_h(t), \sum_{k=h}^{\infty} \alpha_h^k W((x_k(t), v_k(t))\right) \in \mathscr{E}p\, W_M(x_k(t), \cdot) \subset \mathscr{E}p\, W_M(x(t), \cdot) + \varepsilon(B \times B)$$

Hence, by letting $h \to \infty$, we obtain

$$\left(v(t), \liminf_{h \to \infty} \sum_{k=h} \alpha_h^k W(x_k(t), v_k(t))\right) \in cl(\mathscr{E}p\, W_M(x(t), \cdot) + \varepsilon(B \times B)).$$

Since this holds true for all $\varepsilon>0$, and since $W_M(x(t),\cdot)$ is closed, it follows that

(30) $$\left(v(t), \liminf_{h\to\infty} \sum_{k=h}^{\infty} \alpha_h^k W(x_k(t), v_k(t))\right) \in \mathscr{E}p\, W(x(t),\cdot),$$

i.e., that

(31) $$\liminf_{h\to\infty} \sum_{k=h}^{\infty} \alpha_h^k W(x_k(t), v_k(t)) \geq W(x(t), v(t)).$$

We integrate this inequality and we apply Fatou's Lemma, which is possible for W is nonnegative. We obtain

$$\begin{aligned} \underset{\sim}{W}(x, v) &\leq \int_0^{\infty} \liminf_{h\to\infty} \sum_{k=h}^{\infty} \alpha_h^k W(x_k(t), v_k(t))\,dt \\ &\leq \liminf_{h\to\infty} \sum_{k=h}^{\infty} \alpha_h^k \underset{\sim}{W}(x_k, v_k) \leq \lim_{h\to\infty} \left(c+\frac{1}{h}\right) = c. \end{aligned}$$

Hence $\underset{\sim}{W}(x, v) \leq c \doteq \liminf_{k\to\infty} \underset{\sim}{W}(x_k, v_k)$. □

The Time Dependent Case

We shall adapt to the time dependent case the results we proved for the time independent case. We only have to use the classical transformation which amounts to observing that the solutions to the differential inclusion

(32) $$x'(t) \in F(t, x); \quad x(t_0) = x_0$$

are the solutions $\tau \to (t(\tau), x(\tau)) \doteq \hat{x}(\tau)$ of the differential inclusion

(33) $$\hat{x}' \in \hat{F}(\hat{x}), \quad x(0) = (t_0, x_0)$$

where we set

(34) $$\forall (t, x) \doteq \hat{x} \in \operatorname{Dom}(F), \quad \hat{F}(\hat{x}) \doteq (1, F(t, x)).$$

We shall denote by $\hat{K} \doteq \operatorname{Dom} F$ the domain of $\hat{F}$, which is the domain of F. We introduce

(35) $$\begin{cases} \text{i) a nonnegative function } V \text{ from } \operatorname{Dom} F \text{ to } R_+, \\ \text{ii) a nonnegative function } W \text{ from } \operatorname{graph}(F) \text{ to } R_+. \end{cases}$$

Now, the symbol $D_+ V(t, x)(1, v)$ denotes the upper contingent derivative of V at (t, x) in the direction $(1, v)$. We recall that when V is differentiable, we have

(36) $$D_+ V(t, x)(1, v) = \frac{\partial}{\partial t} V(t, x) + \sum_{i=1}^{n} \frac{\partial}{\partial x_i} V(t, x) v_i.$$

Proposition 5. *$\hat{x}(\cdot)$ is a monotone trajectory of the differential inclusion* (33) *with respect to V and W if and only if $x(\cdot)$ is a trajectory of the differential* *inclusion* (32) *which is monotone with respect to* V *and* W *in the sense that*

(37) $$\forall s \geq t, \quad V(s, x(s)) - V(t, x(t)) + \int_0^\infty W(\tau, x(\tau), x'(\tau))d\tau \leq 0. \qquad \blacktriangle$$

Proof. Indeed, $\hat{x}(\cdot)$ is a solution to (33) if and only if $\hat{x}(t) = (t, x(t))$, where $x(\cdot)$ is a solution to (32). Note that $\hat{x}'(t) = (1, x'(t))$. So, condition (37) is equivalent to

$$\forall s \geq t, \quad V(\hat{x}(s)) - V(\hat{x}(t)) + \int_0^\infty W(\hat{x}(\tau), \hat{x}'(\tau))d\tau \leq 0. \qquad \square$$

We introduce now the concept of Liapunov function.

Definition 2. Let F be a set-valued map from $\hat{K} \subset R_+ \times R^n$ to R^n, V be a nonnegative function from $\hat{K}$ to R and W be a nonnegative function from graph(F) to R. We say that V is a Liapunov function for F with respect to W if

(38) $$\forall (t, x) \in \hat{K}, \quad \exists v \in F(t, x) \quad \text{such that } D_+ V(t, x)(1, v) + W(t, x, v) \leq 0 \qquad \blacktriangle$$

We can consider $\hat{K} \subset R \times R^n$ as the graph of a set-valued map $t \to K(t)$. Then monotonicity condition (37) implies in particular that

(39) $$\forall t, \quad x(t) \in K(t)$$

and the Liapunov condition (38) implies in particular that

(40) $$\forall t \geq 0, \quad \forall x \in K(t), \quad \exists v \in DK(t, x)(1)$$

since the latter condition is equivalent to the tangential condition $(l, v) \in T_K(t, x)$. When V is the restriction to $\hat{K}$ of a lower semicontinuous convex function $\hat{V}$, where $0 \in \operatorname{Int}(\hat{K} - \operatorname{Dom} \hat{V})$, the Liapunov condition (36) can be written

(41) $$\begin{cases} \forall t \geq 0, \quad \forall x \in K(t), \quad \exists v \in F(t, x) \cap DK(t, x)(1) \text{ such that} \\ D_+ \hat{V}(t, x)(1, v) + W(t, x, v) \leq 0. \end{cases}$$

When V is the restriction to $\hat{K}$ of a differentiable function $\hat{V}$, the Liapunov condition can be written

(42) $$\begin{gathered} \forall t \geq 0, \quad \forall x \in K(t), \quad \exists v \in F(t, x) \cap DK(t, x)(1) \text{ such that} \\ \frac{\partial}{\partial t} \hat{V}(t, x) + \sum_{i=1}^n \frac{\partial}{\partial x_i} \hat{V}(t, x) v_i + W(t, x, v) \leq 0. \end{gathered}$$

or, equivalently

(43) $$\frac{\partial \hat{V}}{\partial t}(t, x) + \inf_{v \in F(t, x) \cap DK(t, x)(1)} \left(\sum_{i=1}^n \frac{\partial \hat{V}}{\partial x_i}(t, x) v_i + W(t, x, v) \right) \leq 0.$$

This is the *Carathéodory-Hamilton-Jacobi-Bellman equation of optimal control theory.* $\square$

We deduce from Theorem 1 and Proposition 2 the following characterization of existence of monotone trajectories.

Theorem 2. *Let F be bounded upper semicontinuous map from $\hat{K} \subset R_+ \times R^n$ to the compact convex subsets of R^n, V be a nonnegative continuous function from $\hat{K}$ to R and W be a nonnegative continuous function from graph (F) to R which is convex with respect to the last argument. Then the differential inclusion*

$$x'(t) \in F(t, x(t)); \quad x(t_0) = x_0 \tag{44}$$

has a monotone trajectory $x(\cdot) \in \mathscr{C}(t_0, \infty R^n)$ for all $t_0 \geq 0$ and $x_0 \in K(t_0)$ if and only if V is a Liapunov function for F with respect to W. ▲

Remark. Construction of monotone trajectories. The question arises whether V and W being given, there exist monotone trajectories $x(\cdot)$, which, we recall, satisfy

$$\forall t \geq s, \quad V(x(t)) - V(x(s)) + \int_s^t W(x(\tau), x'(\tau) d\tau \leq 0. \tag{45}$$

We can solve this problem by constructing a single-valued map f such that the differential equation $x' = f(x)$ has a monotone trajectory.

In this section, we shall assume that

$$\begin{cases} \text{i) } K \text{ is compact and convex,} \\ \text{ii) } V \text{ is the restriction to } K \text{ of a convex continuous function } \hat{V} \\ \text{iii) } W \text{ is continuous and convex with respect to its second argument.} \end{cases} \tag{46}$$

We recall that a necessary and sufficient condition for f to have monotone trajectories with respect to V and W is that

$$\forall x \in K, \quad D_+ V(x)(f(x)) + W(x, f(x)) \leq 0. \tag{47}$$

Since $D_+ V(x)(v)$ is the restriction to the tangent cone $T_K(x)$ of $D_+ \hat{V}(x)(v)$, we set

$$S(x) \doteq \{v \in X \mid D_+ \hat{V}(x)(v) + W(x, v) \leq 0\}. \tag{48}$$

So, the necessary and sufficient conditions can be written

$$\forall x \in K, \quad f(x) \in S(x) \cap T_K(x). \tag{49}$$

In order to exclude the obvious solution $f \equiv 0$, we introduce the cones

$$\mathring{S}(x) = \{v \in X \mid D_+ \hat{V}(x)(v) + W(x, v) < 0\}. \tag{50}$$

which may be empty. We also set

$$K_0 \doteq \{x \in K \mid \mathring{S}(x) \cap T_K(x) \neq \emptyset\}; \quad K_1 \doteq K \cap \complement K_0. \tag{51}$$

Theorem 3. *Let K be a compact convex subset, V be the restriction to K of a continuous convex function $\hat{V}$ and W be a nonnegative continuous function on $K \times R^n$ which is convex with respect to v. We assume that the subset K_0 defined by (51) is nonempty.*

Then there exists a continuous function f whose set of stationary points is K_1 such that the differential equation $x' = f(x)$, $x(0) = x_0$ has a monotone trajectory with respect to V and W. ▲

Proof. Since $(x, v) \to D_+ \hat{V}(x)(v)$ is upper semicontinuous, the graph of the map $\mathring{S}$,

$$\operatorname{Graph} \mathring{S} = \{(x, v) \in K_0 \times R^n \mid D_+ V(x)(v) + W(x, v) < 0\}$$

is open. Since the set-valued map $x \to T_K(x)$ is lower semicontinuous and has convex values, then $x \to \mathring{S}(x) \cap T_K(x)$ is locally selectionable from K_0 to R^n and its images are convex cones. Hence, by Corollary 1.10.1 there exists a continuous function f from K to R^n satisfying

$$(52) \qquad \forall x \in K_0, \quad f(x) \in T_K(x) \cap \mathring{S}(x) \quad \text{and} \quad \forall x \in K_1, \ f(x) = 0.$$

So, such a function f satisfies the assumption of Theorem 1. Hence there exists monotone trajectories of the differential equation $x' = f(x)$. □

Remark. Feedback Controls Yielding Monotone Trajectories. Let U be a set of controls u and f: $K \times U \to X$ be the map that assigns to each state x and to each control u the velocity $f(x, u)$ of the state.

A feedback control is a map $\underset{\sim}{u}: x \in K \to \underset{\sim}{u}(x) \in U$ associating with each state of the system a control according to a fixed rule for achieving a given purpose. We will require that the trajectories of the differential equation

$$(53) \qquad \begin{cases} \text{i)} \ x'(t) = f(x(t), \underset{\sim}{u}(x(t))), \\ \text{ii)} \ x(0) = 0 \end{cases}$$

exist and satisfy the monotonicity condition:

$$(54) \qquad \forall s \geq t, \qquad V(x(s)) - V(x(t)) + \int_t^s W(x(\tau), x'(\tau)) \, d\tau \leq 0.$$

We assume that V is the restriction to K of a convex continuous function $\hat{V}$.

We introduce the following set-valued map $\mathring{S}$ defined by

$$(55) \qquad \mathring{S}(x) \doteq \{v \in X \mid D_+ \hat{V}(x)(v) + W(x, v) < 0\}$$

and we set

$$(56) \qquad K_0 = \{x \in K \mid \mathring{S}(x) \cap T_K(x) \neq \emptyset\}, \qquad K_1 = K \cap \complement K_0.$$

Theorem 4. *Let us assume that $K \subset X$ and U are both convex compact subsets, that U contains 0 and that f: $K \times U \to X$ is a continuous map that is linear with respect to the controls. We assume that $K_0 \neq \emptyset$ and that there exists $\gamma > 0$*

such that

(57) $$\forall x \in K_0,\ \forall y \in X,\ \|y\| \leq \gamma,\ \exists u \in U \text{ such that } f(x,u) - y \in T_K(x) \cap \mathring{S}(x).$$

Then there exists a feedback control $\underset{\sim}{u} \in \mathscr{C}(K, U)$, vanishing on K_1 that provides monotone trajectories with respect to V and W. ▲

Proof. By Theorem 1, it is sufficient to prove the existence of a feedback control $\underset{\sim}{u} \in \mathscr{C}(K, U)$ such that

(58) $$\begin{aligned} &\text{i)}\ \forall x \in K_0, \quad f(x, \underset{\sim}{u}(x)) \in T_K(x) \cap \mathring{S}(x), \\ &\text{ii)}\ \forall x \in K_1, \quad \underset{\sim}{u}(x) = 0. \end{aligned}$$

Let us set, for $x \in K_0$,

(59) $$G(x) \doteq \{u \in U \mid f(x,u) \in T_K(x) \cap \mathring{S}(x)\}.$$

We already mentioned that $x \to T_K(x) \cap \mathring{S}(x)$ is locally selectionable and thus, lower semicontinuous. Assumption (57) implies that G is lower semicontinuous on K_0 by Theorem 1.2.3.

Michael's theorem states that there exists a continuous selection $\underset{\sim}{v} \in \mathscr{C}(K_0, U)$ of the set-valued map G. We denote by $d_{K_1}(x)$ the distance from x to K_1 and we set:

(60) $$\underset{\sim}{u}(x) = \begin{cases} d_{K_1}(x)\underset{\sim}{v}(x) & \text{if } x \in K_0 \\ 0 & \text{if } x \in K_1. \end{cases}$$

This function $\underset{\sim}{u}$ is continuous on K. It is obviously true when $x \in K_0$. Let us check that it is continuous at $x \in K_1$. Let $\varepsilon > 0$ and $y \in x + \varepsilon B$. Then $\underset{\sim}{u}(x) = 0$ and either $\underset{\sim}{u}(y) = 0$ (when $y \in K_1$) or $\underset{\sim}{u}(y) \leq d_{K_1}(y) M \leq \varepsilon$ where $M = \|U\|$. Then $\|\underset{\sim}{u}(x) - \underset{\sim}{u}(y)\| = \|\underset{\sim}{u}(y)\| \leq \varepsilon M$. Hence $\underset{\sim}{u}$ is continuous. Since $\underset{\sim}{v}(x) \in G(x)$ and since f is linear with respect to the controls, we deduce that

$$f(x, \underset{\sim}{u}(x)) = d_{K_1}(x) f(x, \underset{\sim}{v}(x)) \in T_K(x) \cap \mathring{S}(x)$$

when $x \in K_0$ and that $f(x, \underset{\sim}{u}(x)) = 0$ when $x \in K_1$. □

4. Construction of Liapunov Functions

Given the dynamical system described by the set-valued map $F\colon R^n \to R^n$ and the function W: graph$(F) \to R_+$, the problem arises whether there exists a Liapunov function, i.e., a function V satisfying the property

(1) $$\forall x \in \mathrm{Dom}(F),\ \exists v \in F(x) \text{ such that } D_+ V(x)(v) + W(x, v) \le 0.$$

For this purpose, we denote by $\mathcal{T}_\infty(x_0)$ the set of viable trajectories of the differential inclusion

(2) $$x' \in F(x), \quad x(0) = x_0 \text{ given in } K \doteq \mathrm{Dom}\, F.$$

We define the function V_F from Dom F to $[0, +\infty[$ by

(3) $$\forall x_0 \in K, \quad V_F(x_0) \doteq \inf_{x(\cdot) \in \mathcal{T}_\infty(x_0)} \int_0^\infty W(x(\tau), x'(\tau))\, d\tau.$$

We begin by pointing out the following remark.

Proposition 1. *Let* V: Dom$(F) \to R_+$ *and* W: graph$(F) \to R_+$ *be nonnegative functions.*

a) *If there exists a monotone trajectory* $x(\cdot) \in \mathcal{T}_\infty(x_0)$ *with respect to* V *and* W, *then*

(4) $$0 \le V_F(x_0) \le V(x_0).$$

b) *If* $\bar{x} \in \mathcal{T}_\infty(x_0)$ *is a monotone trajectory with respect to* V_F *and* W *and if* $V_F(x_0)$ *is finite, it achieves the minimum of*

$$x \to \int_0^\infty W(x(\tau), x'(\tau))\, d\tau \quad \text{on } \mathcal{T}_\infty(x_0).$$

c) *Conversely, if* $\bar{x} \in \mathcal{T}_\infty(x_0)$ *achieves the minimum of*

$$x \to \int_0^\infty W(x(\tau), x'(\tau))\, d\tau \quad \text{on } \mathcal{T}_\infty(x_0),$$

then it is a monotone trajectory with respect to V_F *and* W *and furthermore*

(5) $$\forall t \ge 0, \quad V_F(\bar{x}(t)) = \int_t^\infty W(\bar{x}(\tau), \bar{x}'(\tau))\, d\tau.$$ ▲

Remark. Equality (5) is the "*principle of optimality*". It states that if $\bar{x}$ is a solution to the differential inclusion $x' \in F(x)$, $x(0) = x_0$ that minimizes on $\mathcal{T}_\infty(x_0)$ the functional $x \to \int_0^\infty W(x(\tau), x'(\tau))\, d\tau$, then its restriction to $[t, \infty[$ minimizes the functional $x \to \int_t^\infty W(x(\tau), x'(\tau)\, d\tau$ over the set of solutions to the differential inclusion $x' \in F(x)$, $x(t) = \bar{x}(t)$.

Proof. a) Let $x(\cdot)\in\mathcal{T}_\infty(x_0)$ be monotone with respect to V and W. Then

$$\forall t\geq 0, \quad \int_0^t W(x(\tau),x'(\tau))\,d\tau \leq V(x_0)-V(x(t))\leq V(x_0)$$

and, consequently,

$$V_F(x_0)\leq \int_0^\infty W(x(\tau),x'(\tau))\,d\tau \leq V(x_0).$$

b) If $\bar{x}$ is a monotone trajectory with respect to V_F and W, we obtain

$$\int_0^\infty W(\bar{x}(\tau),\bar{x}'(\tau))\,d\tau \leq V_F(x_0) \doteq \inf_{x(\cdot)\in\mathcal{T}_\infty(x_0)} \int_0^\infty W(x(\tau),x'(\tau))\,d\tau.$$

c) Let us prove equation (5). Assume that the restriction of $\bar{x}$ to $[t,\infty[$ is not optimal. Then

$$V_F(\bar{x}(t)) < \int_t^\infty W(\bar{x}(\tau),\bar{x}'(\tau)\,d\tau.$$

Consequently, there exists a solution $y(\cdot)$ to the differential inclusion $y'\in F(y)$, $y(t)=\bar{x}(t)$ such that

$$(6) \qquad \int_t^\infty W(y(\tau),y'(\tau))\,d\tau < \int_t^\infty W(\bar{x}(\tau),\bar{x}'(\tau))\,d\tau.$$

The function $x(\cdot)$ defined by

$$x(\tau)=\begin{cases}\bar{x}(\tau) & \text{if } 0\leq\tau\leq t\\ y(\tau) & \text{if } \tau>t\end{cases}$$

belongs to $\mathcal{T}_\infty(x_0)$. By adding $\int_0^t W(\bar{x}(\tau),\bar{x}'(\tau))\,d\tau$ to both sides of (6), we obtain the contradiction

$$V_F(x_0)\leq \int_0^\infty W(x(\tau),x'(\tau))\,d\tau < \int_0^\infty W(\bar{x}(\tau),\bar{x}'(\tau))\,d\tau = V_F(x_0).$$

We deduce from (5) that when $s>t$,

$$(7) \qquad V_F(\bar{x}(s))-V_F(\bar{x}(t))+\int_t^s W(\bar{x}(\tau),\bar{x}'))\,d\tau=0.$$

Hence the trajectory $\bar{x}(\cdot)$ is monotone with respect to V_F and W. □

We now prove that V_F does satisfy the Liapunov condition for F with respect to W.

Proposition 2. *Let F be a proper upper semicontinuous map with compact convex images and W: $\mathrm{graph}(F)\to R_+$ be a nonnegative lower semicontinuous function that is convex with respect to v. If the minimum in $V_F(x_0)$ is achieved for*

$x_0 \in K$, V_F *satisfies the Liapunov condition*

(8) $$\exists v_0 \in F(x_0) \text{ such that } D_+ V_F(x_0)(v_0) + W(x_0, v_0) \le 0. \qquad \blacktriangle$$

Proof. Let us assume that there exists $\bar{x}(\cdot) \in \mathscr{T}_\infty(x_0)$ such that

$$V_F(x_0) = \int_0^\infty W(\bar{x}(\tau), \bar{x}'(\tau))\, d\tau.$$

Since F is upper semicontinuous, we can associate with any $\varepsilon > 0$ an $\eta > 0$ such that, for all $x \in x_0 + \eta B$, $F(x) \subset F(x_0) + \varepsilon B$. So, for h small enough, $\bar{x}'(\tau) \in F(\bar{x}(\tau)) \subset F(x_0) + \varepsilon B$ for all $\tau \in [0, h]$, hence

$$\frac{\bar{x}(h) - x_0}{h} = \frac{1}{h} \int_0^h \bar{x}'(\tau)\, d\tau$$

belongs to $F(x_0) + \varepsilon B$ by the mean-value theorem, being the latter set convex. So, a subsequence $v_n = \dfrac{\bar{x}(h_n) - x_0}{h_n}$ converges to some $v_0 \in F(x_0)$. On the other hand,

$$\begin{aligned} V_F(\bar{x}(h_n)) &\le \int_{h_n}^\infty W(\bar{x}(\tau), \bar{x}'(\tau))\, d\tau = \int_0^\infty W(\bar{x}(\tau), \bar{x}'(\tau))\, d\tau \\ &\quad - \int_0^{h_n} W(\bar{x}(\tau), \bar{x}'(\tau))\, d\tau \le V_F(x_0) - \int_0^{h_n} W(\bar{x}(\tau), \bar{x}'(\tau))\, d\tau. \end{aligned}$$

Therefore,

(9) $$\frac{V_F(x_0 + h_n v_n) - V_F(x_0)}{h_n} + \frac{1}{h_n} \int_0^{h_n} W(\bar{x}(\tau), \bar{x}'(\tau))\, d\tau \le 0.$$

By the very definition of the upper contingent derivative, we have

$$D_+ V_F(x_0, v_0) = \liminf_{\substack{h \to 0 \\ v \to v_0}} \frac{V_F(x_0 + h v) - V_F(x_0)}{h}$$

and, by Proposition 3.3,

$$W(x_0, v_0) \le \liminf_{\substack{h_n \to 0 \\ v \to v_0}} \frac{1}{h_n} \int_0^{h_n} W(x(\tau), x'(\tau))\, d\tau.$$

So, by taking the limit in inequalities (9), we obtain the Liapunov condition (8). □

Remark. Actually, equality holds in (8) because inequality

(10) $$0 \le \frac{V_F(x(h)) - V_F(x_0)}{h} + \frac{1}{h} \int_0^h W(x(\tau), x'(\tau))\, d\tau$$

holds true for any trajectory $x(\cdot)$ of the differential inclusion.

Indeed, let $x \in \mathcal{T}_\infty(x_0)$ and $y(\cdot)$ be a solution to the inclusion $y' \in F(y)$, $y(h) = x(h)$ satisfying $\int_h^\infty W(y(\tau), y'(\tau))\,d\tau \le V_F(x(h)) + \varepsilon$, where $\varepsilon > 0$ is chosen arbitrarily. Since the function z defined by $z(t) = x(t)$ when $t \in [0, h]$ and $z(t) = y(t)$ when $t > h$ still belongs to $\mathcal{T}_\infty(x_0)$, we deduce that

$$V_F(x_0) \le \int_0^\infty W(z(\tau), z'(\tau))\,d\tau \le \int_0^h W(x(\tau), x'(\tau))\,d\tau + V_F(x(h)) + \varepsilon.$$

It suffices to let ε converge to 0.

Therefore, we can prove that the function V_F is the smallest of the Liapunov functions, whose monotone trajectories are the optimal trajectories.

Theorem 1. *Let F be a bounded upper semicontinuous map from a closed subset $K \subset R^n$ to the compact convex subsets of R^n, satisfying the tangential condition*

$$\forall x \in K, \quad F(x) \cap T_K(x) \neq \emptyset. \tag{11}$$

Let W: graph$(F) \to R_+$ be a nonnegative lower semicontinuous function which is convex with respect to v. If for $x_0 \in K$, $V_F(x_0)$ is finite, then there exists an optimal trajectory $\bar{x}(\cdot) \in \mathcal{T}_\infty(x_0)$; furthermore, $V_F(x_0)$ is lower semicontinuous at x_0 and satisfies

$$\exists v_0 \in F(x_0) \text{ such that } D_+ V_F(x_0) + W(x_0, v_0) \le 0. \tag{12}$$

Consequently, if $V_F(\cdot)$ is finite on K, it is a lower semicontinuous Liapunov function for F with respect to W. This in particular holds when there exists a Liapunov function V; in this case, V_F is the smallest of the Liapunov functions. ▲

Proof. By Theorem 2.2.1, the set-valued map $x_0 \to \mathcal{T}_\infty(x_0)$ is upper semicontinuous from K to the compact subsets of the space of functions $x(\cdot) \in \mathscr{C}(0, \infty; R^n)$ whose derivatives $x'(\cdot)$ belong to $L^\infty(0, \infty, R^n)$, when $\mathscr{C}(0, \infty; R^n)$ is supplied with the topology of uniform convergence on compact intervals and when $L^\infty(0, \infty, R^n)$ is supplied with the weak topology $\sigma(L^\infty, L^1)$. Proposition 3.4 states that the functional

$$x(\cdot) \to \int_0^\infty W(x(\tau), x'(\tau))\,d\tau$$

is lower semicontinuous on that space. Therefore, if V_F is finite at x_0, Theorem 1.2.4 implies that $V_F(\cdot)$ is lower semicontinuous at x_0. Also, the minimum is achieved by an optimal solution $\bar{x}(\cdot)$, which is monotone by statement c) of Proposition 1. Then Proposition 2 implies that V_F is a Liapunov function of F with respect to W, which is the smallest one by statement a) of Proposition 1. □

We translate these results in the time dependent case.

Let F be a set-valued map from $R_+ \times R^n$ to R^n, the domain of which is the graph of a set-valued map $t \to K(t)$ from R_+ to R^n. We introduce a nonnegative function W defined on the graph of F.

We denote by $\mathcal{T}_\infty(t_0, x_0)$ the set of solutions $x(\cdot) \in \mathscr{C}(t_0, \infty; x)$ of the differential inclusion (2). We introduce the function

(13) $$V_F(t_0, x_0) = \inf_{x(\cdot) \in \mathcal{T}_\infty(t_0, x_0)} \int_{t_0}^{\infty} W(\tau, x(\tau), x'(\tau))\, d\tau.$$

Theorem 2. *Let F be a bounded upper semicontinuous map from the closed graph of a set-valued map $K(\cdot)$: $R_+ \to R^n$ to the compact convex subsets of R^n, satisfying*

(14) $$\forall t \geq 0,\ \forall x \in K(t), \quad F(t, x) \cap DK(t, x)(1) \neq \emptyset.$$

Let W: graph$(F) \to R_+$ be a nonnegative lower semicontinuous function which is convex with respect to the last argument. If for all $(t_0, x_0) \in$ graph(K) the function $V_F(t_0, x_0)$ is finite, it is the smallest nonnegative lower semicontinuous Liapunov function for F with respect to W: it satisfies

(15) $$\exists v_0 \in F(t_0, x_0) \text{ such that } D_+ V_F(t_0, x_0)(1, v_0) + W(t_0, x_0, v_0) = 0.$$

The optimal trajectories $\bar{x}(\cdot)$ satisfy

(16) $$\forall t \geq t_0, \quad V_F(t, \bar{x}(t)) = \int_t^{\infty} W(\tau, \bar{x}(\tau), \bar{x}'(\tau))\, d\tau. \qquad \blacktriangle$$

5. Stability and Asymptotic Behavior of Trajectories

According to the assumptions relating V and W, the monotonicity property yields useful informations on the asymptotic behavior of the monotone trajectories of

(1) $$x'(t) \in F(x(t)), \quad x(0) = x_0.$$

Example. *Trajectories with finite length.* Let us consider the case when

(2) $$W(x, v) \doteq \|v\|.$$

Theorem 1. *The trajectories $x(\cdot)$ on $[0, \infty[$ that are monotone with respect to V and W: $(x, v) \to \|v\|$ have finite length $\int_0^{\infty} \|x'(\tau)\|\, d\tau$ and converge to $x_* \in \bar{K}$ when $t \to \infty$. If K is closed and F is upper semicontinuous with compact convex values, then x_* is an equilibrium of F.* $\blacktriangle$

Proof. By Proposition 3.1, $\int_0^\infty \|x'(\tau)\|\, d\tau$, which is the length of the trajectory, is finite. Furthermore, inequality

$$\|x(t)-x(s)\| \leq \int_t^s \|x'(\tau)\|\, d\tau \to 0 \quad \text{when } t, s \to \infty$$

holds true and the Cauchy criterion imply that $\lim_{t\to\infty} x(t) = x_*$ does exist. The following theorem shows that x_* is an equilibrium of F. □

Theorem 2. *Let F be an upper semicontinuous map from a closed subset $K \subset X$ to X with compact convex values and $x(\cdot)$ be a trajectory of the differential inclusion* (1) *that converges to some $x_* \in K$. Then x_* is an equilibrium of F.* ▲

Proof. Assume that $0 \notin F(x_*)$: there exists $\varepsilon > 0$ such that

(3) $$\varepsilon B \cap (F(x_*) + \varepsilon B) = \emptyset$$

(for $F(x_*)$ is a closed subset). Since F is upper semicontinuous, there exists $\delta > 0$ such that $F(y) \subset F(x_*) + \varepsilon B$ whenever $\|y_* - x_*\| \leq \delta$. Hence there exists $T > 0$ such that, $\forall t \geq T$, $\|x(t) - x_*\| \leq \delta$. Consequently:

(4) $$\forall t \geq T, \quad F(x(t)) \subset F(x_*) + \varepsilon B.$$

Since $x'(t) \in F(x(t))$ for almost all $t > 0$, the mean value theorem implies that for all $t \geq T$

(5) $$\frac{x(t)-x(T)}{t-T} = \frac{1}{t-T} \int_T^t x'(\tau)\, d\tau \in \overline{\text{co}}(F(x_*) + \varepsilon B) = F(x_*) + \varepsilon B.$$

Therefore, statements (3) and (5) imply that $\forall t \geq T$, $\left\| \frac{x(t)-x(T)}{t-T} \right\| > \varepsilon$, which is a contradiction to the fact that $x_* = \lim_{t\to\infty} x(t)$, and thus, that $\|x(t) - x(T)\|$ remains bounded. □

We turn our attention to the general case.

Almost Convergence of Monotone Trajectories to Equilibria

We recall that when a Liapunov function V for F (with respect to W) satisfies

(6) V is lower semicontinuous and inf-compact,

there exists x_* and v_* satisfying

(7) $$x_* \in K, \quad v_* \in F(x_*) \quad \text{and} \quad W(x_*, v_*) = 0.$$

So, when the function W satisfies the property

(8) $$\forall x \in K,\ \forall v \neq 0, \quad W(x, v) > 0$$

the existence of a Liapunov function V for F with respect to W and assumptions (6) and (8) imply the existence of equilibria.

We wish to prove that cluster points x_* and v_* of $x(t)$ and $x'(t)$ when $t \to \infty$ do satisfy the property (7). (In this case, property (8) guarantees that such cluster points x_* of $x(t)$ are equilibria of F.)

We already noted that when $W(x, v) \doteq \|v\|$, the trajectory $x(t)$ converges to a limit x_*, which is an equilibrium by Theorem 2. In general, the limit of $x(t)$ when $t \to \infty$ need not exist necessarily: we can only guarantee existence of cluster points.

But a difficulty arises: the derivative $x'(t)$ is only defined almost everywhere, so that we cannot speak of its limit when $t \to \infty$. We shall adapt the concepts of limit and cluster points to measurable functions to prove that measurable functions taking their values in a compact subset do have such "almost" cluster points and that "almost cluster points" x_* and v_* of $x(t)$ and $x'(t)$ satisfy $W(x_*, v_*) = 0$ when $x(\cdot)$ is a monotone trajectory with respect to V and W.

Let $\mu(A)$ denote the Lebesgue measure of a subset $A \subset R$.

Definition 1. Let $x\colon [0, \infty[\to X$ be a measurable function and $x_* \in X$. We say that x_* is *the almost limit of* $x(\cdot)$ *when* $t \to \infty$ (and we write $x_* = a\text{-}\lim_{t \to \infty} x(t)$) if

$$\text{(9)} \qquad \forall \varepsilon > 0,\ \exists T > 0 \text{ such that } \mu\{t \in [T, \infty[\,|\, \|x(t) - x_*\| > \varepsilon\} = 0.$$

We say that x_* is an *almost cluster point* of $x(t)$ when $t \to \infty$ if

$$\text{(10)} \qquad \forall \varepsilon > 0, \quad \mu\{t \in [0, \infty[\,|\, \|x(t) - x_*\| \le \varepsilon\} = \infty. \qquad \blacktriangle$$

These concepts are justified by the following theorem:

Theorem 3. *Let F be an upper semicontinuous map from $K \subset R^n$ to the compact subsets of R^n, W be a nonnegative lower semicontinuous function defined on* graph(F) *and V be a nonnegative lower semicontinuous function defined on K. For any monotone trajectory $x(t)$ of F with respect to V and W, the functions $x(t)$ and $x'(t)$ have almost cluster points x_* and v_* which satisfy*

$$\text{(7)} \qquad x_* \in K, \quad v_* \in F(x_*) \quad \textit{and} \quad W(x_*, v_*) = 0.$$

If W satisfies the condition

$$\text{(8)} \qquad \forall x \in K,\ \forall v \neq 0, \quad W(x, v) > 0$$

then such an almost cluster point x_* is an equilibrium. ▲

The proof of this theorem will be obtained by tying up the following properties of almost convergence. For simplicity, we restrict our study to the case of functions of a real variable. Adaptation for functions defined on a measured space is easy.

We begin by showing that the usual concepts of limit and cluster point are particular cases of almost limit and almost cluster point.

Proposition 1. *Any limit x_* of $x(\cdot)\colon [0,\infty[\to X$ is an almost limit point. If $x(\cdot)$ is uniformly continuous, any cluster point x_* of $x(\cdot)$ is an almost cluster point.* ▲

Proof. a) To say that $x_* = \lim_{t\to\infty} x(t)$ amounts to saying that $\forall \varepsilon > 0$, $\exists T > 0$ such that $[T, \infty[\cap \{t \mid \|x(t) - x_*\| > \varepsilon\} = \emptyset$. Hence the measure of this set is equal to 0.

b) Let x_* be a cluster point of $x(\cdot)$: Since $x(\cdot)$ is uniformly continuous, there exists η such that $|s-t| \le \eta$ implies $\|x(t) - x(s)\| \le \varepsilon/2$. Also, there exists a sequence $t_n \to \infty$ (which satisfies $t_{n+1} - t_n \ge 2\eta$) such that $\|x_* - x(t_n)\| \le \varepsilon/2$ when $n \ge N_\varepsilon$. So, for any $n \ge N_\varepsilon$, the disjoint intervals $[t_n - \eta, t_n + \eta]$ are contained in $\{t \in [0, \infty[\mid \|x(t) - x_*\| \le \varepsilon\}$. Hence

$$\infty = \mu\Big(\bigcup_{n \ge N_\varepsilon} [t_n - \eta, t_n + \eta]\Big) \le \mu(\{t \in [0, \infty[\mid \|x(t) - x_*\| \le \varepsilon\}. \qquad \square$$

The following example justifies the introduction of the concept of almost cluster point.

Proposition 2. *If w is a nonnegative function belonging to $L^1(0,\infty)$, then 0 is an almost cluster point of w when $t \to \infty$.* ▲

Proof. If not, there exists $\varepsilon > 0$ such that the measure of

$$A_\varepsilon \doteq \{t \in [0, \infty[\mid w(t) \le \varepsilon\}$$

is finite. Hence the measure of

$$\complement A_\varepsilon = \{t \in [0, \infty[\mid w(t) > \varepsilon\}$$

is infinite. Therefore:

$$\int_0^\infty w(\tau)\, d\tau = \int_{A_\varepsilon} w(\tau)\, d\tau + \int_{\complement A_\varepsilon} w(\tau)\, d\tau \ge \varepsilon\, \mu(\complement A_\varepsilon) = \infty$$

which is a contradiction. □

Proposition 3. *An almost limit x_* of a measurable function $x(\cdot)\colon [0,\infty[\to X$ is the unique almost cluster point.* ▲

Proof. Let y_* be an almost cluster point different from x_*. We choose $\varepsilon \le \|x_* - y_*\|/2$ and T such that the subset $K \doteq \{t \in [T, \infty[\mid \|x(t) - x_*\| > \varepsilon\}$ has a measure equal to 0.

The subset $L \doteq \{t \in [T, \infty[\mid \|x(t) - y_*\| \le \varepsilon\}$ is obviously contained in K and has an infinite measure for y_* is an almost cluster point. Hence $\infty = \mu(L) \le \mu(K) = 0$, which is impossible. So, $x_* = y_*$. □

Proposition 4. *Let f be a continuous (single valued) map from X to Y and $x(\cdot)$: $[0, \infty[\to X$ be a measurable function. If x_* is the almost limit (resp. an almost cluster point) of $x(\cdot)$, then $f(x_*)$ is the almost limit (resp. an almost cluster point) of $f(x(\cdot))$.* ▲

Theorem 4. *Let K be a compact subspace of X and $x(\cdot)$: $[0, \infty[\to K$ be a measurable function. There exists an almost cluster point $x_* \in K$ of $x(\cdot)$ when $t \to \infty$.* ▲

Proof. We define inductively decreasing sequences of measurable sets $\Delta_n \subset [0, \infty[$ and of closed subsets $E_n \subset K$ such that

$$\Delta_n = x^{-1}(E_n), \quad \mu(\Delta_n) = \infty, \quad \operatorname{diam}(E_n) \leq 1/n. \tag{11}$$

For $n=1$, we cover the compact set K with a finite number of sets B_j^1 of diameter at most 1. Thus the subsets $x^{-1}(B_j^1)$ cover $[0, \infty[$ and, consequently, one of them, denoted Δ_1, has an infinite measure. We set E_1 the corresponding set B_j^1.

Having defined the subsets Δ_k and E_k up to n, we cover the compact set E_n by a finite number of closed subsets B_j^{n+1} of diameter at most $1/n+1$.

Their preimages $x^{-1}(B_j^n)$ form a finite covering of Δ_n. Since $\mu(\Delta_n) = \infty$, at least one of these sets, denoted Δ_{n+1}, has an infinite measure. Call E_{n+1} the corresponding B_j^{n+1}. Hence $\bigcap_{n \geq 0} E_n = \{x_*\}$. It remains to show that x_* is an almost cluster point. Fix $\varepsilon > 0$ and $T > 0$. Then, a neighborhood $N_\varepsilon(x_*)$ contains the subsets E_n for $n \geq n(\varepsilon)$. Consequently, $x^{-1}(N_\varepsilon(x_*)) \supset x^{-1}(E_n) = \Delta_n$ for all $n \geq n(\varepsilon)$. Hence

$$\mu\{t \in [0, \infty[\mid x(t) \in N_\varepsilon(x_*)\} \geq \mu(\Delta_n) = \infty. \qquad \square$$

Proposition 5. *Let W be a nonnegative lower semicontinuous function from $L \subset X$ to R. If $x(\cdot)$ is a measurable function from $[0, \infty[$ to L such that*

$$\int_0^\infty W(x(\tau))\, d\tau < +\infty, \tag{12}$$

then any almost cluster point x_* of $x(t)$ when $t \to \infty$ satisfies the equation $W(x_*) = 0$. ▲

Proof. Let x_* be an almost cluster point of $x(\cdot)$ when $t \to \infty$ and assume that $W(x_*) > 0$. We take $\varepsilon = W(x_*)/2 > 0$. Since W is lower semicontinuous, there exists η such that $W(x_*)/2 \leq W(y)$ when $\|y - x_*\| \leq \eta$. So the subset

$$A_\eta \doteq \{t \in [0, \infty[\mid \|x(t) - x_*\| \leq \eta\},$$

whose measure is infinite, is contained in the set

$$B_\varepsilon = \{t \in [0, \infty[\mid W(x_*)/2 \leq W(x(t))\}.$$

Hence

$$\int_0^\infty W(x(\tau))\,d\tau \ge \int_{B_\varepsilon} W(x(\tau))\,d\tau \ge \frac{W(x_*)}{2}\,\mu(B_\varepsilon) = \infty.$$

This is a contradiction. □

We are ready to prove Theorem 3.

Proof of Theorem 3. Since V is lower semicontinuous and inf-compact, then $x(t)$ remains in the compact subset $\mathcal{Q} \doteq \{x \in K \mid V(x) \le V(x_0)\}$. Because F is upper semicontinuous with compact values, the set

$$F_{\mathcal{Q}} \doteq \{(x, v) \in R^n \times R^n \mid x \in \mathcal{Q},\ v \in F(x)\}$$

which is the graph of the restriction $F|_{\mathcal{Q}}$ of F to $\mathcal{Q}$, is compact. Hence the function $t \to (x(t), x'(t))$ is a measurable function taking its values in the compact set $F_{\mathcal{Q}}$. By Theorem 4, there exists an almost cluster point $(x_*, v_*) \in F_{\mathcal{Q}}$. Since $x(\cdot)$ is a monotone trajectory with respect to V and W, we know that $\int_0^\infty W(x(\tau), x'(\tau))\,d\tau < +\infty$. Hence Proposition 5 implies that $W(x_*, v_*) = 0$. □

Stability and Asymptotic Stability

We consider the case when $W \equiv 0$; in this particular case, we say that V is a *Liapunov function with respect to F* if

(13) $$\forall x \in K,\ \exists v \in F(x) \text{ such that } D_+ V(x)(v) \le 0$$

and that a trajectory $x(\cdot)$ of $x' \in F(x)$ is *monotone* with respect to V if

(14) the function $t \to V(x(t))$ is nonincreasing.

Hence monotone trajectories remain in the "level sets"

(15) $$\{x \in K \mid V(x) \le V(x(t))\}.$$

So, we obtain the following stability property.

Proposition 6. *Let K be a closed subset of R^n and F be a bounded upper semicontinuous map from K to the nonempty compact convex subsets of R^n. Let $V: K \to R_+$ be a continuous inf-compact Liapunov function. Let $P_* \doteq \{x_* \in K \mid V(x_*) = \min_{x \in K} V(x)\}$. Then the following stability property holds:*

(16) *For any open neighborhood M of P_*, there exists a neighborhood $N \subset M$ of P_* such that, for all $x_0 \in N$, there exists a trajectory starting at x_0 and remaining in M.* ▲

Proof. We set $\mathcal{Q} \doteq \{x \in K \mid V(x) \leq \min_{y \in K} V(y) + 1\}$, which is compact. Hence, since M is an open neighborhood of P_*, $\mathcal{Q} \cap \complement M$ is compact and $c \doteq \min_{x \in \mathcal{Q} \cap \complement M} V(x)$ is finite. Thus the subset $N \doteq \{x \in \mathcal{Q} \mid V(x) < c\}$ is contained in M and is a neighborhood of P_* (for V is continuous). Now, if $x_0 \in N$, there exists a trajectory $x(\cdot)$ which is monotone (by Theorem 3.1) and thus, which remains in $N \subset M$ because $V(x(t)) \leq V(x_0) < c$. □

We shall give now conditions implying asymptotic stability when $W \equiv 0$.

Theorem 5. *Let F be an upper semicontinuous map from a closed subset $K \subset R^n$ to the compact convex subsets of R^n and V be the restriction to K of a locally Lipschitzean function $\hat{V}$ which is* inf-compact. *We assume that $\hat{V}$ is a Liapunov function with respect to F in the sense that*

$$(17) \qquad \forall x \in K, \ \exists v \in F(x) \ \text{such that} \ D_+ \hat{V}(x)(v) \leq 0$$

which is strict in the sense that

$$(18) \qquad \begin{array}{l} \forall x \in K, \ \text{if there exists} \ v \in F(x) \ \text{such that} \ D_+ \hat{V}(x)(v) \geq 0, \\ \text{then} \ V(x) = \min_{y \in K} V(y). \end{array}$$

We also assume that

$$(19) \quad \text{the function } (x, v) \in \operatorname{graph}(F) \to D_+ \hat{V}(x)(v) \text{ is upper semicontinuous.}$$

(This is the case when, for instance, $\hat{V}$ is continuously differentiable or convex continuous.) Then any monotone trajectory satisfies

$$(20) \qquad \lim_{t \to \infty} V(x(t)) = \min_{y \in K} V(y). \qquad \blacktriangle$$

Proof. We set $v(t) \doteq \hat{V}(x(t))$. Let us assume that $\alpha \doteq \lim_{t \to \infty} v(t) > \min_{x \in K} V(x) \geq 0$. Let $\mathcal{Q} \doteq \{x \in K \mid V(x) \leq V(x_0)\}$, which is compact because V is inf-compact and lower semicontinuous. Therefore, the graph $\mathcal{F}_{\mathcal{Q}}$ of the restriction to $\mathcal{Q}$ of F is compact, for F is upper semicontinuous from K to the compact subsets of R^n.

Assumption (18) implies that $D_+ \hat{V}(x)(v) < 0$ for all $(x, v) \in \mathcal{F}_{\mathcal{Q}}$; so, we deduce from Assumption (19) that there exists $\mu > 0$ such that

$$(21) \qquad \forall x \in \mathcal{Q}, \quad \sup_{v \in F(x)} D_+ \hat{V}(x)(v) \leq -\mu.$$

Let $x(\cdot) \in \mathscr{C}(0, \infty; R^n)$ be a trajectory such that $x'(\cdot) \in L^\infty(0, \infty; R^n)$. Since $x(t) \in \mathcal{Q}$ for all $t \geq 0$, we deduce that $D_+ \hat{V}(x(t))(x'(t)) \hat{V} \leqq -\mu$. Also, because V is the restriction of a locally Lipschitzean function V, Proposition 1.7 implies that

$$(22) \qquad D_+ V(t) = D_+ \hat{V}(x(t))(x'(t)) \leq -\mu.$$

Therefore, we deduce from Proposition 1.8 that for $T \doteq \frac{v(0)}{\mu}$, we have

$$v(T) - v(0) \le \int_0^T -\mu \, dt = -v(0). \tag{23}$$

Thus $v(T) \le 0 \le \min_{x \in K} V(x)$ and $v(T) \ge \alpha > \min_{x \in K} V(x)$. The theorem follows from this contradiction. □

Estimates of the Function $t \to V(x(t))$

Theorem 6. *Let φ be a continuous bounded function from $[0, \infty[$ to R and let $W(x, v) \doteq \varphi(V(x))$. Let x be a monotone trajectory with respect to V and W and $w(\cdot)$ be a solution to the differential equation*

$$w'(t) + \varphi(w(t)) = 0 \quad \forall t \ge 0, \quad w(0) = V(x(0)). \tag{24}$$

Then, we obtain the following estimate:

$$\forall t \ge 0, \quad V(x(t)) \le w(t). \tag{25}$$ ▲

This statement is an obvious consequence of the following Theorem.

Theorem 7. *Let $\Omega \subset R$ be an open interval and $T < +\infty$. We consider a function φ from $[0, T[\times \Omega$ to R satisfying the following properties.*

$$\begin{cases} \text{i) } \forall t \in [0, T[, & x \to \varphi(t, x) \text{ is continuous,} \\ \text{ii) } \forall x \in \Omega, & t \to \varphi(t, x) \text{ is measurable,} \\ \text{iii) } \exists a \in L^1(0, T) & \text{such that, } \forall (t, x) \in [0, T[\times \Omega, \ |\varphi(t, x)| \le a(t). \end{cases} \tag{26}$$

Let v: $[0, T[\to \Omega$ be a continuous function satisfying

$$\forall s \ge t, \quad v(s) - v(t) + \int_t^s \varphi(\tau, v(\tau)) \, d\tau \le 0. \tag{27}$$

Then, there exists a maximal interval $[0, T_1[$ such that the differential equation $w' + \varphi(w) = 0$, $w(0) = v(0)$ has at least one solution w on $[0, T_1[$ satisfying

$$\forall t \ge 0, \quad v(t) \le w(t). \tag{28}$$ ▲

Proof. a) We introduce the subsets

$$K \doteq \{(t, x) \in [0, T[\times \Omega \text{ such that } x \ge v(t)\},$$
$$L \doteq \{(t, x) \in [0, T[\times \Omega \text{ such that } x \le v(t)\}.$$

Since $K \cup L = [0, T[\times \Omega$ and $K \cap L = \{(t, x) | x = v(t)\}$, we can define a function ψ on $[0, T[\times \Omega$ by

$$\psi(t, x) \doteq \begin{cases} \varphi(t, x) & \text{if } x \geq v(t) \quad (\text{i.e., if } (t, x) \in K), \\ \varphi(t, v(t)) & \text{if } x \leq v(t) \quad (\text{i.e., if } (t, x) \in L). \end{cases} \tag{29}$$

b) The function ψ inherits the properties of φ. To see this, we associate with any $t \in [0, T[$ the subsets

$$K(t) \doteq \{x \in \Omega | x \geq v(t)\}, \quad L(t) \doteq \{x \in \Omega | x \leq v(t)\}.$$

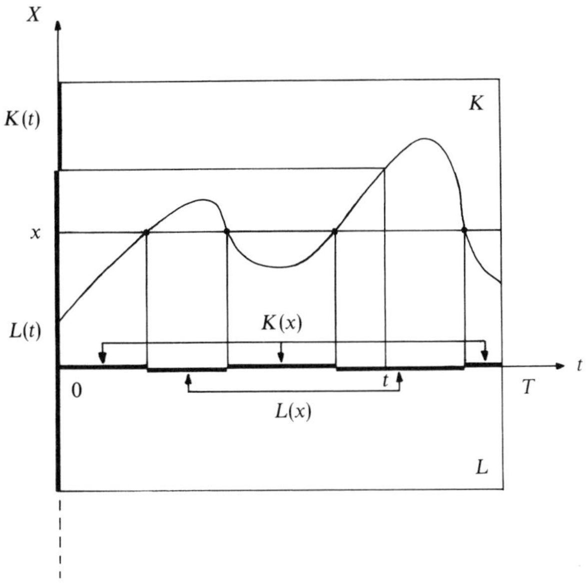

Fig. 29

They are closed and cover Ω. Thus, when $K(t) \neq \emptyset$ (resp. $L(t) \neq \emptyset$), the restriction of $\psi(t, \cdot)$ to $K(t)$ (resp. to $L(t)$) is continuous. Hence we conclude that for all $t \in [0, T[$, $x \to \psi(t, x)$ is continuous. Similarly, we introduce the subsets:

$$K(x) \doteq \{t \in [0, T[| x \geq v(t)\} \quad \text{and} \quad L(x) \doteq \{t \in [0, T[| x \leq v(t)\}.$$

Hence, when $K(x) \neq \emptyset$ (resp. $L(x) \neq \emptyset$), the restriction of $\psi(\cdot, x)$ to $K(x)$ (resp. $L(x)$) is measurable. Consequently, for all $x \in \Omega$, the function $t \to \psi(t, x)$ is measurable. Obviously, $|\psi(t, x)| \leq a(t)$ for all $(t, x) \in [0, T[\times \Omega$.

c) We choose now $T_0 < T$ such that the set of points $(t, x) \in [0, T_0[\times R$ satisfying $|x - x_0| \leq \int_0^t a(\tau) \, d\tau$ is contained in $[0, T[\times \Omega$. Then, by Caratheodory's Theorem, there exists at least a solution $w(\cdot)$: $[0, T_0[\to \Omega$ to the differential equation

$$w'(t)+\psi(t, w(t))=0; \quad w(0)=v(0). \tag{30}$$

d) We assert that for all $t\in[0, T_0]$, $v(t)\leq w(t)$. If not, there would exist $t_2\in]0, T_0]$ such that $v(t_2)>w(t_2)$. Let

$$t_1 \doteq \inf\{s\in[0, T_0] \mid v(t)>w(t) \text{ for all } t\in[s, t_2]\}.$$

By continuity, we have $w(t_1)=v(t_1)$ and $v(t)>w(t)$ for all $t\in]t_1, t_2]$. Since w is a solution to (30), then

$$w(t_2)=w(t_1)-\int_{t_1}^{t_2}\psi(s, w(s))\,ds.$$

Since $v(s)>w(s)$, we deduce that $\psi(s, w(s))=\varphi(s, v(s))$ for all $s\in]t_1, t_2]$. Hence

$$w(t_2)=v(t_1)-\int_{t_1}^{t_2}\varphi(s, v(s))\,ds.$$

By assumption (27), we deduce that $w(t_2)\geq v(t_2)$ (because $t_1<t_2$). This contradicts inequality $v(t_2)>w(t_2)$. Hence $v(t)\leq w(t)$ on $[0, T_0]$. □

Liapunov Functions for U-Monotone Maps

Let $U: K\times K\to R_+$ be a nonnegative continuous function such that $U(y, y)=0$ for all $y\in K$, which plays the role of a semidistance (without having to obey the triangle inequality). We shall associate with U the class of "U-monotone" maps F which enjoy the following property: If we know in advance that $x_*\in K$ is an equilibrium of F, then the "distance function" to x_*: $x\to U(x, x_*)$ is a Liapunov function. When

$$U(x, y)\doteq\frac{1}{2}\sum_{i=1}^{n}|x_i-y_i|^2,$$

the class of U-monotone maps coïncides with the class of monotone maps in the usual sense.

We list some examples of functions U,

$$U_p(x, y)\doteq\frac{1}{p}\sum_{i=1}^{n}|x_i-y_i|^p \quad (1\leq p<+\infty) \tag{31}$$

and, in particular, for $p=2$:

$$U_2(x, y)\doteq\frac{1}{2}\sum_{i=1}^{n}|x_i-y_i|^2. \tag{32}$$

We mention also

$$U_\infty(x, y)\doteq \max_{i=1,\dots,n}|x_i-y_i| \tag{33}$$

and if $K = \mathring{R}^n_+$,

(34) $$U_0(x, y) \doteq \max_{i=1,\ldots,n} x_i/y_i - \min_{i=1,\ldots,n} x_i/y_i.$$

We associate with U the function U' defined on $K \times K \times X$ by

(35) $$U'(x, y)(v) \doteq D_+(x \to U(x, y))(x)(v).$$

Now, we introduce the class of U-monotone set-valued maps.

Definition 2. Let $U\colon K \times K \to R_+$ be a continuous function such that $U(y, y) = 0$ for all $y \in K$. We say that a set-valued map from K to R^n is *U-monotone* if

(36) $$\forall x, y \in K, \quad u \in F(x), \quad v \in F(y), \quad U'(x, y)(v - u) \leq 0.$$

We say that F is *strictly U-monotone* if

(37) $\forall x, y \in K$ such that $U(x, y) > 0$, $\forall u \in F(x)$, $v \in F(y)$, $U'(x, y)(v - u) < 0$

and *strongly U-monotone* if there exists $c > 0$ such that

(38) $$\forall x, y \in K, \quad u \in F(x), \quad v \in F(y), \quad U'(x, y)(v - u) + cU(x, y) \leq 0.$$

Finally, if φ is a bounded continuous map from R_+ to R_+, we say that F is (φ, U)-monotone if

(39) $$\forall x, y \in K, \quad u \in F(x), \quad v \in F(y), \quad U'(x, y)(v - u) + \varphi(U(x, y)) \leq 0.$$

Examples of U-Monotone Operators

When we take $K = R^n_+$ and U_2 as in (32), we see that F is (φ, U_2)-monotone if

(40) $$\forall x, y \in K, \quad u \in F(x), \quad v \in F(y), \quad \langle u - v, x - y \rangle \geq \varphi(U_2(x, y)).$$

So, usual monotone maps are the U_2-monotone maps.

When $1 \leq p < \infty$, we set U_p as in (31) and

(41) $$I_0(x, y) \doteq \{i \mid x_i = y_i\}.$$

Therefore, F is (φ, U_p)-monotone if and only if, for all $x, y \in R^n$, $u \in F(x)$, $v \in F(y)$, we have

(42) $$\sum_{i \notin I_0(x, y)} |x_i - y_i|^{p-2}(x_i - y_i)(u_i - v_i) \geq \varphi(U_p(x, y)).$$

When $p = \infty$, we set U_∞ as in (33) and

(43) $$J(x, y) \doteq \{i \mid |x_i - y_i| = \max_{i=1,\ldots,n} |x_i - y_i|\}.$$

Then F is (φ, U_∞)-monotone if and only if for all $x, y \in R^n$, $u \in F(x)$, $v \in F(y)$, we have

$$\min_{j \in J(x,y)} (u_i - v_i) \geq \varphi(U_\infty(x, y)). \tag{44}$$

Finally, if x and $y \in \mathring{R}^n_+$, we set

$$U_0(x, y) \doteq \max_{i=1,\ldots,n} \frac{x_i}{y_i} - \min_{i=1,\ldots,n} \frac{x_i}{y_i} \tag{45}$$

and

$$K_+(x, y) \doteq \left\{ i \,\middle|\, \frac{x_i}{y_i} = \max_j \frac{x_j}{y_j} \right\}, \quad K_-(x, y) \doteq \left\{ i \,\middle|\, \frac{x_i}{y_i} = \min_j \frac{x_j}{y_j} \right\}. \tag{46}$$

(Note that $U_0(x, y) = 0$ if and only if $\exists \lambda > 0$ such that $x = \lambda y$.)

Then F is (φ, U_0)-monotone if and only if for all $x, y \in \mathring{R}^n_+$, $u \in F(x)$, $v \in F(y)$,

$$\min_{i \in K_+(x,y)} \frac{u_i - v_i}{y_i} - \max_{j \in K_-(x,y)} \frac{u_j - v_j}{y_j} \geq \varphi(U_0(x, y)). \tag{47}$$

Proof. The above characterizations follow obviously from the computation of the contingent derivatives of $x \to U(x, y)$. Since these functions are convex and continuous, the contingent derivatives coïncide with the directional derivatives from the right.

When $1 \leq p < +\infty$, we set $I_+(x, y) \doteq \{i \mid x_i > y_i\}$ and $I_-(x, y) \doteq \{i \mid y_i < x_i\}$. Then it is clear that

$$\begin{aligned} U_p'(x, y)(v) &= \sum_{i \in I_+(x,y)} (x_i - y_i)^{p-1} v_i - \sum_{i \in I_-(x,y)} (y_i - x_i)^{p-1} v_i \\ &= \sum_{i \notin I_0(x,y)} |x_i - y_i|^{p-2} (x_i - y_i) v_i. \end{aligned}$$

If $p = 1$, $U_1'(x, y)(v) = \sum_{i=1}^n v_i$ and, if $p = \infty$, we have

$$U_\infty'(x, y)(v) = \max_{j \in J(x,y)} v_j.$$

Finally, for $p = 0$, we have

$$U_0'(x, y)(v) = \max_{i \in K_+(x,y)} \frac{v_i}{y_i} - \min_{i \in K_-(x,y)} \frac{v_i}{y_i}. \qquad \square$$

The U-monotone maps enjoy the following fundamental properties.

Corollary 1. *Let U be a continuous nonnegative function from $K \times K \to R^n$ such that $U(y, y) = 0$ for all $y \in K$. Let F be a proper bounded upper semicontinuous map from K to the compact convex subsets of R^n. We posit the following assumptions*

$$\begin{cases} \text{i) there exists an equilibrium } x_* \in F \text{ of } F, \\ \text{ii) } -F \text{ is } (\varphi, U) \text{ monotone.} \end{cases} \tag{48}$$

Let $w \in \mathscr{C}(0, \infty; R)$ *be a solution to the differential equation*

$$(49) \qquad w' + \varphi(w) = 0; \quad w(0) = U(x_0, x_*), \quad x_0 \text{ given in } K.$$

Then there exists a solution $x(\cdot)$ *to the differential inclusion* $x' \in F(x)$, $x(0) = x_0$ *such that* $t \to U(x(t), x_*)$ *is nonincreasing, such that* $\int_0^\infty \varphi(U(x(t), x_*))\, dt < +\infty$ *and such that*

$$(50) \qquad \forall t \geq 0, \quad U(x(t), x_*) \leq w(t). \qquad \blacktriangle$$

Proof. We set $V(x) \doteq U(x, x_*)$ and $W(x, v) \doteq \varphi(U(x, x_*))$. Since $0 \in F(x_*)$, then, for all $v \in F(x)$, we get $D_+ V(x)(v) + W(x, v) = U'(x, x_*)(v - 0) + \varphi(U(x, x_*))$. The right-hand side of this inequality is nonpositive since $-F$ is (φ, U)-monotone. Hence V is a Liapunov function for F with respect to W. So, we apply Theorem 7. □

We mention now a result on asymptotic stability. For simplicity, we prove only a special case.

Corollary 2. *Let* $\Omega \subset R^n$ *be an open convex subset of* R^n *and* $U: \Omega \times \Omega \to R_+$ *be a nonnegative continuous function such that*

$$(51) \qquad \begin{cases} \text{i) } \forall y \in \Omega, & U(y, y) = 0, \\ \text{ii) } \forall y \in \Omega, & x \to U(x, y) \text{ is convex.} \end{cases}$$

Let K *be a closed subset of* Ω *and* F *be a bounded upper semicontinuous map from* K *to the compact convex subsets of* R^n. *We assume also that*

$$(52) \qquad \begin{cases} \text{i) there exists an equilibrium } x_* \in K \text{ of } F, \\ \text{ii) } \forall x \in K, \quad F(x) \subset T_K(x), \\ \text{iii) } -F \text{ is strictly } U\text{-monotone.} \end{cases}$$

Then, for any $x_0 \in K$, *there exists a solution to the differential inclusion* $x' \in F(x)$, $x(0) = x_0$ *satisfying*

$$(53) \qquad \lim_{t \to \infty} U(x(t), x_*) = 0. \qquad \blacktriangle$$

Proof. We take for Liapunov function the function $\hat{V}$ defined on Ω by $\hat{V}(x) \doteq U(x, x_*)$, where x_* is an equilibrium of F. Since $\hat{V}$ is convex and continuous, then $D_+ \hat{V}(x)(v)$ is upper semicontinuous with respect to (x, v). It is a Liapunov function since $D_+ \hat{V}(x)(v) = U'(x, x_*)(v - 0) \leq 0$ because $v \in F(x)$, $0 \in F(x_*)$ and $-F$ is U-monotone. Also, let us assume that there exists $v \in F(x)$ such that $D_+ \hat{V}(x)(v) \geq 0$. Then $U'(x, x_*)(v - 0) > 0$ and, since $v \in F(x)$, $0 \in F(x_*)$ and $-F$ is strictly monotone, we deduce from (37) that $V(x) \doteq U(x, x_*) = 0 = \min_{y \in K} V(y)$. Therefore assumptions of Theorem 5 are satisfied and thus, any monotone trajectory satisfies property (36). Such a trajectory does exist by Theorem 3.1. □

Comments

Chapter 1

The definitions and most results of Section 1 are classical. Some related material appears in the book by Berge [1959]. Theorem 1.2 is stated without proof (and without the assumption of completeness of Y) in Choquet [1948]. For the history of the concepts of continuity of set valued maps we refer to the forthcoming book by Rockafellar and Wets. For Theorem 2.2 we refer to the book by Spanier [1966]. Proposition 2.2 is taken from Aubin [1979c], while Proposition 2.3 comes from the book by Ekeland and Teman [1974]. Theorems 2.4 and 2.5 are well known theorems from Berge [1959]. The important results of Section 3 were obtained independently by Robinson [1976a] and Ursescu [1975].

Several partial versions of the convergence theorem of Section 4 were used essentially by Filippov [1959], more explicitly by Lasota and Opial [1965] who actually constructed the coefficients of the convex combinations and received an abstract formulation by Castaing [1966].

Example 1 of Section 6 appears in Filippov [1967]. Theorem 2 of Section 7 is of Ekeland-Valadier [1971]. An analogue for lipschitzean maps admitting a parameterization $f(x, u)$ lipschitzean in x appears in Le Donne and Marchi [1980]. The construction of Section 9 is derived from Lojasiewicz, Plis and Suares [1979]. Section 10 is due to Cornet (private communication). The selection theorem of Section 11 was proved by Michael [1965a]. Several extensions and refinements appear in Michael [1956b; 1957; 1959]. The approximate selection theorem and the proof of Kakutani's Theorem appear in Cellina [1969a; b; c]. The useful concept of σ-selectionable maps as well as Theorem 13.1 are due to Haddad and Lasry [1983].

The first measurable selection theorem became known as Filippov's Lemma from its use in Filippov [1959]. Several mathematicians contributed in extending the validity of this theorem to more general situations. The construction we present is that of Kuratowski and Ryll-Nardzewski [1965]. Two excellent surveys exist on the subject of measurable selections, one by Wagner [1977] and the other, concerned with the Russian literature, by Ioffe [1978].

Additional results about selections concern the link about the theorems of Sections 11 and 14. In Cellina [1976] it is proved that a convex valued map $F(t, x)$, separately upper semicontinuous in t and lower semicontinuous in x admits a selection separately measurable in t and continuous in x. This result was extended by Castaing [1976] to maps measurable in t and continuous in x under the assumption of joint measurability in the product (t, x).

Not considered in this chapter is the problem of the existence of an extension of a map F defined on some subset A of a metric space X i.e. the analogue for set valued maps of the theorems of Tietze and of Dugundji [1951]. Theorems about the existence of an extension of an upper semicontinuous convex valued map were given by Cellina [1969c] and Ma [1972].

A theorem about existence of an extension for a continuous but not necessarily convex valued map was proved by Antosiewicz and Cellina [1977].

Chapter 2

In two papers, by Marchaud [1934] and Zaremba [1936], equations of the kind $Dx(t) \subset F(t, x)$ were investigated, where D was interpreted as the contingent or paratingent derivative, i.e. the set of cluster points of the quotient ratios. Results on the existence of solutions of these equations were presented under the assumption of convexity and of continuity in the paper of Marchaud, of upper semicontinuity in the paper of Zaremba. A solution to a contingent equation was meant to be a continuous $x(\cdot)$ such that for every t in an interval I, $Dx(t) \subset F(t, x(t))$. Wazewski [1961a, b] proved that when F is compact and convex valued and upper semicontinuous jointly in (t, x), $x(\cdot)$ is a solution to the contingent equation if and only if $x(\cdot)$ is absolutely continuous and for almost all t, $x'(t) \in F(t, x(t))$. Meanwhile, differential inclusions with right hand sides had been considered by Filippov in [1959] in the context of a problem of optimal control and in [1960] as a tool for investigating equations with discontinuous right hand side. The theory was extended to cover the Caratheodory conditions, i.e., a) the map $(t, x) \to F(t, x)$ is upper semicontinuous in x for every fixed t and measurable in t for every fixed x and b) there exists an integrable function $\gamma(t)$ bounding the norm of $F(t, x)$ for all x, by Plis [1965], Lasota and Opial [1965], Castaing [1966]. All these proofs used in a way or another a result of the kind of the Convergence Theorem provided in Chapter 1. The idea of passing to the limit from results known from ordinary differential equations appears in Cellina [1970a]; the integral representation for the solutions was used by Ghouila-Houri [1965]. The differential inclusion appearing in the section on the regularization of a differential equation with discontinuous right hand side was introduced by Filippov [1960] and solutions to (1.9) are called solutions of (6) in the sense of Filippov. Since the synthesis of optimal controls is a main source of differential equations with discontinuous right hand side, a natural question that arose was whether the time optimal solutions of a linear control problem were solutions of the corresponding optimal synthesis in the sense of Filippov. Counterexamples are provided in Hermes [1965] and Brunovski [1974]. For further results in this direction see Kurzweil [1977].

The closure of the set of trajectories for a differential inclusion with upper semicontinuous convex valued right hand side was known since Filippov [1959]. That the map $\xi \to A_T(\xi)$ enjoys all the main properties of maps with convex images was pointed out by Cellina [1969a]. The fixed point property was proved in Cellina [1971b]. A proof of the connectedness of the set of trajectories appears in Cellina [1971a]. Aronszajn [1942] proved that the set of solutions to an ordinary differential equation without uniqueness is homeomorphic to the intersection of a decreasing sequence of absolute retracts. Theorem 2.3 extends and makes precise this statement to differential inclusion and appears in Haddad [1981a]. The proof of Hukuhara's theorem presented in Section 2 is derived from a reasoning of Davy [1972].

Theorem 1 of Section 3 is in a fundamental paper of Filippov [1971]. Theorem 2 appears in Kikuchi-Tomita [1971] and, with a somewhat different definition in Hermes [1971]. Theorem 1 of this section was extended to the Caratheodory conditions by Kaczynski and Olech [1974] and by Antosiewicz and Cellina [1975]. An existence theorem under Lipschitz-Caratheodory conditions is presented in Himmelberg and Van Vleck [1973]. Theorem 3 is in Filippov [1976]. Olech [1975] proved the following theorem: let $F(t, x)$ be measurable in t for every fixed x, and upper semicontinuous in x for every fixed t, and let there exist an integrable function λ such that $\|F(t, x)\| \leq \lambda(t)$ for all x. Assume moreover that F is continuous in x at x^* for every fixed (t^*, x^*) such that $F(t^*, x^*)$ is not convex. Then the differential inclusion $x'(t) \in F(t, x(t))$ admits at least one solution.

The investigation of the relations between the set of solutions of a differential inclusion and of its convexified appears in Wazewski [1962]. In it a map $x(\cdot)$, uniform limit of a sequence of functions $k_i(\cdot)$, such that $d(k_i'(t), F(t, k_i(t))) \to 0$ for almost all t, is called a quasitrajectory. It is proved that whenever $F(t, x)$ is *continuous*, every solution of $x'(t) \in F(t, x(t))$ is the limit of a sequence of quasitrajectories of $x'(t) \in \operatorname{co} F(t, x(t))$. Note that the above result is not equivalent to Theorem 2, since no estimate is provided about the distance of a quasitrajectory to a true solution; and this estimate could not be provided for a continuous F. Under Lipschitz conditions the missing estimate is what we called the Gronwall inequality for differential inclusion, Theorem 1, and was provided by Filippov [1967]. The example that concludes Section 4 is of Plis [1962]. A generalization of Theorem 2 is given by Pianigiani [1977].

The relation between the set of solutions to a differential inclusion and the set of solutions to its convexified analogue was further investigated by Cellina [1980] who proved that the set of solution to $x' \in \{-1, +1\}$ is a subset of the second category (in the metric of $C(I)$) of the set of solution to

$x' \in [-1, +1]$. This idea was used as a tool by De Blasi and Pianigiani [1981] to yield existence for a particular class of inclusions in an infinite dimensional Banach space.

Kakutani's Theorem was used for the first time to prove existence for convex valued right hand sides (Theorem 1 of Section 4) by Ghouila-Houri [1965] and Lasota and Opial [1965]. Attouch and Danlamian [1972] by the same theorem proved existence for a right hand side sum of a maximal monotone and of a convex valued upper semicontinuous map. The properties of the integral operator for the convex valued case were further studied by Lasota and Olech [1966] who proved that whenever $(x_n, x_n) \in \operatorname{graph} J$; $x_n \to x$ in $C(I)$ and $y_n \to y$ in the sense of distributions, then $(x, y) \in \operatorname{graph} J$.

The definition of a continuous partition of an interval and Theorem 4.2 on the existence of a selection is due to Antosiewicz and Cellina [1975]. Pianigiani [1977] uses the same approach to investigate some qualitative properties of the set of solution without convexity assumptions.

Extensions of this approach to cover the lower semicontinuous case are given by Bressan [1980] and Lojasiewicz [1980] with proofs somewhat different from ours. The same selection approach but using a construction that yields a selection continuous from L^1 into L^1 to cover differential inclusions whose right hand side in the sum of a maximal monotone map and of a non convex perturbation is used by Cellina and Marchi [1982]. Kisielewicz [1980] uses the same construction for studying functional differential inclusions.

Chapter 3

The concept of monotone map was introduced by Zarantonello [1960] and the concept of maximal monotone map by Minty [1962], who proved Theorem 3.1.1. Existence and uniqueness of the solution to the differential inclusion $x'(t) \in -Ax(t)$, $x(0) = x$ where A is maximal monotone was obtained by Crandall and Pazy [1969] after a pioneering work of Kimura [1967]. A detailed exposition of this subject can be found in the book by Brézis [1973].

The recent developments in the ergodic theory of non linear mappings started with the result of Baillon [1975] for the case of Cesaro means (first statement of Corollary 3.1.1) followed soon after by the extension to the continuous case by Baillon and Brézis [1976]. Extensions to more general averages were given by Reich [1977]. We follow the exposition by Pazy [1981] of these results in a simple unified way.

Theorem 3.4.3 is due to Brézis and Ekeland [1975]. The treatment of gradient methods of Section 3.5 was proposed in Aubin [1981 b]. The application to the convergent of some dynamical processes to Pareto minima is taken from [1981 d].

Chapter 4

In Bouligand [1932] the author introduced the concept of contingent to a set, which spans what we call the Bouligand contigent cone defined in Section 4.1.

It was Nagumo who, in 1942, proved the necessary and sufficient condition for a differential equation to have viable trajectories. It was later observed that his condition could be rephrased in terms of the contingent cones. Again, this justified a further study of Bouligand cones and its connections with Clarke's tangent cones (see Clarke [1975], Aubin-Clarke [1977], Aubin [1981a], and Cornet [1981b]).

Nagumo's theorem was forgotten and rediscovered several times. Let us mention Crandall [1972], Martin [1973], Hartman [1972] and Yorke [1967] in the case of continuous right-hand side, by Brézis [1970], Bony [1969] and Redheffer [1972], in the case of lipschitzean right-hand side.

The first extension of the Nagumo theorem for upper semicontinuous maps with compact convex images was stated with no proof in Yorke [1969]. Proofs appeared in connection with monotone evolution in Aubin-Cellina-Nohel [1976] under convexity of the preordering and in Haddad [1981a] in the general case. The sufficient condition of viability was proved independently by Gautier [1976]. The idea of using monotone trajectories as a mathematical metaphor for darwinian evolution was proposed in Aubin-Nohel [1976]. We follow the exposition of Haddad's papers in Section 4.2. The concept of contingent derivative of a set-valued map was introduced in Aubin [1981a], Sections 4.4 and 4.5 are taken from this paper. Ekeland's theorem, which plays an important role in many domains, was proved in 1974. A simple proof can be found in Aubin [1977a], Aubin-Siegel [1980] and Brézis-Browder [1976b]. The general Newton's method was studied in Smale [1976c]. Theorem 6.1 is due to Aubin-Clarke [1977] and Theorem 6.2 to Clarke [1975]. An extension to the lower semicontinuous case appears in Bressan [1982]. The rest of this chapter is devoted to differential inclusion with memory. We follow the papers of Haddad [1981a, b and c] who proved all the results we present.

Chapter 5

Section 5.1 on tangent cones to convex sets is taken from Aubin [1979c]. Theorem 5.2.1 on the existence of an equilibrium is a formulation due to Cornet of earlier results of Ky Fan [1972] and Browder [1962]. The proof used Ky Fan's inequality (see Ky Fan [1972]), which plays a fundamental role in nonlinear analysis.

The fixed point theorem for strongly inward selectionable maps in Fréchet space is due to Haddad-Lasry [1983], as well as its application to the existence of periodic solutions of differential inclusions.

The application to game theory can be found in Aubin [1980a], Section 5.5 on dynamical price decentralization is taken from Aubin [1981b]. The existence of solutions to projected differential inclusion is due to the pioneering work of Henry [1973] which was motivated by planning procedures in economics (see also Henry [1972] and [1974]). The connection with differential variational inequalities (Proposition 5.6.2) was observed by Cornet [1981a]. Theorem 5.6.1 on existence of solutions to differential variational inequalities is a particular case of the theorems proved by Attouch-Danlamian [1972] and [1975]. A result with no convexity assumptions of the images of F appears in Cellina-Marchi [1982].

The results on slow solution as well as their application to the construction of planning procedures are due to Cornet [1981a and d]. Theorem 5.7.2 is due to Cornet-Lasry [1976].

Chapter 6

Most of the material presented in this chapter appeared already in Aubin [1981a]. We refer to the books by Yoshizawa [1966] and [1975] for the classical theory of Liapunov second method. Proposition 6.2.1 is due to Moreau [1978]. The proof of Theorem 6.5.7 we provide is due to Antoziewicz (private communication).

Bibliography

Antosiewicz, H.A.

1969 Set valued mappings and differential equations, in *Advances in Differential and Integral equations*. Studies in Appl. Math. n. 5 SIAM, Philadelphia

1971 Stability theory: an overview, in: Transactions of the seventeenth Conference of Army Mathematicians, 335–350

Antosiewicz, H.A., Cellina, A.

1975 a) Continuous selections and differential relations, J. Diff. Eq. *19,* 386–398

1977 a) Continuous extension of multifunctions, Ann. Polonici Mat. *34,* (1977) 107–111

b) Continuous extensions, their construction and their application in the theory of differential equations. International Conference on Differential Equations, Academic Press, New York, 1977

Aronszajn N.

1942 Le correspondent topologique de l'unicité dans la théorie des équations différentielles, Ann. of Math. *43,* 730–738

Arrow, K., Debreu, G.

Existence of an equilibrium for a competitive economy. Econometrica *22,* 265–290

Arrow, K.J., Hahn, F.M.

1971 *General competitive analysis,* Holden-Day, San Francisco

Artstein, Z.

1977 Continuous dependence of solutions of operator equations I. Trans. AMS *231,* 143–166

1978 a) Uniform asymptotic stability via the limiting equations. J. Diff. Eq. *27,* 172–189

b) Relaxed controls and the dynamics of control systems. SIAM J. Opt. and Cont.*16,* 689–701

d'Aspremont, C., Dreze, J.H.

1979 On the stability of dynamic processes in economic theory. *47,* 733–737

d'Aspremont, C., Tulkens, H.

1980 Commodity exchanges as gradient processes. Econometrica *48,* 387–399

Attouch, H.

1979 Famille d'opérateurs maximaux monotones et mesurabilité. Ann. di Mat. pura appl, *120,* 35–111

Attouch, H., Damlamian, A.

1972 On multivalued evolution equations in Hilbert spaces. Israel J. Math. *12,* 373–390

1975 Problèmes d'évolution dans les Hilberts et application. J. Math. pures et appl. *54,* 53–74

Aubin, J.P.

1977 a) *Applied Abstract Analysis.* Wiley-Interscience

b) Evolution monotone d'allocations de biens disponibles. C.R.A.S. 293–296

1978 a) Analyse fonctionnelle non linéaire et applications à l'équilibre économique, Ann. Sc. Math. Québec *2,* 5–47

b) Gradients généralisés de Clarke. Ann. Sc. Math. Québec *2,* 197–252

1979 a) *Applied Functional Analysis.* Wiley-Interscience

b) *Mathematical methods of game and economic theory.* North-Holland

c) Cônes tangents à un sous-ensemble convexe fermé. Ann. Sc. Math. Québec *3,* 63–80

d) Monotone evolution of resource allocation. J. Math. Economics *6,* 43–62

1980 a) Formation of coalitions in a dynamical model where agents act on the environment. Economie et Société, 1583–1594

b) Monotone trajectories of differential inclusions: a Darwinian approach. Methods of Operations Research *37,* 19–40

1981 a) Contingent derivatives of set-valued maps and existence of solutions to nonlinear inclusions and differential inclusions. Advances in Mathematics. Supplementary studies, Ed.L., Nachbin, Academic Press, 160–232

b) A dynamical, pure exchange economy with feedback pricing. J. Economic Behavior and Organizations *2,* 95–127

Aubin, J.P., Cellina, A., Nohel, J.

1977 Monotone trajectories of multivalued dynamical systems. Annali di Matematica pura e applicata. *115,* 99–117

Aubin, J.P., Clarke, F.H.

1977 Monotone invariant solutions to differential inclusions. J. London Math. Soc. *16,* 357–366

1979 Shadow prices and duality for a class of optimal control problems. SIAM J. Opt. and Control *17,* 567–586

Aubin, J.P., Day, R.H.

1980 Homeostatic trajectories for a class of adaptative economic systems. J. Econ. Dyn. Control *2,* 185–203

Aubin, J.P., Ekeland, J.

1980 Second order evolution equation with convex Hamiltonian. Can. Bull. Math. *23,* 81–94

1983 *Applied Nonlinear Analysis.* Wiley-Interscience

Aubin, J.P., Frankowska, H.

1984 Heavy viable trajectories of controlled systems. Cahiers de Math. de la Decision, Université de Paris-Dauphine

Aubin, J.P., Nohel, J.

1976 Existence de trajectories monotones de systèmes dynamiques discrets C.R.A.S. *282,* 267–270

Aubin, J.P., Siegel, J.

1980 Fixed points and stationary points of dissipative multivalued maps. Proceedings of Am. Math. Soc. *78,* 391–398

Aumann, R.J.

1965 Integrals of set-valued functions. J. Math. An. Appl. *12,* 1–12

Baillon, J.B.

1975 Un théorème de type ergodique pour les contractions non linéaires dans une espace de Hilbert. C.R.A.S. *280,* 1511–1514

1976 Quelques proprietés de convergence asymptotique pour les semigroupes de contraction impaires. C.R.A.S. *283,* 75–78 & 587–590

1978 Un exemple concernant le comportement asymptotique de la solution du problème $0 \in \frac{du}{dt} + \partial\varphi(u)$ J. Funct. Anal. *28,* 369–376

Baillon, J.B., Brezis, H.

1976 Une remarque sur le comportement asymptotique des semigroupes non linéaires. Houston J. Math. *2,* 5–7

Baillon, J.B., Haddad, G.

1977 Quelques proprietés des opérateurs angle-bornés et n-cycliquement monotones. Israel J. Math. *26,* 137–150

Baillon, J.B., Lions, P.L.

Convergence de suites de contractions dans un espace de Hilbert. (in: Thèse d'Etat, Baillon, Paris 6, 1978)

Baiocchi, C., Capelo, A.

1978 a) Disequazioni variazionali e quasivariazionali. Applicazioni a problemi di frontiera libera. Vol. I. Problemi Variazionali. Pitagora Editrice, Bologna

b) Vol. II Problemi quasivariazionali. Pitagora Ed., Bologna

Banks, H.T., Jacobs, M.Q.

1970 A differential calculus for multi-functions, J. Math. An. Appl. *29,* 246–272

Barbu, V.
1976 *Nonlinear semigroups and differential equations in Banach spaces,* Noordhoff International Publ., Leiden
Bebernes, J., Schurr, J.
1970 The Ważewski topological method for contingent equations, Ann. Mat. Pura Appl. *87,* 271–280
Berge, C.
1959 *Espaces topologiques et fonctions multivoques.* Dunod, Paris
Blagodatskih, V.I.
1973 a) Sufficient conditions of optimality for contingent equations, in: Lecture Notes in Comp. Sc. *3,* 319–328. Springer-Verlag, Berlin Heidelberg New York
b) Local controllability of differential inclusions, Differ. Uravn. *9,* 361–362
c) The differentiability of solutions with respect to the initial conditions, Differ. Uravn. *9,* 2136–2140
1974 Sufficient optimality conditions for differential inclusions. Izv. Akad. Nauk SSSR *38,* 615–624
1979 Theory of differential inclusions. Part I, Editions of Moskow Univ, Moskow
Blaquière, A., Leitman, G.
1967 On the geometry of optimal processes, in: *Topics in Optimization.* Academic Press, New York, 263–371
Bony J.M.
1969 Principe du maximum, inégalité de Harnack, et unicité du problème de Cauchy pour les opérateurs elliptiques dégénerés. Ann. Inst. Fourier *19,* 277–304
Bouligand, G.
1932 *Introduction à la géométrie infinitesimale directe.* Gauthier-Villars, Paris
Bourgin
1963 *Modern algebraic topology*. MacMillan, New York
Bressan, A.
1960 On differential relations with lower semicontinuous right hand side, J. Diff. Eq. *37,* 89–97
1982 Integrals of lower semicontinuous orientor fields on closed sets, in print
Brézis, H.
1970 On a characterization of flow invariant sets. Comm. Pure Appl. Math. *23,* 261–263
1973 *Opérateurs maximaux monotones et semigroupes de contractions dans les espaces de Hilbert.* North Holland, Amsterdam
1979 Asymptotic behavior of some evolution systems. *Non linear evolution Equations,* Ed. Crandall M.G., Academic Press
Brézis, H., Browder, F.
1976 a) Nonlinear egodic theorems. Bull. AMS *82,* 959–961
b) A general principle on ordered sets in non linear functional analysis. Adv. in Math. *21,* 355–369
Brézis, H., Ekeland, J.
1976 Un principe variationnel associé à certaines équations paraboliques. CRAS *282,* 971–974
Brézis, H., Lions, P.L.
1978 Produits infinis de résolvantes. Israel J. Math. *29,* 329–345
Brézis, H., Nirenberg, L., Stampacchia, G.
1972 A remark on Ky Fan's minimax principle, Boll. Un. Mat. Ital. (4) *6,* 293–300
Bridgland Jr., T.F.
1969 a) Contributions to the theory of generalized differential equations I. Math. Systems Theory *3,* 17–50
b) Contributions to the theory of generalized differential equations II. Math. Systems Theory *3,* 156–165
Browder, F.
1968 The fixed point theory of multivalued mappings in topological vector spaces, Math. Ann. *177,* 183–301

1975 Non linear operators and non linear equations of evolution in Banach Spaces. In: *Nonlinear Functional Analysis,* Proc. Symp. Pure Math. Vol. 118, Part 2, American Math. Soc. Providence

Brunovski, P.

1973 On two conjectures about the closed loop time optimal control. Lect. Notes Comp. Sci. *3,* 341–344

Caristi, J.

1972 Fixed point theorems for mappings satisfying inwardness conditions, Trans. Am. Math. Soc. *78,* 186–197

Castaing, C.

1966 Sur les équations différentielles multivoques. C.R. Acad. Sc. Paris *263,* 63–66

1975 Rafle par un convexe aléatoire à variation continue à droite, Séminaire d'analyse convexe, Montpellier

1976 A propos de l'existence des sélections séparément mesurables et séparément continues d'une multi-application séparément mesurable et séparément semicontinue inférieurement. Séminaire d'analyse convexe, Montpellier

1978 Equation différentielle multivoque avec contraite sur l'état dans les espaces de Banach. Séminaire d'analyse convexe, Montpellier

Castaing, C., Valadier, M.

1968 Equations différentielles multivoques dans les espaces vectoriels localement convexes. C.R. Acad. Sc. *266,* 985–987

1969 Equations différentielles multivoques dans les espaces vectoriels localement convexes. Revue Franc. Inf. Rech. Oper. *16,* 3–16

1977 *Convex Analysis and measurable multifunctions.* Lecture Notes in Mathematics *580.* Springer-Verlag, Berlin Heidelberg New York

Cellina, A.

1969 a) Multivalued Functions and Multivalued Flows, Univ. of Maryland Tech. Note BN 615

b) Approximation of set valued functions and fixed point theorems, Annali di Mat. Pura Appl. *82,* 17–24

c) A theorem on the approximation of compact multivalued mappings Rend. Acc. Naz. Lincei *47,* 429–433

1970 a) Multivalued differential equations and ordinary differential equations. SIAM J. Appl. Math. *18,* 533–538

b) A further result on the approximation of set valued mappings, Rend. Acc. Naz. Lincei *48,* 230–234

1971 a) The role of approximation in the theory of multivalued mappings in *Differential games and related topics,* H.W. Kuhn and G.P. Szegö eds. North-Holland, Amsterdam

b) On mappings defined by differential equations. Zeszyty nauk. Uniw. Jagiellonski *252,* Prace Mat, *15,* 17–19

1976 A selection theorem. Rend. Sem. Univ. Padova *55,* 143–149

1980 On the differential inclusion $x' \in [-1+1]$, Rend. Acc. Naz. Lincei 69, 1-6

Cellina, A., Lasota, A.

1969 A new approach to the definition of topological degree for multivalued mappings. Rend. Acc. Naz. Lincei *47,* 434–440

Cellina, A., Marchi, M.V.

1982 Nonconvex perturbations of maximal monotone differential inclusions. Cahiers de Math. de la Décision, Université de Paris-Dauphine

Champsaur, P.

1975 How to share the cost of a public good? J. of Game Theory *4,* 113–129

1976 Neutrality of planning procedures in an economy with public goods. Review of Econ. Stud. *43,* 273–294

Champsaur, P., Dreze, J.H., Henry, A.

1977 Stability theorems with economic applications, Econometrica *45,* 133–150

Choquet, G.
1948 Convergences, Ann. Univ. Grenoble, Sect. Sc. Math. Phys. *23*, 57–112
Clarke, F.H.
1975 a) Generalized gradients and applications. Trans. of A.M.S. *205*, 247–262
b) The Euler-Lagrange differential inclusion. J. Diff. Eq., *19*, 80–90
1976 a) Optimal solutions to differential inclusions. J. Opt. Th. Appl. *19*, 469–478
b) The maximum principle under minimal hypotheses. SIAM J. Contr. Opt. *14*, 1078–1091
1981 Generalized gradients of Lipschitz functional. Advances in Math. *40*, 52–67
1983 *Optimization and non smooth analysis*, Wiley-Interscience, New York
Conti, R.
1974 *Problemi di controllo e di controllo ottimale*, UTET, Torino
1976 *Linear differential equations and control*, Institutiones Math. n. 1. Academic Press, New York
Conti, R., Sansone, G
1964 *Nonlinear differential equations*. Revised edition, Pergamon Press, New York
Cornet, B.
1977 a) Sur la neutralité d'une procédure de planification, Cahiers du Séminaire d'Econométrie, CNRS, Paris, *19*, 71–81
b) An abstract theorem for planning procedures, Lectures Notes in Economics and Mathematical Systems *144*, 53–59
c) Accessibilité des optima de Pareto par des processus monotones. C.R.A.S., 17 Octobre 1977, Série A, 641–644
1979 Monotone planning procedures and accessibility of Pareto optima, in: *New Trends in Dynamic system theory and economics*, Aoki and Marzollo, Ed., Academic Press, 337–349
1981 a) Existence of flows solutions for a class of differential inclusions, Journal of Math. Analysis and Applications
b) Regular properties of tangent and normal cones. Cahiers de Math. de la Décision, Université de Paris-Dauphine 8130
c) Neutrality of planning procedures. Cahiers de Math. de la Décision, 8132 – Université de Paris-Dauphine
d) Contributions à la théorie mathématique des mécanismes dynamiques d'allocation des ressources. Thèse de doctorat d'etat. Université de Paris-Dauphine
Cornet, B., Lasry, J.M.
1976 Un theorème de surjectivité pour une procedure de planification. CRAS, *282*, 1375–1378
Crandall, M.G.
1970 Differential equations on convex sets. J. Math. Soc. Japan, *22*. 443–455
1972 A generalization of Peano's existence theorem and flowinvariance. Proc. A.M.S., *36*, 151–155
Crandall, M.G., Pazy, A.
1969 Semigroups of nonlinear contractions and dissipative sets. J. Funct. Analysis *3*, 376–418
Crandall, M.G. (Ed.)
1979 *Nonlinear Evolution Equations*. Academic Press, New York
Da Prato, G.
Applications croissantes et équations d'évolutions dans le espaces de Banach. Institutiones Math. Vol 2, Academic Press, New York
Davy, J.L.
1972 Properties of the solution set of a generalized differential equation. Bull. Austral. Math. Soc. *6*, 379–398
Day, R.K., Groves, T. (Eds.)
1975 *Adaptative economic models*. Academic Press
De Blasi, F.S.
1976 a) Existence and stability of solutions for autonomous multivalued differential equations in Banach spaces, Rend. Acc. Naz. Lincei *60*, 767–774
b) On the differentiability of multifunctions, Pacific. J. Math. *66*, 67–81

De Blasi, F.S., Iervolino, F.
1968 Equazioni differenziali con soluzioni a valore compatto convesso. Boll. Un. Mat. Ital. (4) *2*, 491–501; errata corrige ibid (4) *3*, 699
De Blasi, F.S., Lasota, A.
1968 Daniell's method in the theory of the Aumann-Hukahara integral of set-valued functions, Rend. Acc. Naz. Lincei *45*, 252–256
1969 Characterization of the integral of set valued functions. Rend. Acc. Naz. Lincei, *46*, 154–157
De Blasi, F.S., Lopes Pinto, A.J., Iervolino, F.
1970 Uniqueness and existence theorems for differential equations with compact convex valued solutions. Boll. Un. Mat. Ital. (4) *3*, 47–54
De Blasi, F.S., Pianigiani, G.
1982 A category approach to the existence of solutions of multivalued differential equations in Banach spaces. Nota Tecn. Ist. Mat. "U. Dini"
Debreu, G.
1959 *Theory of value*. Wiley New York
1971 A tatonnement process for public goods. Rev. of Econ. Stud. *37*, 133–150
Dugundji, J.
1951 An extension of Tietze's theorem. Pacific J. Math. *1*, 353–367
Dunford, N., Schwartz, J.T.
1958 *Linear operators*. Wiley-Interscience. New York
Edelstein, M.
1972 The construction of an asymptotic center with a fixed point property. Bull. Ann. Math. Soc. *78*, 206–208
Eggleston, H.C.
1958 *Convexity*. Cambridge Tracts in Mathematics *47*. Cambridge University Press, London
Eilenberg, S., Montgomery, D.
1966 Fixed point theorems for multivalued transformations. Am. J. Math. *58*, 214–222
Ekeland, I.
1968 Paramétrisation en théorie de la commande. C.R.A.S. *267*, 98–100
1970 a) On the variational principle. J. Math. Anal. Appl. *47*, 324–353
b) *La théorie des jeux et ses applications à l'économie mathématique*. Presses Univ. France, Paris
1979 *Eléments d'économie mathématique*. Hermann, Paris
Ekeland, J., Temam, R.
1974 *Analyse convexe et problèmes variationnels*. Dunod, Paris
Ekeland, J., Temam, R
1974 Analyse convexe et problèmes variationnels. Dunod, Paris
Ekeland, I., Valadier, M.
1971 Representation of set-valued mappings. J. Math. An. Appl. *35*, 621–629
Fan Ky
1972 A minimax inequality and applications, in: Inequalities III, O Sisha Ed., Academic Press, 103–113
Filippov, A.F.
1959 On certain questions in the theory of optimal control. Vestnik Moskov. Univ. Ser. mat. Mech. Astr. *2*, 25–32. (English Translation: SIAM J. Control *1*, (1962) 76–84)
1960 Differential equations with discontinuous right hand side. Math. Sbornik *51*, 99–128 (English Translation: Transl. Ann. Math. Soc. *42*, (1964) 199–232)
1967 Classical solutions of differential equations with multivalued right hand side. Vestnik, Moskov. Univ. Ser. Mat. Mech. Astr. *22*, 16–26 (English translation: SIAM J. Control, *5* (1967) 609–621)
1971 On the existence of solutions of multivalued differential equations. Mat. Zametki *10*, 307–313
Frankowska, H.
1983 Inclusions adjointes associées aux trajectoires minimales d'inclusions différentielles. C.R.A.S. *297* (461–464)

1984 a) The first order necessary conditions for nonomooth variational and control problems. SIAM I. Cont.
b) Necessary conditions for Bolza problems. Math. Op. Res.
c) A viability approach to Skorohod problem

Frankowska, H., Olech, C.
1982 a) R. Convexity of integral of a set-valued function. Am. Math. J., 117–129
b) Boundary solutions of a differential inclusion. J. Diff. Eq.

Fraser, R.B.
1971 On continuous and measurable selections and the existence of solutions of generalized differential equations. Proc. Amer. Math. Soc. *29,* No. 3, 535–542

Gale, D.
1960 *The theory of linear economic models.* McGraw Hill, New York

Gautier, S.
1976 Equations différentielles multivoques sur un fermé. Publ. Math. de Pau
1978 Différentiabilité des multiapplications. Publ. Math. de Pau

Glodde, B., Niepage, H.D.
1982 Einführung in die mengenwertige Analysis und die Theorie der Kontingentgleichungen. Seminarberichte. Humboldt-Universität zu Berlin

Haddad, G.
1981 a) Monotone trajectories of differential inclusions and functional differential inclusion with memory. Israel J. of Math. *39,* 83–100
b) Monotone viable trajectories for functional differential inclusions. J. Diff. Eq. *42,* 1–24
c) Topological properties of the set of solutions for functional differential inclusions. Nonlinear analysis, Theory, methods, appl. *5,* 1349–1366

Haddad, G., Lasry, J.M.
1983 Periodic solutions of functional differential inclusions and fixed points of σ-Selectionable correspondences. J. Math. Anal. Appl.

Hale, J.
1977 *Theory of functional differential equations.* Springer-Verlag, New York Berlin Heidelberg

Hartman, P.
1965 Generalized Liapunov functions and functional equations. Annali di Mat. Pura Appl. *69,* 305–320
1972 On invariant sets and on a theorem of Ważewski. Proc. Am. Math. Soc. *32,* 511–520
1973 *Ordinary differential equations.* Corrected reprint. S.H. Hartman, Baltimore

Heal, G.
1973 *The theory of economic planning.* North Holland, Amsterdam

Henry Cl.
1972 Differential equations with discontinuous righ-hand side for planning procedures. J. Econ. Theory, *4,* 545–551
1973 An existence theorem for a class of differential equations with multivalued right-hand side. J. Math. Anal. Appl. *41,* 179–186
1974 Problèmes d'éxistence et de stabilité pour des processus dynamiques considerés en économie mathématique. CRAS, *278,* 97–100

Hermes, H.
1965 Discontinuous vector fields and feedback control. Brown Univ. tech. Report. 65–8
1971 On continuous and measurable selections and the existence of solutions of generalized differential equations. Proc. Am. Math. Soc. *29,* 535–542

Himmelberg, C.J.
1972 Fixed points of compact multifunctions. Indiana Univ. Math. J. *22,* 719–729
1975 Measurable relations. Fund. Math. *87,* 53–72

Himmelberg, C.J., Jacobs, M.Q., Van Vleck, F.S.
1969 Measurable multifunctions, selectors and Filippov's implicit function Lemma. J. Math. Anal. Appl. *25,* 276–284

Himmelberg, C.J., van Vleck, F.S.
1971 Selection and implicit function theorems for multifunctions with Souslin graph. Bull. Acad. Pol. Sc. *19,* 911–916
1972 a) Fixed points of semi-condensing multifunctions. Boll. Un. Mat. Ital. (4) *5,* 187–194
b) Lipschitzian generalized differential equations. Rend. Sem. Mat. Padova *48,* 159–169
1976 An extension of Brunovski's Scorza Dragoni type theorem for unbounded set-valued functions. Math. Slovaca *26,* 47–52
Hiriart-Urruty, J.B.
1979 Tangent cones, generalized gradients and mathematical programming in Banach spaces. Math. of Oper. Res. *4,* 79–97
Hukuhara, M.
1930 Sur les systèmes d'équations différentielles ordinaires, II., Jap. J. of Math. *6,* 269–299
1967 a) Sur l'application semi-continue dont la valeur est un compact convexe, Funkcialaj Ekvacioj. *10,* 43–66
b) Intégration des applications mesurables dont la valeur est un compact convexe, Funkcialaj Ekvacioj. *10,* 205–223
Ioffe, A.D.
1978 Survey of measurable selection theorems: Russian literature supplement. SIAM J. Contr. Opt. *16,* 728–723
Ioffe, A.D., Tichomirov
1974 *Theory of extremal problems* (English Translation: North-Holland, Amsterdam, 1978)
Itoh, S., Takahashi, W.
1977 Single-valued mappings, multivalued mappings and fixed-point theorems, J. Math. Anal. Appl. *59,* 514–521
Jacobs, M.Q.
1967 Remarks on some recent extension of Filippov's implicit function Lemma. SIAM J. Control *5,* 622–627
1968 Measurable multivalued mappings and Lusin's theorem. Trans. Am. Math. Soc. *134,* 471–481
1969 On the approximation of integrals of multivalued functions, SIAM J. Control *7,* 158–177
Jarnik, J., Kurzweil, J.
1977 Reducing differential inclusions, in: *Internationale Konferenz über nichtlineare Schwingungen,* Akademie-Verlag, Berlin
Kaczynski, H., Olech, C.
1974 Existence of solutions of orientor fields with non-convex right-hand side. Ann. Polon. Math. *29,* 61–66
Kato, T.
1967 Nonlinear semigroups and evolution equations. J. Math. Soc. Japan *19,* 508–520
Kikuchi, N.
1967 On some fundamental theorems of contingent equations in connection with the control problems. Publ. RIMS Kyoto Univ. Ser. A *3,* 177–201
Kikuchi, N., Tomita
1971 On the absolute continuity of multifunctions and orientor fields, Funkc. Ekvacioj *14,* 161–170
Kisielewicz M.
1980 Existence theorem for generalized differential equations of neutral type. J. Math. An. Appl. *78,* 173–182
Komura, Y.
1967 Nonlinear semigroups in Hilbert spaces. J. Math. Soc. Japan *19,* 498–507
Kuratowski, K.
1958 Topologie, Vol. I and II, 4th. edition corrected. Panstowowe Wyd. Nauk, Warszawa
Kuratowski, K., Ryll-Nardzewski, C.
1965 A general theorem on selectors. Bull. Acad. Pol. Sc. *13,* 397–403

Kurzanskii, A.B.
1970 The existence of solutions of equations with aftereffect. Diff. Urav. *6,* 1800–1809
Kurzweil, J.
1977 On differential relations and on Filippov's concept of differential equations; in: *Diff. Equations,* Almqvist & Wiksel, Stockholm
Laborde, P.
1981 a) Sur un problème d'évolution non monotone C.R.A.S. *292,* 319–322
b) Processus de rafle non convexe. C.R.A.S.
1982 Approximation en viscoplasticité. J. de Math. Pures et Appl.
Lasota, A., Olech, C.
1966 On the closedness of the set of trajectories of a control system. Bull. Acad. Pol. Sc. *14,* 615–621
1968 On Cesari's semicontinuity condition for set valued mappings, Bull. Acad. Pol. Sc. *16,* 711–716
Lasota, A., Opial, Z.
1965 An application for the Kakutani-KyFan theorem in the theory of ordinary differential equations. Bull. Acad. Pol. Sc. *13,* 781–786
1968 Fixed point theorems for multivalued mappings and optimal control problems. Bull. Acad. Pol. Sc. *16,* 645–649
Lasry, J.M., Robert, R.
1976 a) Acyclicité de l'ensemble des solutions de certaines équations fonctionnelles. C.R.A.S. *282,* 1283–1286
b) Analyse nonlinéaire multivoque, Cahiers de Math. de la Décision 7611, Université de Paris-Dauphine
Le Donne, A., Marchi, M.V.
1980 Representation of Lipschitzean compact convex valued mappings. Rend. Acc. Naz. Lincei *68,* 278–280
Lee, E.B., Markus, L.
1967 Foundations of optimal control theory. Wiley, New York
Leela, S., Mauro
1978 Existence of solutions in a closed set for delay differential equations in Banach spaces. J. Nonlinear Analysis TMA *2,* 391–423
Lions, J.L.
1968 *Contrôle optimal de systèmes gouvernés par des équations aux derivées partielles.* Dunod, Paris
1969 *Quelques méthodes de résolution de problèmes non linéaires.* Dunod, Paris
Lions, P.L.
1977 Approximation de point fixe de contractions. C.R.A.S. *284,* 1357–1359
1978 a) Products infinis de resolvantes. Israel J. of Math. *29,* 329–345
b) Une méthode iterative de résolution d'une inéquation variationnelle. Israel J. Maths. *31,* 204–208
Lions, P.L., Mercier, B.
1979 Splitting algorithms for the sum of two nonlinear operators. SIAM J. Num. Anal. *16,* 964–979
Lojasiewicz Jr., S.
1980 The existence of solutions for lower semicontinuous orientor fields. Bull. Acad. Pol. Sc. *28,* 483–487
Lojasiewicz Jr., S., Plis, A., Suarez, R.
1979 Necessary conditions for nonlinear control systems. Inst. of Math. PAN Preprint 139
Lyapunov, A.
1910 Problème général de la stabilité du mouvement. Annales de la Faculté des Sciences de l'Université de Toulouse. *9,* 27–474

Ma, T.W.

1972 Topological degree of set valued compact fields in locally convex spaces. Dissertationes Math. (Rozprawy Mat.) *92,* 1–47

Malinvaud, E.

1970 Procédures pour la détermination d'un programme de consommation collective. European Econ. Rev. *2,* 187–217

1971 A planning approach to the public goods problem. Swedish J. Econ. *73,* 96–112

1972 Prices for individual consumption, quantity indicators for collective consumption. Rev. Econ. Stud. *39,* 385–406

Marchaud, H.

1938 Sur les champs de demi-cônes et les équations différentielles du premier ordre. Bull. Sc. Math. *62,* 1–38

Markin, J.T.

1973 Continuous dependence of fixed point sets, Proc. Amer. Math. Soc. *38,* 545–547

Martin, R.M.

1973 Differential equations on closed subsets of a Banach space. Trans. AMS *179,* 399–414

1976 *Nonlinear operators and differential equations in Banach spaces.* Wiley Interscience, New York

Michael, E.

1956 a) Continuous selections I. Annals of Math. *63,* 361–381

b) Continuous selections II. Annals of Math. *64,* 562–580

1957 Continuous selections III. Annals of Math. *65,* 375–390

1959 A theorem on semicontinuous set valued functions. Duke Math. J. *26,* 647–651

Minty, G.

1962 Monotone (nonlinear) operators in a Hilbert space. Duke Math. J *29,* 341–348

1964 On the monotonicity of the gradient of a convex function. Pacific J. Math. *14,* 243–247

1965 A theorem on maximal monotone sets in Hilbert space. J. Math. Anal. Appl. *11,* 434–439

Moreau, J.J.

1971 Rafle par un convexe variable I. Sem. Analyse Convexe Montpellier. Ex. n. 15

1972 Rafle par un convexe variable II. Sem. Analyse Convexe Montpellier. Ex. n. 3

1977 Evolution problem associated with a moving convex set in a Hilbert space. J. Diff. Eq. *26,* 347–374

1978 Un cas de convergence des iterés d'une contraction d'un espace Hilbertien. C.R.A.S. *286,* 143–144

Mosco, U.

1969 Convergence of convex sets and of solutions of variational inequalities. Adv. in Math. *3,* 510–585

Nadler Jr., S.B.

1970 Some results on multi-valued contraction mappings, Lect Notes Math. *171,* 64–69

Nagumo, M.

1942 Über die Lage der Integralkurven gewöhnlicher Differentialgleichungen. Proc. Phys. Math. Soc. Japan *24,* 551–559

Negishi, T.

1962 The stability of a competitive economy: A survey article. Econometrica *30,* 635–669

Nikaido, H.

1968 *Convex structures and economic theory.* Academic Press, New York

Olech, C.

1965 a) A note concerning set-valued measurable functions. Bull. Acad. Pol. Sc. *13,* 317–321

b) A note concerning extremal points of a convex set. Bull. Acad. Pol. Sc. *13,* 347–351

1967 Lexicographical order, range of integrals and "bang-bang" principle, in: *Mathematical Theory of Control.* Academic Press. New York, 35–45

1968 Approximation of set-valued functions by continuous functions. Colloq. Math. *19,* 285–293

1969 Existence theorems for optimal control problems involving multiple integrals. J. Diff. Eq. *6,* 512–526
1975 Existence of solutions of non convex orientor fields. Boll. Un. Mat. It. (4) *11,* 189–197
Pazy, A.
1977 On the asymptotic behavior of iterates of nonexpensive mappings in Hilbert spaces. Israel J. Math. *26,* 197–204
1978 On the asymptotic behavior of semigroups of nonlinear contractions in Hilbert spaces. J. Func. Anal. *27,* 293–307
1979 Remarks on nonlinear ergodic theory in Hilbert spaces. Nonlinear Analysis TMA *3,* 863–871
Pchenitchny, B.W.
1980 *Convex analysis and extremal problems* (in russian). Nauka, Moscow
Penot, J.P.
1981 A characterization of tangential regularity. J. Nonlinear Analysis. T.A.M. *5,* 625–643
Pianigiani, G.
1977 On the fundamental theory of multivalued differential equations. J. Diff. Eq. *25,* 30–38
Plis, A.
1962 Trajectories and quasi-trajectories of an orientor field. Bull. Acad. Pol. Sc. *10,* 529–531
1965 Measurable orientor fields. Bull. Acad. Pol. Sc. *13,* 565–569
Pontryagin, L.S. et al.
1962 *The mathematical theory of optimal processes* (transl. Wiley Interscience, New York)
Reder, C.
1982 Familles de convexes invariantes et équations de diffusion-réaction. Annales de l'Inst. Fourier
Redheffer, R.M.
1972 The theorems of Bony and Brézis on flow invariant set. Am. Math. Month. *79,* 790–797
Reich, S.
1977 Nonlinear evolution equations and non linear ergodic theorems. Nonlinear Analysis TMA *1,* 319–330
Robinson, S.
1976 a) Regularity and stability for convex multivalued functions. Math. Op. Res. *1,* 130–143
b) Stability theory for systems of inequalities Part II: differentiable nonlinear systems. SIAM J. Num. Analysis. *13,* 497–513
1980 Generalized equations and their solutions. Part II: applications to nonlinear programming. Cahiers de Math. de la Décision 8006, Université de Paris-Dauphine
Rockafellar, R.T.
1966 Characterization of the subdifferentials of convex functions. Pacific J. Math. *17,* 497–510
1969 Convex functions, monotone operators and variational inequalities, in: *Theory and applications of monotone operators.* Oderisi, Gubbio, 35–65
1970 a) *Convex Analysis.* Princeton Math. Series. n. 28, Princeton University Press, Princeton
b) On the maximal monotonicity of subdifferential mappings. Pacific J. Math. *33,* 209–216
c) Generalized Hamiltonian equation for convex problems of Lagrange. Pacific J. Math. *33,* 411–427
Rockafellar, R.T., Wets, R.J.-B.
1983 *Extended real analysis*
Sansone, G.
1941 *Equazioni differenziali nel campo reale.* Zanichelli, Bologna
Schecter, S.
1977 Accessibility of Pareto optima in pure exchange economies. J. Math. Econ. *4,* 197–216
Seifert, G.
1976 Positively invariant closed sets for systems of delay differential equations. J. Diff. Eq. *22,* 292–304
Smale, S.
1976 a) Dynamics in general equilibrium theory. American Econ. Rev. *66,* 288–294
b) Exchange processes with price adjustment. J. Math. Econ. *3,* 211–216
c) A convergent process of price adjustment and global Newton method

Spanier, E.
1966 *Algebraic topology*. McGraw Hill, New York

Tolstonogov, A.A.
1979 Differential inclusions in Banach spaces and continuous selectors. Dokl. Akad. Nauk SSSR *244*, 1088–1092

Treves, F.
1967 *Topological vector spaces, distributions and kernels*. Academic Press, New York

Tulkens, H.
1978 Dynamic processes for public goods. J. Public Economy *9*, 163–201

Tulkens, H., Zamir, S.
Local games in dynamic exchange processes. Rev. of Econ. Stud.

Turowicz, A.
1962 Sur les trajectoires et quasitrajectoires des systèmes de commande nonlinéaires, Bull. Acad. Polon. Sci., Ser. Sci. Math. Astron. Phys. *10*, 529–531

Ursescu, C.
1975 Multifunctions with closed convex graph. Czecos. Math. J. *25*, 438–441

Valadier, M.
1971 a) Multi applications mesurables à valeurs convexes compactes. J. Math. Pures Appl. *50*, 265–297
b) Existence globale pour les équations différentielles mutivoques. *272*, 474–477

Vitali, G., Sansone, G.
1943 *Moderna teoria delle funzioni di variabile reale*. Zanichelli, Bologna

Wagner, D.M.
1977 Survey of measurable selection theorems. SIAM J. Contr. Opt. *15*, 859–903

Walras, L.
1974 *Eléments d'économie politique pure*. Corbez, Lausanne

Wazewski, T.
1961 a) Systèmes de commande et équations au contingent. Bull. Acad. Pol. Sc. *9*, 151–155
b) Sur une condition équivalente à l'équation au contingent. Bull. Acad. Pol. Sc. *9*, 865–867
c) Sur la semicontinuité inférieure du "Tendeur" d'un ensemble compact variant d'une façon continue. Bull. Acad. Pol. Sc. *9*, 869–872
1962 Sur une généralization de la notion des solutions d'une équation au contingent. Bull. Acad. Pol. Sc. *10*, 11–15
1964 On an optimal control problem. Proc. Conference "Differential equations and their applications". Prague 1964, 229–242

Yorke, J.A.
1967 Invariance for ordinary differential equations. Math. Systems Theory *1*, 353–372
1969 Invariance for contingent equations, in: Lecture Notes in Operations Research and Math. Economics *12*, 379–381. Springer-Verlag, Berlin Heidelberg New York
1970 Differential inequalities and non-Lipschitz scalar functions. Math. Systems Theory *4*, 140–153

Yoshizawa, T.
1966 *Stability theory by Liapunov's second method*. The Mathematical Society of Japan, Tokyo
1975 *Stability theory and the existence of periodic solutions and almost periodic solutions*. Springer-Verlag, New York Heidelberg Berlin

Yosida, K.
1974 *Functional Analysis*. Springer-Verlag, New York Heidelberg Berlin

Zarantonello, E.
1960 Solving fonctional equations by contraction averaging. Math. Res. Center Rep. n. 160. University of Wisconsin, Madison

Zaremba, S.C.
1936 Sur les équations au paratingent. Bull. Sc. Math. *60*, 139–160

Index

Absolutely continuous, function 10
– –, set valued map 115
Alaoglu 13
Allocation 246
Almost cluster point 311
– limit 311
Approximate selection 84–86
– Selection Theorem 84–85, 105
Arrow-Debreu 248
Ascoli-Arzelà 13
Asymptotic center 15–17
– stability 314
Attainable set 95, 104

Banach's Open Mapping Principle 55
Barrier cone 30
Barycenter 77
Best Approximation 21–28
Bouligand 174
Bouligand's Contingent Cone 176–179
Browder 218, 232

Carathéodory 20, 93
– -Hamilton-Jacobi-Bellman 301
Caristi's Fixed Point Theorem 290
Chebischev center 74
– selection 73–76
Coalition of players 216
Commodity 251, 257
Compactness 13–15
Cone-valued map 50
Conjugate functions 29
Connected, $\mathscr{A}_T$ 106
–, $\mathscr{T}_T(x)$ 109, 209
–, image of a map 47–48
Constrained minimization problems 163
Contingent Derivative of a Map 188–192
Convergence Theorem 60–61, 101
Convex body 77
– hull 18–21

Decentralized system 215
Demand function 246, 251
Differential variational inequality 217, 265–271
Dini Derivative 282
Discontinuous right-hand side 101–103
Discounted prices 262
Domain of a map 39
Duality pairing 59

Ekeland's Theorem 195, 287
Ergodic Theorem 140, 153–156
– – for products of resolvents 156–158
Epigraph 22
Equilibrium 213
Equioscillating, family 14
Explicit scheme 228

Feed-back control map 38, 49–50
– map in game theory 244
Fenchel inequality 29
Filippov 95
Functional Differential Inclusions 204–209
Fuzzy coalition 243

Gateau differentiable function 31
Gradient 31
– inclusions 158–171
Graph of a map 40

Hartman 97
Hausdorff continuous map 67
– topology 65–67
Hemicontinuous map 59–64
Henry 217
Hukuhara 95, 109

Implicit Differential Inclusions 192–194
– scheme 228
Indicator 22
Inf-compact 292
Integral representation 98–99

Invariance 203, 233–234
Inverse of a map 40

Kakutani's Fixed Point Theorem 85, 213, 232
Kuratowski 90
Kuratowski's index 46
Ky Fan-Rogalski 232
Ky Fan's Inequality 230–233

Liapunov function 283
– – and equilibria 310
– –, construction 305–309
Lipschitzean map 44
– selection 77–80
Lions 218
Locally compact 40, 84
– selectionable map 80
Logistic equation 274–276
Lotka 275
Lower contingent derivative 287
– semicontinuous in the ε-sense 45
– –, map 43

Marginal map 38, 51–54
Maximal monotone map 139, 140–144
Mean Value Theorem 21
Measurable Selections 90–92
Michael's Selection Theorem 82–83, 130
Monotone Trajectories 179–185, 293–300

Negative polar cone 26
Newton's Method 195–198
Normal cone 33
– – to a convex set 218–228

Orthogonal Projector 25
Oscillation 14

Paracompact, space 10
Parametrization problem 72
Parametrized, map 37, 46–49
Pareto minimum 168, 169, 265, 271
Partition of an interval 130
– – unity 10
Peano 97
Periodic solutions 236–238
Planning procedures 216
Players 243
Polar cone 26
Price regulation 253–257
– system 252

Principle of optimality 305
Process 40
Projected differential inclusion 266–267
Projector 24
Proper, function 22

Range of a map 40
Rate equations 274–280
Regularization of a differential equation 101–103
Regulation through viability 238–242
Relaxation Theorem 94
Robinson-Ursescu 38, 54–56
Ryll-Nardzewski 90

Selections 68–69
Selection, barycentric 77–80
–, Lipschitzean 77–86
–, minimal 70
Slow solution 95, 139, 264, 268
Smith, Adam 245, 246, 247, 249
Stampacchia 218
Strict, map 40
Sub-differential 31–36
Supply 246
Support function 30–31, 59
– of a function 10
σ-Selectionable map 86–89, 109

Tangent cones to convex sets 218–228
Tangential condition 228
– –, strong 229, 234
Tatonnement 248

U-Monotone map 318–320
Upper contingent derivative 282, 284–290
– hemicontinuous map 59–64
– semicontinuous in the ε-sense 45
– – map 41–43

Variational principle 161
Verhurst-Pearl 274
Viability without convexity 198–202
Vitali 94

Walras, Léon 246, 249
– equilibrium 247, 253
– law 247, 250
– –, instantaneous 254
– –, – coalitional 260

Yosida approximation 139, 144–146
– – of the subdifferential 161–163

Die Grundlehren der mathematischen Wissenschaften

A Series of Comprehensive Studies in Mathematics

A Selection

114. MacLane: Homology
115. Hewitt/Ross: Abstract Harmonic Analysis I
123. Yosida: Functional Analysis
127. Hermes: Enumerability, Decidability, Computability
131. Hirzebruch: Topological Methods in Algebraic Geometry
132. Kato: Perturbation Theory for Linear Operators
143. Schur: Vorlesungen über Invariantentheorie
144. Weil: Basic Number Theory
145. Butzer/Berens: Semi-Groups of Operators and Approximation
146. Treves: Locally Convex Spaces and Linear Partial Differential Equations
148. Chandrasekharan: Introduction to Analytic Number Theory
152. Hewitt/Ross: Abstract Harmonic Analysis. Vol. 2: Structure and Analysis for Compact Groups. Analysis on Locally Compact Abelian Groups
153. Federer: Geometric Measure Theory
154. Singer: Bases in Banach Spaces I
155. Müller: Foundations of the Mathematical Theory of Electromagnetic Waves
156. van der Waerden: Mathematical Statistics
157. Prohorov/Rozanov: Probability Theory. Basic Concepts. Limit Theorems. Random Processes
158. Constantinescu/Cornea: Potential Theory on Harmonic Spaces
159. Köthe: Topological Vector Spaces I
160. Agrest/Maksimov: Theory of Incomplete Cylindrical Functions and their Applications
162. Nevanlinna: Analytic Functions
163. Stoer/Witzgall: Convexity and Optimization in Finite Dimensions I
164. Sario/Nakai: Classification Theory of Riemann Surfaces
165. Mitrinović: Analytic Inequalities
166. Grothendieck/Dieudonné: Eléments de Géométrie Algébrique I
167. Chandrasekharan: Arithmetical Functions
168. Palamodov: Linear Differential Operators with Constant Coefficients
170. Lions: Optimal Control of Systems Governed by Partial Differential Equations
171. Singer: Best Approximation in Normed Linear Spaces by Elements of Linear Subspaces
172. Bühlmann: Mathematical Methods in Risk Theory
173. Maeda/Maeda: Theory of Symmetric Lattices
174. Stiefel/Scheifele: Linear and Regular Celestial Mechanics. Perturbed Two-body Motion - Numerical Methods - Canonical Theory
175. Larsen: An Introduction to the Theory of Multipliers
176. Grauert/Remmert: Analytische Stellenalgebren
177. Flügge: Practical Quantum Mechanics I
178. Flügge: Practical Quantum Mechanics II
179. Giraud: Cohomologie non abélienne
180. Landkof: Foundations of Modern Potential Theory
181. Lions/Magenes: Non-Homogeneous Boundary Value Problems and Applications I
182. Lions/Magenes: Non-Homogeneous Boundary Value Problems and Applications II
183. Lions/Magenes: Non-Homogeneous Boundary Value Problems and Applications III
184. Rosenblatt: Markov Processes. Structure and Asymptotic Behavior
185. Rubinowicz: Sommerfeldsche Polynommethode
186. Handbook for Automatic Computation. Vol. 2 Wilkinson/Reinsch: Linear Algebra
187. Siegel/Moser: Lectures on Celestial Mechanics
188. Warner: Harmonic Analysis on Semi-Simple Lie Groups I
189. Warner: Harmonic Analysis on Semi-Simple Lie Groups II
190. Faith: Algebra I: Rings, Modules, and Categories
191. Faith: Algebra II: Ring Theory
192. Mal'cev: Algebraic Systems
193. Pólya/Szegö: Problems and Theorems in Analysis I
194. Igusa: Theta Functions
195. Berberian: Baer*-Rings
196. Athreya/Ney: Branching Processes
197. Benz: Vorlesungen über Geometrie der Algebren
198. Gaal: Linear Analysis and Representation Theory
199. Nitsche: Vorlesungen über Minimalflächen

200. Dold: Lectures on Agebraic Topology
201. Beck: Continuous Flows in the Plane
202. Schmetterer: Introduction to Mathematical Statistics
203. Schoeneberg: Elliptic Modular Functions
204. Popov: Hyperstability of Control Systems
205. Nikol'skii: Approximation of Functions of Several Variables and Imbedding Theorems
206. André: Homologie des algébres commutatives
207. Donoghue: Monotone Matrix Functions and Analytic Continuation
208. Lacey: The Isometric Theory of Classical Banach Spaces
209. Ringel: Map Color Theorem
210. Gihman/Skorohod: The Theory of Stochastic Processes I
211. Comfort/Negrepontis: The Theory of Ultrafilters
212. Switzer: Algebraic Topology – Homotopy and Homology
213. Shafarevich: Basic Algebraic Geometry
214. van der Waerden: Group Theory and Quantum Mechanics
215. Schaefer: Banach Lattices and Positive Operators
216. Pólya/Szegö: Problems and Theorems in Analysis II
217. Stenström: Rings of Quotients
218. Gihman/Skorohod: The Theory of Stochastic Processes II
219. Duvaut/Lions: Inequalities in Mechanics and Physics
220. Kirillov: Elements of the Theory of Representations
221. Mumford: Algebraic Geometry I: Complex Projective Varieties
222. Lang: Introduction to Modular Forms
223. Bergh/Löfström: Interpolation Spaces. An Introduction
224. Gilbarg/Trudinger: Elliptic Partial Differential Equations of Second Order
225. Schütte: Proof Theory
226. Karoubi: K-Theory. An Introduction
227. Grauert/Remmert: Theorie der Steinschen Räume
228. Segal/Kunze: Integrals and Operators
229. Hasse: Number Theory
230. Klingenberg: Lectures on Closed Geodesics
231. Lang: Elliptic Curves: Diophantine Analysis
232. Gihman/Skorohod: The Theory of Stochastic Processes III
233. Stroock/Varadhan: Multi-Dimensional Diffusion Processes
234. Aigner: Combinatorial Theory
235. Dynkin/Yushkevich: Controlled Markov Processes
236. Grauert/Remmert: Theory of Stein Spaces
235. Köthe: Topological Vector Spaces II
238. Graham/McGehee: Essays in Commutative Harmonic Analysis
239. Elliott: Probabilistic Number Theory I
240. Elliott: Probabilistic Number Theory II
241. Rudin: Function Theory in the Unit Ball of $\mathbb{C}^n$
242. Huppert/Blackburn: Finite Groups II
243. Huppert/Blackburn: Finite Groups III
244. Kubert/Lang: Modular Units
245. Cornfeld/Fomin/Sinai: Ergodic Theory
246. Naimark/Štern: Theory of Group Representations
247. Suzuki: Group Theory I
249. Chung: Lectures from Markov Processes to Brownian Motion
250. Arnold: Geometrical Methods in the Theory of Ordinary Differential Equations
251. Chow/Hale: Methods of Bifurcation Theory
252. Aubin: Nonlinear Analysis on Manifolds. Monge-Ampère Equations
253. Dwork: Lectures on p-adic Differential Equations
254. Freitag: Siegelsche Modulfunktionen
256. Hörmander: The Analysis of Linear Partial Differential Operators I
257. Hörmander: The Analysis of Linear Partial Differential Operators II
258. Smoller: Shock Waves and Reaction-Diffusion Equations
259. Duren: Univalent Functions
260. Freidlin/Wentzell: Random Perturbations of Dynamical Systems
263. Krasnosel'skĭ/Zabreĭko: Geometrical Methods of Nonlinear Analysis

Springer-Verlag Berlin Heidelberg New York Tokyo